AF179567

A Geometric Approach to Free Boundary Problems

Luis Caffarelli
Sandro Salsa

Graduate Studies
in Mathematics
Volume 68

American Mathematical Society
Providence, Rhode Island

Editorial Board

Walter Craig
Nikolai Ivanov
Steven G. Krantz
David Saltman (Chair)

2000 *Mathematics Subject Classification.* Primary 35–01, 35R35; Secondary 35J25, 35K20.

For additional information and updates on this book, visit
www.ams.org/bookpages/gsm-68

Library of Congress Cataloging-in-Publication Data

Caffarelli, Luis A.
A geometric approach to free boundary problems / Luis Caffarelli, Sandro Salsa.
p. cm. — (Graduate studies in mathematics, ISSN 1065-7339 ; v. 68)
Includes bibliographical references and index.
ISBN 0-8218-3784-2 (alk. paper)
1. Boundary value problems. 2. Lipschitz spaces. I. Salsa, S. II. Title. III. Series.

QA379.C34 2005
515′.35—dc22 2005041181

Copying and reprinting. Individual readers of this publication, and nonprofit libraries acting for them, are permitted to make fair use of the material, such as to copy a chapter for use in teaching or research. Permission is granted to quote brief passages from this publication in reviews, provided the customary acknowledgment of the source is given.

Republication, systematic copying, or multiple reproduction of any material in this publication is permitted only under license from the American Mathematical Society. Requests for such permission should be addressed to the Acquisitions Department, American Mathematical Society, 201 Charles Street, Providence, Rhode Island 02904-2294, USA. Requests can also be made by e-mail to `reprint-permission@ams.org`.

© 2005 by the American Mathematical Society. All rights reserved.
The American Mathematical Society retains all rights
except those granted to the United States Government.
Printed in the United States of America.

∞ The paper used in this book is acid-free and falls within the guidelines
established to ensure permanence and durability.
Visit the AMS home page at `http://www.ams.org/`

10 9 8 7 6 5 4 3 2 1 10 09 08 07 06 05

Contents

Introduction

Free or moving boundary problems appear in many areas of mathematics and science in general. Typical examples are shape optimization (least area for fixed volume, optimal insulation, minimal capacity potential at prescribed volume), phase transitions (melting of a solid, Cahn-Hilliard), fluid dynamics (incompressible or compressible flow in porous media, cavitation, flame propagation), probability and statistics (optimal stopping time, hypothesis testing, financial mathematics), among other areas.

They are also an important mathematical tool for proving the existence of solutions in nonlinear problems, homogenization limits in random and periodic media, etc.

A typical example of a free boundary problem is the evolution in time of a solid-liquid configuration: suppose that we have a container D filled with a material that is in solid state in some region $\Omega_0 \subset D$ and liquid in $\Lambda_0 = D \setminus \Omega_0$.

We know its initial temperature distribution $T_0(x)$ and we can control what happens on ∂D at all times (perfect insulation, constant temperature, etc.). Then from this knowledge we should be able to reconstruct the solid-liquid configuration, Ω_t, Λ_t, and the temperature distribution $T(x,t)$ for all times $t > 0$.

Roughly, on Ω_t, Λ_t the temperature should satisfy some type of diffusion equation, while across the transition surface, we should have some "balance" conditions that express the dynamics of the melting process.

The separation surface $\partial\Omega_t$ between solid and liquid is thus determined implicitly by these "balance conditions". In attempting to construct solutions to such a problem, one is thus confronted with a choice. We could try to build "classical solutions", that is, configurations Ω_t, Λ_t, $T(x,t)$ where the separation surface $F = \partial\Omega_t$ is smooth, the function T is smooth up to F from both sides, and the interphase conditions on T, ∇T, ... are satisfied pointwise. But this is in general not possible, except in the case of low dimensions (when F is a curve) or very special configurations.

The other option is to construct solutions of the problem by integrating the transition condition into a "weak formulation" of the equation, be it through the conservation laws that in many cases define them, or by a Perron-like supersolution method since the expectation that transition processes be "organized" and "smooth" is usually linked to some sort of "ellipticity" of the transition conditions.

The challenging issue is then, of course, to fill the bridge between weak and classical solutions

A comparison is in order with calculus of variations and the theory of minimal surfaces, one of the most beautiful and successful pursuits of the last fifty years.

In the theory of minimal surfaces one builds weak solutions as the boundary of sets of finite perimeters (weak limits of polygonals of uniformly bounded perimeter) or currents (measures supported in countable unions of Lipschitz graphs) and ends up proving that such objects are indeed smooth hypersurfaces except for some unavoidable singular set perfectly described.

This is achieved by different methods:

(i) by exploiting the invariance of minimal surface under dilations and reducing the problem of local regularity to global profiles (monotonicity formulas, classifications of minimal cones),

(ii) by exploiting the fact that the minimal surface equation linearizes into the Laplacian (improvement of flatness),

(iii) by, maybe the most versatile approach, the DeGiorgi "oscillation decay" method, which says that under very general conditions a Lipschitz surface that satisfies a "translation invariant elliptic equation" improves its Lipschitz norm as we shrink geometrically into a point.

We will see these three themes appearing time and again in these notes. In fact, we consider here a particular family of free boundary problems accessible to this approach: those problems in which the transition occurs when a "dependent variable u" (a temperature, a density, the expected value of a random variable) crosses or reaches a prescribed threshold value $\varphi(x)$.

In the same way that zero curvature forces regularity on a minimal surface, the interplay of both functions $(u-\varphi)^{\pm}$ at each side of $F = \partial\Omega$ and the transition conditions (typically relating the speed of F with $(u-\varphi)^{\pm}_{\nu}$) force regularity on F, although in a much more tenuous way.

To reproduce the general framework of the methods described above for minimal surfaces, it is then necessary to understand the interplay between harmonic and caloric measures in both sides of a domain, the Hausdorff measure of the free boundary and the growth properties of the solutions (monotonicity formulas, boundary Harnack principles).

These are important, deep tools developed in the last thirty years, which we have included in Part 3 of this book. We choose in this book to restrict ourselves to two specific free boundary problems, one elliptic and one parabolic, to present the main ideas and techniques in their simplest form.

Let us mention two other problems of interest that admit a similar treatment: the obstacle problem (see the notes [C5]) and the theory of flow through porous media.

In this book, we have restricted ourselves to the problem of going from weak solutions to classical solutions.

The issue of showing that classical solutions exhibit higher regularity has been treated extensively and forms another body of work with different techniques, more in the spirit of Schauder and other a priori estimates.

There are of course many other problems of great interest: elliptic or parabolic systems, hyperbolic equations, random perturbations of the transition surface, etc.

Although the issues become very complicated very fast, we hope that the techniques and ideas presented in this book contribute to the development of more complex methods for treating free boundary problems or, more generally, those problems where, through differential relations, manifolds and their boundaries interact.

We would like to thank our wives, Anna and Irene, who supported and encouraged us so much, and our institutions, the Politecnico di Milano and the University of Texas at Austin, that hosted each other during the many years of our collaboration. Finally, we are specially thankful to Margaret Combs for her generosity, dedication and support that made this book possible.

Austin, Texas, and Milano, Italy
December 2004

Part 1

Elliptic Problems

Chapter 1

An Introductory Problem

1.1. Introduction and heuristic considerations

In this part we consider free boundary problems of the following type.

In the ball $B_1 = B_1(0)$ we have a continuous function u satisfying

(a) $\Delta u = 0$ in $\Omega^+(u) := \{u > 0\}$ and in $\Omega^-(u) := \{u \leq 0\}^0$,

(b) the flux balance

$$G(u_\nu^+, u_\nu^-) = 0 \tag{1.1}$$

across $F(u) := \partial\Omega^+(u) \cap B_1$, the free boundary.

In (1.1), u_ν^+ and u_ν^- denote the normal derivatives in the inward direction to $\Omega^+(u)$ and $\Omega^-(u)$, respectively, so that $u_\nu^\pm$ are both nonnegative.

The simplest example of this type of problems arises in the minimization of the variational integral

$$J_0(u) = \int_{B_1} \left(|\nabla u|^2 + \chi_{\{u>0\}}\right) dx \tag{1.2}$$

that appears in many applications (e.g., in jet flows, see [AC], [ACF1], [ACF2], [ACF3]).

Suppose u is a local minimizer and assume that the free boundary is differentiable (say) at the origin.

Since

$$u_\lambda(x) = \frac{1}{\lambda} u(\lambda x)$$

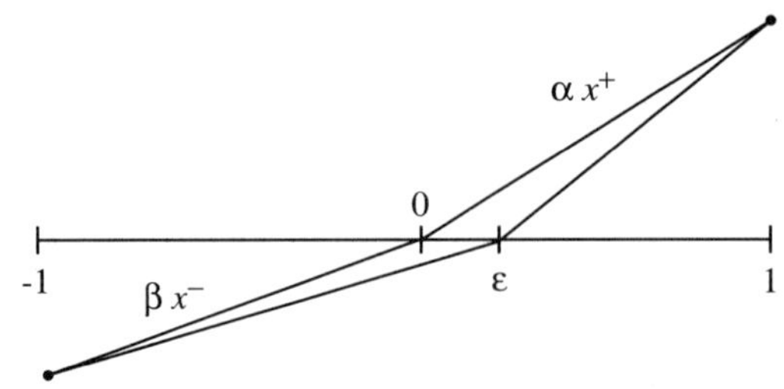

Figure 1.1

is again a local minimizer, that is, the problem is invariant under "elliptic" dilations if we let λ go to zero, we could expect to "guess" the free boundary condition by examining one-dimensional, linear solutions of the type

$$u(x) = \alpha x^+ - \beta x^- \qquad (\alpha, \beta > 0)$$

in $[-1, 1]$.

If we perturb the origin by ε ($\varepsilon \gtrless 0$), the Dirichlet integral

$$\int_{-1}^{1} (u')^2 \, dx$$

changes from $\alpha^2 + \beta^2$ to

$$\frac{\alpha^2}{(1-\varepsilon)^2}(1-\varepsilon) + \frac{\beta^2}{(1+\varepsilon)^2}(1+\varepsilon) = \frac{\alpha^2}{1-\varepsilon} + \frac{\beta^2}{1+\varepsilon} \approx \alpha^2 + \beta^2 + \varepsilon(\alpha^2 - \beta^2)$$

while the volume integral $\int_{-1}^{1} \chi_{\{u>0\}} \, dx$ changes from 1 to $1-\varepsilon$. The "Euler equation" at the origin is therefore

$$\alpha^2 - \beta^2 - 1 = 0;$$

i.e., in this case we expect

$$G(u_\nu^+, u_\nu^-) = (u_\nu^+)^2 - (u_\nu^-)^2 = 1 \, . \tag{1.3}$$

Another way to recover, at least formally, the free boundary condition (1.3) is to use the classical Hadamard formula. Indeed, if we assume that $F(u)$ is smooth and perturb it inward w.r.t. $\Omega^+(u)$ around a point $x \in F(u)$, so that $|\Omega^+(u)|$ decreases by $\delta(\mathrm{Vol})$, then, from Hadamard's formula we get that $\int_{B_1} |\nabla u^+|^2$ increases by an amount $(u_\nu^+)\delta(\mathrm{Vol})$ while $\int_{B_1} |\nabla u^-|^2$ decreases by $(u_\nu^-)^2\delta(\mathrm{Vol})$.

Thus, the minimization condition implies

$$(u_\nu^+)^2 - (u_\nu^-)^2 \geq 1 \, .$$

An inward perturbation w.r.t. $\Omega^-(u)$ would give the opposite inequality so that, on $F(u)$, the "Euler equation" for J_0 is exactly (1.3).

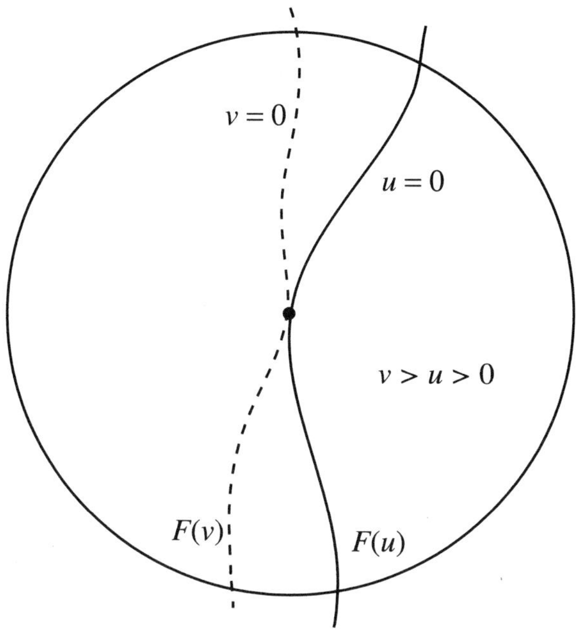

Figure 1.2

The above considerations show a sort of "stability" property of the free boundary relation: if we perturb it towards the positive (negative) region, u_ν^+ tends to increase (decrease), so that u deviates from being a solution.

Referring to the general free boundary condition (1.1), this behavior is reflected in an "ellipticity" condition on G that can be motivated in terms of a comparison principle, like maximum principles.

Consider, for instance, the equation

$$H(D^2u) = 0 .$$

One way of requiring its ellipticity is by asking for a *strict comparison principle*: what is the natural condition on H that prevents two (smooth) solutions u and v ($H(D^2u) = H(D^2v) = 0$) from becoming into contact, i.e.,

$$u \geq v \quad \text{and} \quad u(x_0) = v(x_0) \ ?$$

At x_0 we have $D^2u(x_0) \geq D^2v(x_0)$, but, suppose (poetic license...)

$$D^2u(x_0) > D^2v(x_0) .$$

Thus, if we ask that H be strictly increasing as a function of symmetric matrices, we conclude

$$H(D^2u(x_0)) > H(D^2v(x_0)) ,$$

a contradiction. In other words, monotonicity in H with respect to D^2u is "an ellipticity condition".

Suppose now that u and v are solutions of the same free boundary problem, with $u \geq v$. Since away from their free boundaries $F(u)$ and $F(v)$, respectively, u and v are harmonic and hence cannot touch, the analog of the question above is: what will exclude the possibility that $F(u)$ and $F(v)$ can touch at a free boundary point (say $x_0 = 0$)?

The Hopf maximum principle gives, at $x_0 = 0$,

$$u_\nu^+ > v_\nu^+ \quad \text{and} \quad u_\nu^- < v_\nu^- \ .$$

Therefore, if $G = G(a, b)$ is strictly increasing w.r.t. a and strictly decreasing w.r.t. b, the possibility that both

$$G(u_\nu^+, u_\nu^-) = G(v_\nu^+, v_\nu^-) = 0$$

is excluded.

Definition 1.1. The free boundary relation (1.1) is *elliptic* if $G = G(a, b)$ is strictly increasing w.r.t. a and strictly decreasing w.r.t. b.

The one-dimensional computation done above indicates also an important difference between one-phase and two-phase problems. In one-phase problems it is possible to get universal interior bounds, in the sense that, if u is a solution in a ball $B_1(0)$ and $0 \in F(u)$, then $|\nabla u|$ or even D^2u (as in the case of the obstacle and the one-phase Stefan problem) is bounded in $B_{1/2}(0)$ by a universal constant, no matter what the boundary data arc.

In two-phase problems this is, in general, impossible. For instance, in the one-dimensional minimization problem, by raising the boundary data, one can have a large gradient of the solution near the origin.

1.2. A one-phase singular perturbation problem

Let us go back to the minimization of the functional (1.2). If we put boundary data

$$u_{|\partial B_1} = f \geq 0,$$

then the solution u will be nonnegative and we are dealing with a one-phase problem. We will discuss this problem in full detail in order to introduce some of the main ideas and techniques, useful also in a more general context.

We shall consider minimizers constructed as limits of singular perturbations because in this case the technique can be somewhat simplified. Observe that the problem has no uniqueness so there could exist other types of solutions (see [AC]). All the theory can be developed anyway for any minimizer of J_0 ([AC]).

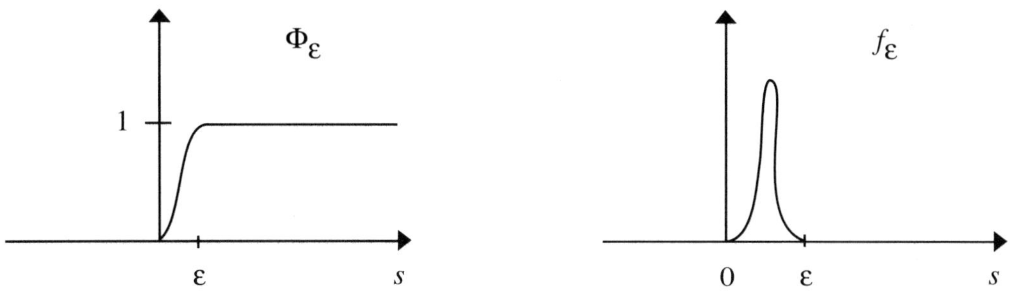

Figure 1.3

So we start studying the minimizers v_ε of the variational integral

$$J_\varepsilon(u) = \int_{B_1} \{|\nabla u|^2 + \Phi_\varepsilon(u)\}\, dx$$

where Φ_1 is a smooth nondecreasing function on the real line with $\Phi_1(s) = 0$ for $s \leq 0$, $\Phi_1(s) = 1$ for $s \geq 1$, and $\Phi_\varepsilon(s) = \Phi_1(s/\varepsilon)$. Therefore,

$$f_\varepsilon(s) = \Phi_\varepsilon'(s) = \frac{1}{\varepsilon}\Phi_1'\Big(\frac{s}{\varepsilon}\Big) = \frac{1}{\varepsilon} f_1\Big(\frac{s}{\varepsilon}\Big)$$

is an approximation of the Dirac measure.

Proposition 1.1. *Given $g \in H^1(B_1)$, there exists a minimizer $u \in H^1(B_1)$ with $u - g \in H_0^1(B_1)$.*

A minimizer u_ε satisfies the Euler equation

$$2\Delta u = f_\varepsilon(u) \tag{1.4}$$

and so, since f_ε is smooth, it is a smooth solution. Since $g_{|\partial B_1} \geq 0$ and f_ε is supported in $0 < u < \varepsilon$, by the maximum principle, u_ε is nonnegative.

We start now with optimal regularity and nondegeneracy. Given the jump conditions along the free boundary, the optimal global regularity we can expect from the limiting solution is Lipschitz continuity and this is what we prove for u_ε.

Theorem 1.2. *Let $\varepsilon < 1/3$ and suppose $u_\varepsilon(0) = \varepsilon$. Then there exists a (universal) constant C_0 such that*

$$\|\nabla u_\varepsilon\|_{L^\infty(B_{1/2}(0))} \leq C_0.$$

At this stage it is important to emphasize a renormalization property of the problem. If u is a solution of equation (1.4) in $B_R(x_0)$, then the renormalization of u given by

$$w(y) = \frac{1}{\varepsilon} u(x_0 + \varepsilon y) \tag{1.5}$$

satisfies the equation

$$\Delta w = \frac{1}{2} f_1(w) \quad \text{in} \quad B_{R/\varepsilon}(0)$$

where, notice, $0 \le f_1(w) \le 1$.

Moreover, observe that $\nabla w(0) = \nabla u(x_0)$. Thus, proving that u is Lipschitz in $B_1(x_0)$ is equivalent to proving that w is Lipschitz in $B_{R/\varepsilon}(0)$. When a property is invariant under the rescaling (1.5), we say that it is a "renormalization property". So, Lipschitz continuity is a renormalization property in this sense.

The next two lemmas show two of these renormalization properties. Since they express general facts, useful in many other situations, we state them in a renormalized form. Theorem 1.2 is then an immediate consequence.

Lemma 1.3. *Let $w \ge 0$ be a solution of*

$$|\Delta w| \le C \quad \text{in} \quad B_2(0).$$

If $w(0) \le 1$, then

$$|\nabla w(0)| \le C_0.$$

Proof. Interior Shauder estimates and Harnack's inequality give

$$|\nabla w(0)| \le \bar{C}(C + \|w\|_{L^\infty(B_1(0))}) \le \bar{C}(C+1) \equiv C_0. \qquad \square$$

Corollary 1.4. *If $x \in B_{1/2}$ and $0 \le u_\varepsilon(x) \le \varepsilon$, then*

$$|\nabla u_\varepsilon(x)| \le C \ .$$

Corollary 1.4 shows Lipschitz continuity in the region where $0 \le u_\varepsilon \le \varepsilon$. Applying the next lemma to $v = u - \varepsilon$, we take care of the region Ω_ε where $u > \varepsilon$.

Lemma 1.5. *Let Ω be an open set with $0 \in \partial\Omega$ and let v be harmonic and nonnegative in $B_1 \cap \Omega$. Suppose that on $\Gamma \equiv \partial\Omega \cap B_1$, $v = 0$ and $|\nabla v|$ is bounded. Then for every $x \in B_{1/2}$*

(a) $v(x) \le C\, d(x, \partial\Omega) \sup_\Gamma |\nabla v|$,

(b) $\|\nabla v\|_{L^\infty(B_{1/2}\cap\Omega)} \le C \sup_\Gamma |\nabla v|$.

Proof. Let $x_0 \in B_{1/2} \cap \Omega$, $h = d(x_0, \partial\Omega)$ and $A = \sup_\Gamma |\nabla v|$. Suppose

$$v(x_0) = \lambda h \ .$$

We want to show that $\lambda \le CA$. Rescale v in $B_h(x_0)$ by putting

$$w(y) = \frac{1}{h} v(x_0 + hy) \ .$$

Then w is harmonic and nonnegative in $B_1(0)$ with $w(0) = \lambda$. Since $h = d(x_0, \partial\Omega)$, there exists $y_1 \in \partial B_1(0)$ at which $w(y_1) = 0$. Moreover, since $\nabla w(y) = \nabla v(x_0+hy)$, we have $|\nabla w(y_1)| \leq A$. Now, by Harnack's inequality, in $B_{1/2}(0)$,

$$w \geq c\lambda \ .$$

In the ring $B_1 \setminus \bar{B}_{1/2}$, compare w with the harmonic function

$$z(y) = \frac{c\lambda}{2^{n-2}-1}(|y|^{2-n} - 1) \ .$$

z vanishes on ∂B_1 and equals $c\lambda$ on $\partial B_{1/2}$; therefore, by the maximum principle

$$w \geq z \ \text{ in } \ B_1 \setminus B_{1/2} \ .$$

If ν is the inward unit normal to ∂B_1 at y_1, we have

$$A \geq |\nabla w(y_1)| \geq w_\nu \geq \frac{n-2}{2^{n-2}-1} c\lambda \ .$$

This proves (a).

For (b), use interior Shauder estimates and Harnack's inequality to get for any point $y \in B_{1/2} \cap \Omega$

$$|\nabla w(y)| \leq C \frac{w(y)}{d(y, \Gamma)}$$

and using (a), we end the proof. □

Remark. In this proof we only use the behavior of v_ν at those points with an inner tangent ball.

The next theorem is a linear growth result: if we stay a fixed amount away from $\partial\Omega_\varepsilon$, then u_ε starts growing linearly.

Theorem 1.6. *There exist (universal) constants c_1, c_2 such that if $x_0 \in B_{1/2}$ and $u_\varepsilon(x_0) = \lambda \geq c_1\varepsilon$ ($\varepsilon < \frac{1}{3}$), then*

$$d(x_0, \partial\Omega_\varepsilon) \leq c_2\lambda \ .$$

From Theorems 1.2 and 1.6 we immediately obtain

Corollary 1.7. *In $\Omega_{c_1\varepsilon} = \{u > c_1\varepsilon\}$,*

$$C^{-1}\operatorname{dist}(x, \partial\Omega_\varepsilon) \leq u_\varepsilon(x) \leq C \operatorname{dist}(x, \partial\Omega_\varepsilon) \ . \tag{1.6}$$

Remark. Observe that also (1.6) is a renormalization property.

Proof of Theorem 1.6. Put $d_0 = d(x_0, \partial\Omega_\varepsilon)$ and suppose $\lambda = \alpha d_0$. We want to show that $\alpha \geq c > 0$. Rescale u in $B_{d_0}(x_0)$ by setting

$$w(y) = \frac{1}{d_0} u_\varepsilon(d_0 y + x_0) \ .$$

Then $\Delta w = 0$ and $w \geq 0$ in B_1 with $w(0) = \alpha$. By Harnack's inequality, in $B_{1/2}$, $\underline{c}\alpha \leq w \leq \bar{c}\alpha$. Now let ψ be a radial cut-off function, $\psi \equiv 0$ in $B_{1/4}$, $\psi \equiv 1$ outside $B_{1/2}$ and define

$$z = \begin{cases} \min\{w, \bar{c}\alpha\psi\} & \text{in } B_{1/2}, \\ w & \text{in } B_1 \setminus B_{1/2}. \end{cases}$$

Then since w is a minimizer of the functional

$$\tilde{J}(v) = \int_{B_1} \left\{ |\nabla v|^2 + \Phi_\varepsilon(d_0 v) \right\} dy$$

among all $v \in w + H_0^1(B_1)$ and since z is an admissible function, we must have

$$\tilde{J}(w) \leq J(z).$$

On the other hand,

$$\int_{B_1} |\nabla z|^2 - \int_{B_1} |\nabla w|^2 \leq c\alpha$$

while

$$\int_{B_1} \Phi_\varepsilon(d_0 v)\, dy - \int_{B_1} \Phi_\varepsilon(d_0 v)\, dy \geq c > 0$$

since, in $B_{1/4}$, $z = 0$ and $\frac{d_0}{\varepsilon} w \geq \frac{d_0}{\varepsilon} \underline{c}\alpha \geq \underline{c}c_1$. The assertion follows easily. $\square$

Remark. Regularity uses only the "weak equation", while linear growth needs the minimization property.

Our next purpose is to estimate the Hausdorff measure of the sets $\partial\Omega_{c\varepsilon}$. To this aim we need the strong nondegeneracy expressed in the following theorem.

Theorem 1.8. *There exist two (universal) constants c, c_1 such that if $x_0 \in B_{1/2}$ and $u_\varepsilon(x_0) \geq c_1\varepsilon$, then*

$$\sup_{B_\rho(x_0)} u_\varepsilon \geq c\rho\ .$$

If $w_\varepsilon(y) = \frac{1}{\rho} u_\varepsilon(x_0 + \rho y)$, this is equivalent to saying that if $\delta = \varepsilon/\rho$ and $w_\varepsilon(0) \geq c_1\delta$, then

$$\sup_{B_1(0)} w_\varepsilon \geq c\ .$$

Once again, this is a consequence of a renormalization result expressed in the following

Theorem 1.9. *Let Ω be an open set with $0 \in \partial\Omega$ and $w \geq 0$, Lipschitz in $B_2(0)$ and harmonic in $\Omega \cap B_2$. Suppose*

(i) $w(x_0) = \delta > 0$,

(ii) *in the region* $\{w \geq \delta/c_1\}$, $c_1 > 1$,

$$w(x) \sim \operatorname{dist}(x, \partial\Omega) \ .$$

Then

$$\sup_{B_1(x_0)} w \geq c \ .$$

First, we give a lemma.

Lemma 1.10. *Under the condition of Theorem 1.9 there exist positive constants* c, C, γ *such that*

$$C\delta \geq \sup_{B_{c\delta}(x_0)} w \geq (1+\gamma)\delta \ .$$

Proof. Let $B_\rho(x_0)$ be the largest ball contained in $\{w > \delta/c_1\}$. From (ii) in Theorem 1.9, $\rho \sim c\delta$. Let $y_0 \in \partial B_\rho(x_0)$ with $w(y_0) = \delta/c_1$. By Lipschitz continuity, for a suitable positive h, $w \leq \delta/2c_1$ in $B_{h\rho}(y_0)$ and therefore also in a fixed fraction of $\partial B_\rho(x_0)$. Since w is harmonic in $B_\rho(x_0)$,

$$\delta \leq w(x_0) = \fint_{\partial B_\rho(x_0)} u$$

so that there must be a point $x_1 \in \partial B_\rho(x_0)$ with $w(x_1) \geq (1+\gamma)\delta$. This gives the second inequality. The first one follows from Lipschitz continuity. □

Proof of Theorem 1.9. We construct a polygonal along which w grows linearly, starting from x_0. From the proof of Lemma 1.10, there exists a point x_1 such that $|x_1 - x_0| = \operatorname{dist}(x_0, \partial\Omega_\varepsilon)$ and

$$cw(x_0) \geq w(x_1) \geq (1+\gamma)w(x_0) \geq w(x_0) + \gamma'|x_1 - x_0| \ .$$

Starting now from x_1 and iterating the application of Lemma 1.10, we construct a sequence $\{x_j\}_{j\geq 1}$ such that $|x_j - x_{j-1}| = \operatorname{dist}(x_{j-1}, \partial\Omega_\varepsilon)$ and

$$cw(x_{j-1}) \geq w(x_j) \geq (1+\gamma)w(x_{j-1}) \geq w(x_{j-1}) + \gamma'|x_j - x_{j-1}| \ .$$

Therefore

(a) $c^j w(x_0) \geq w(x_j) \geq (1+\gamma)^j w(x_0)$,

(b) $w(x_j) - w(x_{j-1}) \geq \gamma'|x_j - x_{j-1}|$,

and in particular,

$$w(x_j) - w(x_0) \geq \gamma' \sum_{k=1}^{j} |x_k - x_{k-1}| \geq \gamma'|x_j - x_0| \ .$$

From (a) we deduce that after a finite number of steps, x_j exits from $B_1(x_0)$. Let k be such that $x_k \in B_1(x_0)$ and $x_{k+1} \notin B_1(x_0)$. Then

$$|x_k - x_0| \geq c > 0 \ .$$

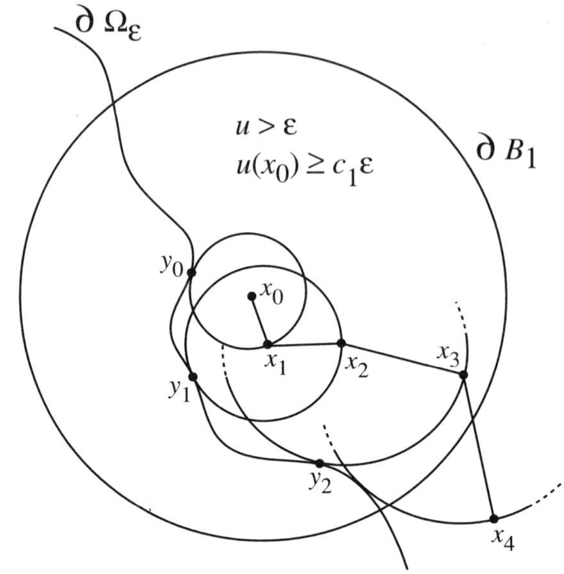

Figure 1.4. The polygonal constructed in the proof of Theorem 1.9

In fact, if $|x_k - x_0| = \alpha$, from (b) and Lipschitz continuity,

$$\gamma'|x_k - x_{k-1}| \le \gamma' \sum_{j=1}^{k} |x_j - x_{j-1}| \le w(x_k) - w(x_0) \le c_0|x_k - x_0| = c_0\alpha$$

so that

$$\operatorname{dist}(x_k, \partial\Omega_\varepsilon) \le 2|x_k - x_{k-1}| \le \frac{2c_0}{\gamma'}\alpha\ .$$

Thus

$$\begin{aligned} 1 \le |x_{k+1} - x_0| &\le |x_{k+1} - x_k| + |x_k - x_0| \le \operatorname{dist}(x_k, \partial\Omega_\varepsilon) + \alpha \\ &\le \left(2\frac{c_0}{\gamma'} + 1\right)\alpha \end{aligned}$$

or $|x_k - x_0| \ge (2\frac{c_0}{\gamma'} + 1)^{-1}$.

From (b)

$$\sup_{B_1(x_0)} w \ge w(x_k) \ge w(x_0) + \gamma'|x_k - x_0| \ge c\ . \qquad \square$$

A first consequence of Theorem 1.8 is the following result, which we call the *uniform positive density* of $\Omega_{c\varepsilon}$ along $\partial\Omega_{c\varepsilon}$, $c \gg 1$.

Corollary 1.11. *Let $x_0 \in B_{1/2}$. There exist (universal) constants c_1, c_2, c_3 such that if $x_0 \in B_{1/2}$, $u(x_0) = \lambda \ge c_1\varepsilon$ and $\rho \ge c_2\lambda$, then*

$$|B_\rho(x_0) \cap \{u_\varepsilon > \lambda\}| \ge c_3\rho^n\ .$$

Proof. Let $u_\varepsilon(y) = \sup_{B_{\rho/2}(x_0)} u_\varepsilon = c\rho$ (Theorem 1.8). Then $d(y, \partial\Omega_\varepsilon) \geq c_1\rho$, by Lipschitz continuity (or Corollary 1.7). By Harnack's inequality, for $\bar{c}$ small and c_2 large,

$$u_\varepsilon(x) \geq \frac{c\rho}{2} \geq \frac{cc_2\lambda}{2} > \lambda \text{ in } B_{\bar{c}\rho}(y)\ .$$

Thus, $B_{\bar{c}\rho}(y) \subset B_\rho(x_0) \cap \{u_\varepsilon > \lambda\}$. □

Remark. The proof of Theorem 1.9 uses the conclusions of Theorem 1.6 as its hypotheses but we make no use of the variational properties of w in its proof.

We are now in a position to estimate the Hausdorff measure at the λ scale of the level sets $\Omega_{c\varepsilon}$, for $\lambda > 3c\varepsilon$, where c is a large universal constant. This is a consequence of the following theorem, where $\mathcal{N}_\delta(E)$ denotes a δ-neighborhood of the set E.

Theorem 1.12. *If $x_0 \in \partial\Omega_{c\varepsilon}$, $\lambda > 3c\varepsilon$, $R \geq c_1\lambda$, $B_R = B_R(x_0)$, then*

$$|\mathcal{N}_\lambda(\partial\Omega_{c\varepsilon}) \cap B_R| \leq c_3\lambda R^{n-1}$$

where all the constants are universal.

We split the proof into several lemmas.

Lemma 1.13. *Under the hypotheses of Theorem 1.12,*

$$|\mathcal{N}_\lambda(\partial\Omega_{c\varepsilon}) \cap B_R| \sim |\mathcal{N}_\lambda(\partial\Omega_{c\varepsilon}) \cap \Omega_{c\varepsilon} \cap B_R|$$

and

$$\begin{aligned}\big[\{c\varepsilon < u_\varepsilon < c^{-1}\lambda\} \cap B_R\big] \ &\subset \ [\mathcal{N}_\lambda(\partial\Omega_{c\varepsilon}) \cap \Omega_{c\varepsilon} \cap B_R] \\ &\subset \ \big[\{c\varepsilon < u_\varepsilon < C\lambda\} \cap B_R\big]\ .\end{aligned}$$

Proof. It follows from Lipschitz continuity and nondegeneracy. □

Lemma 1.14.

$$\int_{\{c\varepsilon < u_\varepsilon < \lambda\} \cap B_R} |\nabla u_\varepsilon|^2 \leq c\lambda R^{n-1}.$$

Proof. By the Gauss formula in $\Omega_{c\varepsilon} \cap B_R$, if $w = \min\{(u_\varepsilon - c\varepsilon)^+, \lambda - c\varepsilon\}$, we have

$$\int_{\{c\varepsilon < u_\varepsilon < \lambda\} \cap B_R} \nabla(u_\varepsilon - c\varepsilon)\nabla w + \int_{\{c\varepsilon < u_\varepsilon < \lambda\} \cap B_R} w\Delta(u_\varepsilon - c\varepsilon) = \int_{\partial[\Omega_{c\varepsilon} \cap B_R]} w(u_\varepsilon - c\varepsilon)^+_\nu \, d\sigma$$

or

$$\int_{\{c\varepsilon < u_\varepsilon < \lambda\} \cap B_R} |\nabla u_\varepsilon|^2 \leq c_0\lambda R^{n-1}\ . \quad \square$$

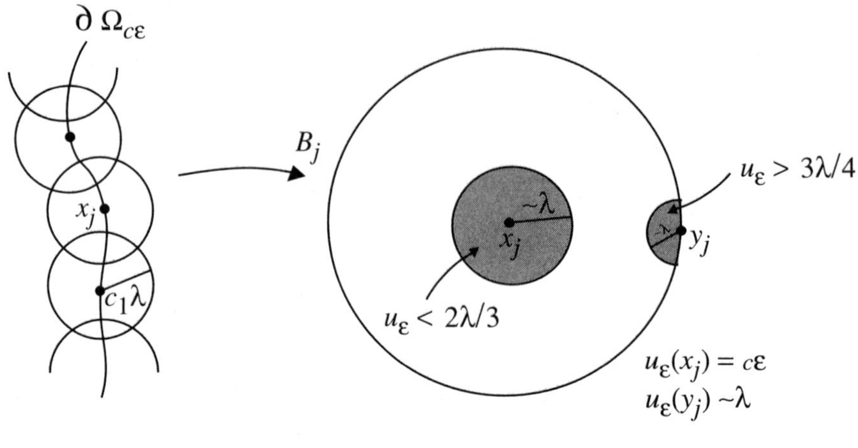

Figure 1.5

It remains to relate the Dirichlet integral of u_ε with the measure of the set $\{c\varepsilon < u_\varepsilon < \lambda\} \cap B_R$. The next lemma completes the proof of Theorem 1.12.

Lemma 1.15. *If* $x_0 \in \partial\Omega_{c\varepsilon}$, $\lambda > 3c\varepsilon$, $R \geq c_1\lambda$, $B_R = B_R(x_0)$, *then*

$$\int_{\{c\varepsilon < u_\varepsilon < \lambda\} \cap B_R} |\nabla u_\varepsilon|^2 \sim |\{c\varepsilon < u_\varepsilon < \lambda\} \cap B_R|.$$

Proof. From Lipschitz continuity the "$\leq$" inequality follows immediately. On the other hand, let $\{B_j\}$ be a finite overlapping covering of $\partial\Omega_{c\varepsilon}$ by balls of radius $c_1\lambda$ and center on $\partial\Omega_{c\varepsilon}$. In every B_j there are subballs B_j^1 and B_j^2 of radius r_j of order λ, where

$$u_\varepsilon \geq \frac{3}{4}\lambda \quad \text{and} \quad u_\varepsilon \leq \frac{2}{3}\lambda,$$

respectively. Therefore, if $m_j = ⨍_{B_j} u_\varepsilon$, then $|u - m_j| \geq c\lambda$ on at least one of the two subballs (c universal). Thus, by Poincaré inequality,

$$c\lambda^2 \leq ⨍_{B_j} (u_\varepsilon - m_j)^2 \leq \bar{c}r_j^2 ⨍_{B_j} |\nabla_\varepsilon u|^2$$

so that

$$\int_{B_j} |\nabla_\varepsilon u|^2 \geq c|B_j|.$$

For c_1 large enough, $\{c\varepsilon < u_\varepsilon < \lambda\} \subset UB_j$ and the proof is complete. □

We are now ready to pass to the limit as $\varepsilon \to 0$. Since $\{u_\varepsilon\}$ is a bounded set in $H^1(B_1)$ and uniformly locally Lipschitz, there exists a sequence $u_k = u_{\varepsilon_k}$, converging to u_0, strongly in $L^2(B_1)$, weakly in $H^1(B_1)$ and uniformly in every compact subset of B_1, as $\varepsilon_k \to 0$.

Theorem 1.16. *u_0 is a local minimizer of J_0.*

Let us first record the main properties of u_0, $\Omega_0 = \{u_0 > 0\}$ and $F(u_0) = \partial\Omega_0 \cap B_1$ in the following

Lemma 1.17. *If $\Omega_0 = \{u_0 > 0\}$, then the following hold.*

(a) *u_0 is locally Lipschitz in B_1, harmonic in Ω_0 and nondegenerate away from $\partial\Omega_0 \cap B_{1/2}$, that is, $\sup_{B_\rho(x)} u_0 \geq c\rho$.*

(b) *Ω_0 is the limit in the Hausdorff distance of $\Omega_k = \{u_k > c\varepsilon_k\}$; i.e., given $\delta > 0$, for c and k large enough,*

$$B_{1/2} \cap \Omega_k \subset N_\delta(\Omega_0) \cap B_{1/2}$$

and

$$B_{1/2} \cap \Omega_0 \subset N_\delta(\Omega_k) \cap B_{1/2}.$$

(c) *$|\mathcal{N}_\delta(\partial\Omega_0) \cap B_R| \leq c\delta R^{n-1}$ for every $\delta > 0$, in particular*

$$|H_{n-1}(\partial\Omega_0) \cap B_R| \leq cR^{n-1} \; . \tag{1.7}$$

Proof. (a) It is clear that u_0 is locally Lipschitz and harmonic in Ω_0. To see it is also nondegenerate, let $x_0 \in \Omega_0 \cap B_{1/2}$. Then there must exist a sequence $\{x_j\}$, with $x_j \to x_0$, $x_j \in \Omega_j \cap B_{2/3}$. By nondegeneracy, $u_j(x_j) \geq c\varepsilon_j$ implies, for any $\rho > 0$ small,

$$u_j(y_j) = \sup_{B_{\rho/4}(x_j)} u_j \geq c\rho \qquad (y_j \in \partial B_{\rho/4}(x_j)) \; .$$

When $|x_j - x_0| \leq \rho/4$, $B_{\rho/4}(x_j) \subset B_\rho(x_0)$ and y_j (or a subsequence) converges to some $y^* \in B_\rho(x_0)$. Since $c\rho \leq u_j(y_j) \to u_0(y^*)$, we conclude that

$$\sup_{B_\rho(x_0)} u_0 \geq c\rho \; .$$

(b) Suppose the first inclusion is false. Then there exists a sequence $\{x_k\}$ such that, for some δ,

- $\operatorname{dist}\{x_k, \Omega_0\} \geq \delta$,
- $x_k \in \Omega_k \cap B_{1/2}$,
- $x_k \to x_0$ with $\operatorname{dist}(x_0, \Omega_0) \geq \delta$.

Therefore, $u_0(x_0) = 0$ while $u_k(x_k) \geq c\varepsilon_k$. By nondegeneracy,

$$u_k(y_k) = \sup_{B_\rho(x_k)} u_k \geq c\rho \; .$$

When $|x_k - x_0| < \delta/8$, for $\rho = \delta/8$, $B_\rho(x_k) \subset B_{\delta/2}(x_0)$ and (a subsequence of) $y_k \to y^* \in B_\delta(x_0)$. Since $u_k(y_k) \to u_0(y^*)$, we conclude that

$$0 = \sup_{B_{\delta/2}(x_0)} u_0 \geq c\delta,$$

a contradiction.

If the second inclusion is false, there exists a sequence $\{x_k\} \subset \Omega_0 \cap B_{1/2}$ such that $\operatorname{dist}(x_k, \Omega_k \cap B_{1/2}) \geq \delta$. This means that $u_k(x) \leq c\varepsilon_k$ for $x \in B_{\delta/2}(x_k)$. Suppose $x_k \to x^*$; then when $|x_k - x^*| < \delta/8$, $B_{\delta/8}(x^*) \subset B_{\delta/2}(x_k)$ and therefore $B_{\delta/8}(x^*) \subset B_{1/2} \setminus \Omega_0$. Contradiction.

(c) It is a consequence of (b) and Theorem 1.12. □

End of the proof of Theorem 1.16. We want to show that u_0 is a local minimizer. Assume not. Then in some ball $B_r = B_r(x_0) \subset\subset B_1$, there exist $a > 0$ and $v \in H^1(B_1)$ such that $v = u_0$ on ∂B_r and

$$\int_{B_r} \{|\nabla v|^2 + \chi_{\{v>0\}}\} \leq \int_{B_r} \{|\nabla u_0|^2 + \chi_{\{u_0>0\}}\} - a \,.$$

Fix $h > 0$ small, and radially interpolate in a linear fashion between u_0 and u_k in the ring $B_{r+h} \setminus B_r$; i.e., define

$$v_{h,k} = \begin{cases} u_0 + \dfrac{|x| - r}{h}(u_k - u_0) & \text{in } B_{r+h} \setminus B_r, \\ v & \text{in } B_r. \end{cases}$$

Then, on $B_{r+h} \subset\subset B_1$,

$$\begin{aligned} J_{r+h,k}(v_{h,k}) &= \int_{B_{r+h}} \{|\nabla v_{h,k}|^2 + F_{\varepsilon_k}(v_{h,k})\} \\ &\leq chr^{n-1} + \frac{2}{h^2} \int_{B_{r+h} \setminus B_r} (u_k - u_0)^2 + J_{r,0}(v) \end{aligned}$$

where

$$J_{r,0}(v) = \int_{B_r} \Big\{|\nabla v|^2 + F(v)\Big\}$$

since

$$\int_{B_r} F_{\varepsilon_k}(v_{h,k}) = \int_{B_r} F_{\varepsilon_k}(v) \leq |\chi_{\{v>0\}}| \,.$$

Thus

$$\overline{\lim_{k\to\infty}}\, J_{r+h,k}(v_{h,k}) \leq chr^{n-1} + J_{r,0}(v) \,.$$

On the other hand,

$$J_{r+h,k}(v_{h,k}) \geq J_{r,k}(v_{h,k}) \geq J_{\varepsilon_k}(u_k)$$

and

$$J_{r,0}(u_0) \leq \underline{\lim_{k\to\infty}}\, J_{\varepsilon_k}(u_k) \,.$$

This follows from

$$\int_{B_r} |\nabla v_0|^2 \leq \underline{\lim_{k\to\infty}} \int_{B_r} |\nabla u_k|^2$$

by the weak convergence of u_k to u_0 and (by Lemma 1.17 applied to B_r instead of $B_{1/2}$ and Theorem 1.12) from

$$|\Omega_0 \cap B_r| \leq |\mathcal{N}_h(\Omega_k) \cap B_r| \leq chr^{n-1} + |\Omega_k \cap B_r| \qquad (h \gg \varepsilon_k)\ .$$

Thus

$$J_{r,0}(u_0) \leq chr^{n-1} + J_{r,0}(v) \leq J_{r,0}(v) + chr^{n-1} - a,$$

a contradiction for $h < \frac{a}{2c} r^{1-n}$. □

1.3. The free boundary condition

At this point we have constructed a local minimizer u_0 of the functional J_0; u_0 is Lipschitz and nondegenerate inside B_1 and for any ball $B_R(x_0) \subset\subset B_1$ and any $\delta > 0$,

$$|\mathcal{N}_\delta(\partial\Omega_0) \cap B_R| \leq c\delta R^{n-1} \tag{1.8}$$

which in particular means that the $(n-1)$-dimensional Hausdorff measure of $\partial\Omega_0$ is locally finite.

Differing from the obstacle problem, the structure of the free boundary $\partial\Omega_0 \cap B_1 = F(u_0)$ is somewhat nicer since cusps cannot occur. This is expressed in the next theorem.

Theorem 1.18. *Let $x_0 \in \partial\Omega_0 \cap B_{1/2}$, $r \leq 1/4$. Then both*

$$|\mathcal{C}\Omega_0 \cap B_r(x_0)| \sim cr^n \quad \text{and} \quad |\Omega_0 \cap B_r(x_0)| \sim r^n. \tag{1.9}$$

As a consequence, from the isoperimetric inequality and (1.7)

$$H_{n-1}(\partial\Omega_0 \cap B_r(x_0)) \sim r^{n-1}\ . \tag{1.10}$$

Proof. $|\Omega_0 \cap B_r(x_0)| \sim r^n$ follows from Lemma 1.17(a). We have to prove that

$$|\{u_0 = 0\} \cap B_r(x_0)| \geq cr^n\ .$$

Let v be the harmonic function in $B_r(x_0)$ with $u_0 = v$ on $\partial B_r(x_0)$. Then since $v > 0$ in $B_r(x_0)$,

$$\begin{aligned}\int_{B_r(x_0)} \left(|\nabla u_0|^2 + \chi_{\{u_0>0\}}\right) &\leq \int_{B_r(x_0)} \left(|\nabla v|^2 + \chi_{\{v>0\}}\right) \\ &= \int_{B_r(x_0)} |\nabla v|^2 + |B_r(x_0)|\ .\end{aligned}$$

Therefore

$$\int_{B_r} \left(|\nabla u_0|^2 - |\nabla v|^2\right) \leq |\{u = 0\} \cap B_r(x_0)|\ .$$

On the other hand, by Poincaré inequality

$$\int_{B_r(x_0)} \left(|\nabla u_0|^2 - |\nabla v|^2\right) = \int_{B_r(x_0)} |\nabla(u_0 - v)|^2 \geq \frac{c}{r^2} \int_{B_r(x_0)} (u_0 - v)^2$$

so that

$$\int_{B_r} (u_0 - v)^2 \leq cr^2 |\{u_0 = 0\} \cap B_r(x_0)| \ .$$

Now,

$$v(x_0) = \fint_{\partial B_r(x_0)} u \geq cr$$

by nondegeneracy, and

$$v(y) \geq cr$$

on $B_{r/2}(x_0)$. Since, by Lipschitz continuity,

$$u(y) \leq chr$$

in $B_{hr}(x_0)$, we conclude that

$$v - u_0 \geq cr \quad \text{on} \quad B_{hr}(x_0)$$

if h is small enough.

Therefore

$$|\{u = 0\} \cap B_r(x_0)| \geq \frac{c}{r^2} \int_{B_r(x_0)} (u_0 - v)^2 \geq \frac{c}{r^2} \int_{B_{hr}(x_0)} (u_0 - v)^2 \geq cr^2 \ . \quad \square$$

Theorem 1.18 says that both Ω_0 and its complement $\mathcal{C}\Omega_0$ have uniform density along the free boundary $F(u_0)$ and that Ω_0 is a set of finite perimeter. But the main and most challenging mathematical question is the regularity of $F(u_0)$.

Before addressing this problem, we must ask another basic question: precisely, in which sense are the free boundary conditions satisfied by u_0? Recall that if we know that $F(u_0)$ is smooth, Hadamard's classical formula gives

$$\partial_\nu u_0^+ = 1 \ . \tag{1.11}$$

There are several ways to interpret in a weak sense the condition (1.11). One way is suggested by (1.10) and by the fact that Δu_0 is a nonnegative measure whose total mass in a ball B_r centered on $F(u_0)$ is equivalent to r^{n-1}. Precisely, we have

Theorem 1.19. *Let $x_0 \in F(u_0)$ and put $\mu = \Delta u_0$. Then μ is a nonnegative measure supported on $F(u_0)$ and for any $r > 0$*

$$\int_{\partial B_r(x_0)} \partial_\nu u_0^+ \, dH_{n-1} = \int_{B_r(x_0)} d\mu \sim r^{n-1} \ . \tag{1.12}$$

Proof. Since (1.12) expresses a renormalization property, it is enough to check it for $r = 1$. The inequality

$$\int_{\partial B_1(x_0)} \partial_\nu u_0^+ \, dH_{n-1} \leq c$$

follows just from Lipschitz continuity.

To prove the opposite inequality, let w be harmonic in $B_1(x_0)$ with $w = u_0$ on $\partial B_1(x_0)$. Then $\Delta(w - u_0) = -\mu$, $w \geq u_0$ and

$$w(y) - u_0(y) = \int_{B_1(x_0)} G(y, z) \, d\mu(z) \ \text{ in } \ B_1$$

where G is the Green's function for B_1. By nondegeneracy and Lipschitz continuity there exists a point $y \in B_h(x_0)$, h small, with $u_0(y) \sim ch$ and consequently

$$u_0 > 0 \ \text{ in } \ B_{ch}(y) \ .$$

Thus $d\mu = 0$ on $B_{ch}(y)$ and

$$w(y) - u_0(y) = \int_{B_1(x_0) \setminus B_{ch}(y)} G(y, z) \, d\mu(z) \leq c \int_{B_1(x_0)} d\mu \ . \tag{1.13}$$

On the other hand, from nondegeneracy, if $p > 1$,

$$⨍_{B_{1/2}(x_0)} w^p \geq ⨍_{B_{1/2}(x_0)} u_0^p \geq c$$

and, from Harnack's inequality,

$$w(y) \geq \left(⨍_{B_{1/2}(x_0)} w^p \right)^{1/p} \geq c$$

so that, if h is small enough

$$w(y) - u_0(y) \geq c - ch \geq \frac{c}{2}$$

and we conclude the proof from (1.13). □

From Theorem 1.19, we obtain immediately the following representation theorem.

Theorem 1.20. *There exists a H_{n-1}-measurable function g on $F(u_0) \cap B_{1/2}$ such that, in $B_{1/2}$*

(i) $0 < c \leq g \leq C$,

(ii) $\Delta u_0 = g H_{n-1} \lfloor_{F(u_0)}$.

Remark. $g = \frac{d\mu}{dH_{n-1}}$ in the Radon-Nikodym sense.

Since, heuristically, $d\mu$ represents $\partial_\nu u_0^+ \, dH_{n-1}$ on $F(u_0)$, we expect $g \equiv 1$ so that, in conclusion, the free boundary condition $\partial_\nu u_0^+ = 1$ should be interpreted as

$$\Delta u_0 = H_{n-1} \quad \text{on} \quad F(u_0) \tag{1.14}$$

in the sense of measure. We can better understand the free boundary condition if we replace $F(u_0)$ with its *reduced* part $F^*(u_0)$, that is, the set of points x at which a generalized (interior to $\Omega_0(u_0)$) normal

$$\nu(x) = \lim_{r \to 0} \frac{\nabla \chi_{\Omega_0}(B_r)}{|\nabla \chi_{\Omega_0}(B_r)|} \tag{1.15}$$

exists with $|\nu(x)| = 1$. In (1.15), $|\nabla \chi_{\Omega_0}(B_r)|$ denotes the total variation in B_r of the measure $\nabla \chi_{\Omega_0}$.

Thus, let $0 \in F^*(u_0)$ and consider the "blow-up" sequence

$$u_k(x) = k u_0 \left(\frac{x}{k} \right) \qquad \text{in} \quad B_k(0) \ .$$

Now let $k \to +\infty$. Let us record the main properties of "blow-up limits".

First of all, since u_0 is locally Lipschitz continuous, for a subsequence

$$u_k \to u_\infty \quad \text{in} \quad C^\alpha_{\text{loc}}(\mathbb{R}^2) \quad (\text{for every } \alpha < 1),$$
$$\nabla u_k \to \nabla u_\infty \quad \text{weakly star in} \quad L^\infty_{\text{loc}}(\mathbb{R}^n) \ .$$

Clearly u_∞ is nonnegative, harmonic in $\{u_\infty > 0\}$ and globally Lipschitz. Moreover,

Lemma 1.21.

(a) $F(u_k) \to F(u_\infty)$ *locally, in the Hausdorff distance,*

(b) $\chi_{\{u_k > 0\}} \to \chi_{\{u_\infty > 0\}}$ *in* $L^1_{\text{loc}}(\mathbb{R}^n)$,

(c) $\nabla u_k \to \nabla u_\infty$ *a.e. in* $\mathbb{R}^n$.

Proof. (a) It follows from the uniform convergence of u_k to u_∞ and the uniform nondegeneracy of u_k.

(b) u_∞ is nondegenerate along $F(u_\infty)$. Indeed, if $x \in F(u_\infty)$, then there exists a sequence $y_k \in F(u_k)$ such that $y_k \to x$, with

$$\fint_{\partial B_r(y_k)} u_k \geq cr$$

and therefore, also

$$\fint_{\partial B_r(x)} u_\infty \geq cr \ .$$

This implies that

$$|\{u_\infty > 0\} \cap B_r(x)| \geq cr^n$$

and therefore that $|F(u_\infty)| = 0$. Using (a), (b) follows.

(c) It is enough to show that $|\nabla u_k| \to |\nabla u_\infty|$ a.e. in $\{u_\infty = 0\}$. Now, a.e. points in $\{u_\infty = 0\}$ are 1-density point. Let S denote the set of such points and let x_0 be one of these. We claim that

(1.16) $$u_\infty(x) = o(|x - x_0|) \text{ near } x_0 .$$

Suppose not. Then there exists a sequence $x_m \to x_0$ such that

$$\frac{u_\infty(x_m)}{|x_m - x_0|} \geq c > 0 .$$

By Lipschitz continuity,

$$u_\infty(x) \geq c|x_m - x_0| \equiv cr_m$$

in $B_{hr_m}(x_m)$, h small. Therefore, for m large enough and $r \gg r_m$, $B_r(x_0)$ contains the ball $B_{hr_m}(x_m)$, which is a contradiction to the 1-density of x_0.

From (1.16) we get $\nabla u_\infty(x_0) = 0$ and, in particular, that, given $\varepsilon > 0$,

$$\frac{u_k(x)}{\delta} < 2\varepsilon \text{ in } B_\delta(x_0)$$

provided k is large enough, say $k \geq k_0(\varepsilon, \delta)$. Then, by nondegeneracy, $u_k \equiv 0$ in $B_{\delta_{\varepsilon/2}}(x_0)$ and consequently, $u_0 = 0$ in a neighborhood of x_0 and S is open. The above argument shows that $u_k = u_0$ in any compact subset of S if k is large enough. This completes the proof of (c). □

We now identify u_∞.

Theorem 1.22. *Let $0 \in F^*(u_0) \cap B_{1/2}$. Then u_∞ is a local minimizer in $\mathbb{R}^n$ and*

$$u_\infty(x) = \langle x, \nu(0)\rangle^+.$$

Corollary 1.23.

(1.17) $$\Delta u_0 = H_{n-1}\lfloor_{F^*(u_0)} .$$

Proof. One can prove that u_∞ is a local minimizer in $\mathbb{R}^n$ by using the same technique of Theorem 1.16.

We may suppose $\nu(0) = e_n$. A well-known property of sets of finite perimeter says that the blow-up limits of Ω_0 and of $\mathcal{C}\Omega_0$ are, respectively, the half planes

$$x_n > 0 \text{ and } x_n < 0 .$$

Thus $u_\infty(x)$ is positive if $x_n > 0$ and equal to zero if $x_n \leq 0$ and in particular on the hyperplane $x_n = 0$. Reflecting u_∞ in an odd way with respect to this hyperplane, we get a function $\tilde{u}_\infty$, harmonic in $\mathbb{R}^n$. Since u_∞ is globally Lipschitz, it follows (Liouville theorem) that $\tilde{u}_\infty$ is a linear function and therefore $u_\infty(x) = \alpha x_n^+$ for some positive α.

We want now to show that $\alpha = 1$; as a consequence we obtain $g \equiv 1$ proving Corollary 1.23:

$$\Delta u_0 = H_{n-1} \quad \text{on} \quad F^*(u_0) \ . \tag{1.18}$$

This gives a possible interpretation of the free boundary condition.

Suppose the problem is 1-dimensional. Then

$$u_\infty(x) = \alpha x^+ \ .$$

Make a perturbation inside $(-1, 1)$ by taking

$$w(x) = \frac{\alpha}{1-b}(x_n - b) \qquad (|b| < 1) \ .$$

Then the minimization condition gives, as in Section 1.1,

$$\alpha^2(b + o(b)) - b \geq 0$$

from which $\alpha = 1$.

In the n-dimensional case we make a similar perturbation inside a strip $|x_n| < 1$, $|x'| = |(x_1, \ldots, x_{n-1})| < M$ with a large M. Let $\psi = \psi(x')$ be a cut-off function such that $\psi \equiv 1$ for $|x'| \geq M+1$, $0 \leq \psi \leq 1$, and $\psi \equiv 0$ in $B'_M = \{|x'| < M\}$. Define

$$w(x) = \max\left\{\frac{\alpha}{1-b}(x_n - b)^+\big(1 - \psi(x')\big) + \alpha x_n^+ \psi(x'), 0\right\} .$$

Then it is not difficult to check that

$$\int_{B_M} (|\nabla w|^2 - |\nabla u_\infty|^2) \leq \alpha^2(b + o(b))|B'_{M+1}| + C|b|\,|B'_{M+1} \setminus B'_M|$$

and

$$|\{w > 0\}| - |\{u_\infty > 0\}| \leq b|B'_{M+1}| + c|b|\,|B'_{M+1} \setminus B'_M| \ .$$

The minimization condition gives

$$(\alpha^2 - 1)\big(b + o(b)\big) + c|b|/M \geq 0$$

and letting $M \to \infty$, we get again $\alpha = 1$. □

There are other ways to interpret the free boundary condition. Let us go back to one of the key lemmas, Lemma 1.5, where the Lipschitz continuity of the minimizers u_ε (and consequently of u_0) is proved.

In that lemma, to the nonnegative harmonic function v we required having bounded gradient along its zero level set. In fact, we only used that v had some bounded linear behavior at points of the free boundary where there exists a touching ball from inside $\{v > 0\}$, i.e., a ball $B \subset \{v > 0\}$ such that $B \cap \{v > 0\} = \{y\}$.

This would suggest taking into consideration, as far as the free boundary condition is concerned, only the points at which there exists a touching ball from one side or the other of $F(u)$. (We will call these points *regular points*.)

This leads naturally towards notions of the free boundary condition in a viscosity sense, which we will formalize in the next chapter.

We point out that these notions of solutions are considerably weaker than the one in the measure-theoretical sense, described above: H_{n-1}-a.e. points on the free boundary have a generalized normal, since $H_{n-1}(F(u)\setminus F^*(u)) = 0$, while, in principle, the set of regular points can be very small.

Clearly, a careful balance is required in constructing the definitions if one looks for both existence of a solution and regularity of the free boundary. For minimizers in variational problems this need is less essential, since the minimization process conveys some "stability" both to the solution and its free boundary. It is not so, for instance, in evolution free boundary problems, where the two requirements could strongly compete.

We end this section by briefly examining what happens to the free boundary conditions $u_\nu^+ = 1$ in the one-phase minimization problem if we adopt this new point of view. Suppose 0 is a regular point of $F(u_0)$ with a touching ball B (from either one of the two sides of $F(u_0)$), whose inward normal at 0 is e_n.

Then from Lemma 11.17 and the remark after it, u_0 has the linear behavior

$$u_0(x) = \alpha x_n^+ + o(|x|) \tag{1.19}$$

near zero if $B \subset \mathcal{C}\Omega_0$, near zero and $x_n \geq 0$ if $B \subset \Omega_0$. In (1.19), $0 < c_1 \leq \alpha \leq c_2$, by Lipschitz continuity and nondegeneracy.

Look again at the blow-up sequence $u_k = ku_0(x/k)$ and at its limit u_∞. From (1.19) we get

$$u_\infty(x) = \alpha x_n^+ \tag{1.20}$$

for $x_n \geq 0$ if $B \subset \Omega_0$ or for all x if $B \subset \mathcal{C}\Omega_0$.

From the monotonicity formula (Theorem 12.3) and the uniform density estimate of $\mathcal{C}\Omega_0$ (or from Lemma 12.8) we deduce that (1.19) holds for every $x \in \mathbb{R}^n$, also if $B \subset \Omega_0$.

We can now proceed as before to prove that $\alpha = 1$.

The free boundary condition is therefore to be interpreted in the following sense: near any regular point $x_0 \in F(u_0)$, u_0 has the linear behavior

$$u_0(x) = \langle x - x_0, \nu(x_0)\rangle^+ + o(|x - x_0|)$$

where $\nu(x_0)$ is the normal to ∂B at x_0, inward to Ω_0.

Let us summarize. In this chapter we have shown the main general properties of a solution to our introductory free boundary problem:

optimal regularity (Lipschitz),

nondegeneracy and linear growth,

uniform density of Ω and $\mathcal{C}\Omega$,

finite Hausdorff measure of the free boundary.

As we will see, this will provide us with the model approach to studying more general free boundary problems.

Chapter 2

Viscosity Solutions and Their Asymptotic Developments

2.1. The notion of viscosity solution

The notion of viscosity solution was introduced by M. Crandall and P. L. Lions (see [CL]) in the context of Hamilton-Jacobi equations and in the last two decades has become the central notion in the theory of fully nonlinear parabolic and elliptic equations.

Let us see, for instance, how viscosity harmonic functions are defined. The key idea is to switch the action of the Laplace operator to smooth test functions, using the comparison principle.

Suppose D is a domain in $\mathbb{R}^n$ and $\varphi \in C^2(D)$. If u is a classical subharmonic function in D and φ touches u (locally) from above at x_0, i.e.,

$$\varphi \geq u \quad \text{near } x_0 \text{ and } \quad \varphi(x_0) = u(x_0), \tag{2.1}$$

then clearly

$$\Delta\varphi(x_0) \geq 0. \tag{2.2}$$

Analogously, if u is a classical superharmonic function in D,

$$\varphi \leq u \quad \text{near } x_0 \text{ and } \varphi(x_0) = u(x_0), \tag{2.1$'$}$$

then

$$\Delta\varphi(x_0) \leq 0\ . \tag{2.2$'$}$$

Notice that conditions (2.2) and (2.2′) do not require derivatives of u, while (2.1) and (2.1′) make sense with u continuous.

Definition 2.1. A function $u \in C(D)$ is a viscosity subharmonic (superharmonic) function if for every $\varphi \in C^2(D)$ satisfying (2.1) (resp. (2.1′)), (2.2) (resp. (2.2′)) holds; u is a viscosity harmonic function if it is both a viscosity subharmonic and superharmonic function.

In other words u is subharmonic if the only way a smooth function may touch it from above is by being itself subharmonic.

Definition 2.1 is perfectly consistent with the classical definition: u is a viscosity harmonic function if and only if it is harmonic in the classical sense. Indeed, it is easy to prove that if $u \in C(D)$ is a viscosity harmonic function, then $u \in C^2(D)$ and $\Delta u = 0$ in D.

Observe that we could equivalently have defined u subharmonic (superharmonic) in the viscosity sense by requiring that if $\varphi \in C^2(\Omega)$ and $\Delta\varphi \leq 0$ (resp. $\Delta\varphi \geq 0$) in Ω, then $\varphi - u$ cannot have a local minimum (resp. maximum). We leave the proof of the equivalence of the two definitions as an exercise.

As we see, the "leitmotiv" of the definition is to prevent a subsolution (a supersolution) from touching a solution from below (above). This is exactly what we require for a viscosity solution of a free boundary problem. First, define a C^k-classical subsolution and supersolution ($1 \leq k \leq \infty$) as follows:

Definition 2.2. A function $v \in C(D)$ is called a C^k-classical subsolution of our free boundary problem if

(i) $v \in C^k(\bar{\Omega}^+(v)) \cap C^k(\bar{\Omega}^-(v))$,

(ii) $\Delta v \geq 0$ in $\Omega^+(v) = \{v > 0\} \cap D$ and $\Omega^-(v) = \{v \leq 0\}^0 \cap D$,

(iii) the free boundary $F(v) = \partial\Omega^+(v) \cap D$ is a C^k-surface, $|\nabla u^+| > 0$ on $F(v)$ and

$$G(v_\nu^+, v_\nu^-) \geq 0 \quad \text{on} \quad F(v) \tag{2.3}$$

where $\nu = \frac{\nabla u^+}{|\nabla u^+|}$.

If the inequality in (2.3) is strict, we call v the *strict subsolution*. A C^k-classical supersolution is defined by reversing the inequalities in Definition 2.2, while a C^k-classical solution is both a C^k-classical subsolution and supersolution.

Notice that in the one-phase minimization problem of Section 1.1, a subsolution of the problem satisfies the condition $u_\nu^+ \geq 1$.

We now use classical strict subsolutions and supersolutions as test functions. From now on we choose $k = 2$ to fix the ideas, but other choices may be more convenient.

Definition 2.3. A function $u \in C(D)$ is a viscosity subsolution if for every classical strict supersolution v, v cannot touch u from above at a free boundary point.

Notice that if $v \in C^2(\bar{\Omega}(v))$ is a superharmonic function in $\Omega^{\pm}(v)$ touching u from above at a free boundary point x_0, i.e., $v(x) \geq u(x)$ near x_0 and $x_0 \in F(v) \cap F(u)$, then necessarily

$$G\big(v_\nu^+(x_0), v_\nu^-(x_0)\big) \geq 0 \; . \tag{2.4}$$

Viscosity supersolutions are defined reversing the inequalities in Definition 2.3 and interchanging sub with super. A viscosity solution is both a viscosity subsolution and supersolution. So, no classical strict subsolution (supersolution) can touch a solution from below (above).

It is not difficult to prove that a C^2-classical subsolution (supersolution, solution) is also a viscosity subsolution (supersolution, solution) and that a viscosity solution (subsolution, supersolution) of class C^2, with its free boundary, in $\bar{\Omega}^+(w)$ and $\bar{\Omega}^-(w)$ is a classical solution (subsolution, supersolution).

2.2. Asymptotic developments

The definitions of C^2-viscosity subsolutions, supersolutions and solutions can be restated in terms of asymptotic linear behavior near a regular point of the free boundary.

Indeed, let u be a C^2-viscosity solution and suppose that a classical strict subsolution v touches u from below at 0, say. Then the origin is a regular point from the right, i.e., there is a ball B touching $F(u)$ at zero, inside $\Omega^+(w)$. By Lemma 11.17, if ν denotes the unit normal to ∂B inward to $\Omega^+(u)$, we have, near zero, in B

$$u^+(x) = \alpha\langle x, \nu\rangle^+ + o(|x|)$$

with $\alpha > 0$ and in $\mathcal{C}B$

$$u^-(x) = \beta\langle x, \nu\rangle^- + o(|x|)$$

with $\beta \geq 0$, finite. The free boundary condition translates into a relation between α and β. Which is the correct one? Since v is a classical strict subsolution, we have

$$G(v_\nu^+(0), v_\nu^-(0)) > 0 \; .$$

On the other hand, the Hopf maximum principle gives

$$v_\nu^+(0) < \alpha \quad \text{and} \quad v_\nu^-(0) > \beta$$

so that from the ellipticity of G

$$0 < G(v_\nu^+(0), v_\nu^-(0)) < G(\alpha, \beta) \ .$$

We reach a contradiction if $G(\alpha,\beta) \le 0$, which is, therefore, the correct inequality. Analogously, if v is a supersolution touching u from above at zero, there exists a ball B touching $F(u)$ at zero, inside $\Omega^-(u) = \{u \le 0\}^o$. Then if ν denotes the unit normal to ∂B inward to $\Omega^+(u)$, we have, near zero, in B

$$u^-(x) = \beta\langle x, \nu\rangle^- + o(|x|)$$

with $\beta > 0$, while in $\mathcal{C}B$

$$u^+(x) = \alpha\langle x, \nu\rangle^+ + o(|x|)$$

with $\alpha \ge 0$.

This time we have

$$0 > G(v_\nu^+(0), v_\nu^-(0)) > G(\alpha, \beta) \ ,$$

a contradiction if $G(\alpha,\beta) \ge 0$.

Substantially, we require that u satisfy a supersolution condition at regular points from the right (touching ball inside $\Omega^+(w)$) and a subsolution condition at regular points from the left (touching ball inside $\Omega^-(u)$).

The strict condition $G(\bar\alpha, \bar\beta) > 0$ compared to $G(\alpha,\beta) = 0$ indicates that $\bar\alpha > \alpha$ and/or $\bar\beta < \beta$ and therefore a supersolution is "more concave" than a solution at a common point of the free boundary. This clearly prevents a supersolution from touching a solution from above and it is perfectly analogous to the fact that a superharmonic function is more concave than a harmonic function at a common point of their graphs.

We summarize the conclusions by giving an alternative definition.

Definition 2.4. A continuous function u is a viscosity solution in D if the following hold.

(i) It is harmonic in $\Omega^+(u)$ and $\Omega^-(u)$.

(ii) Along $F(u)$, u satisfies the free boundary condition in the following sense:

(a) If at x_0 there is a ball $B \subset \Omega^+(u)$, $B \cap \Omega^+(u) = \{x_0\}$ and near x_0,

in B

$$u^+(x) \ge \alpha\langle x - x_0, \nu\rangle^+ + o(|x - x_0|) \qquad (\alpha > 0), \tag{2.5}$$

in $\mathcal{C}B$

$$u^-(x) \le \beta\langle x - x_0, \nu\rangle^- + o(|x - x_0|) \qquad (\beta \ge 0) \tag{2.6}$$

with equality along every nontangential domain in both cases, then

$$G(\alpha, \beta) \le 0. \tag{2.7}$$

(b) If at x_0 there is a ball $B \subset \Omega^-(u)$, $B \cap \Omega^-(u) = \{x_0\}$ and near x_0, in B

$$u^-(x) \ge \beta\langle x - x_0, \nu\rangle^- + o(|x - x_0|) \qquad (\beta > 0), \tag{2.8}$$

in $\mathcal{C}B$

$$u^+(x) \le \alpha\langle x - x_0, \nu\rangle^+ + o(|x - x_0|) \qquad (\alpha \ge 0) \tag{2.9}$$

with equality along every nontangential domain in both cases, then

$$G(\alpha, \beta) \ge 0\ . \tag{2.10}$$

We leave it as an exercise to prove that Definitions 2.3 and 2.4 are equivalent. (Hint: if (2.5) and (2.6) hold and $G(\alpha, \beta) > 0$, construct a strict subsolution touching u from below at x_0.)

Definition 2.4 is based on the asymptotic behavior of u. We get an equivalent definition involving the linear behavior of classical test functions touching the free boundary at regular points (see Definition 2.3) replacing condition (ii) by the following.

(ii)* (a) If x_0 has a touching ball B inside $\Omega^+(u)$ and in B, near x_0

$$u^+(x) \ge \bar{\alpha}\langle x - x_0, \nu\rangle^+ + o(|x - x_0|) \qquad (\alpha \ge 0), \tag{2.11}$$

then in $\mathcal{C}B$, near x_0

$$u^-(x) \ge \bar{\beta}\langle x - x_0, \nu\rangle^- + o(|x - x_0|) \qquad (\beta \ge 0) \tag{2.12}$$

for any $\bar{\beta}$ such that

$$G(\,\bar{\alpha}, \bar{\beta}\,) > 0\ .$$

(b) If x_0 has a touching ball B inside $\Omega^-(u)$ and in B, near x_0

$$u^-(x) \ge \bar{\beta}\langle x - x_0, \nu\rangle^+ + o(|x - x_0|) \qquad (\bar{\beta} \ge 0), \tag{2.13}$$

then in $\mathcal{C}B$, near x_0

$$u^+(x) \ge \bar{\alpha}\langle x - x_0, \nu\rangle^+ + o(|x - x_0|) \qquad (\bar{\alpha} \ge 0) \tag{2.14}$$

for any $\bar{\alpha}$ such that

$$G(\,\bar{\alpha}, \bar{\beta}\,) < 0\ . \tag{2.15}$$

Let us check that conditions (ii) and (ii)* are equivalent. Assume (ii)(b) holds. Then if (2.13) holds, we have $\beta \ge \bar{\beta}$. If $\bar{\alpha}$ is such that $G(\bar{\alpha}, \bar{\beta}) < 0$, since $G(\alpha, \bar{\beta}) \geq G(\alpha, \beta) \geq 0$, it must be that $\alpha \geq \bar{\alpha}$. Since equality holds in (2.9) along nontangential domains, (2.14) follows.

Assume now (ii)*(b). Let (2.8) and (2.9) hold; we want to show that $G(\alpha, \beta) \geq 0$. If not, $G(\alpha, \beta) < 0$ and, for a small $\varepsilon > 0$, $G(\alpha + 2\varepsilon, \beta) < 0$. Then (2.9) gives

$$u^+(x) \leq (\alpha + \varepsilon)\langle x - x_0, \nu\rangle^+ + o(|x - x_0|)$$

while (2.13) and (2.14) with $\bar{\alpha} = \alpha + 2\varepsilon$ and $\bar{\beta} = \beta$ give

$$u^+(x) \geq (\alpha + 2\varepsilon)\langle x - x_0, \nu\rangle^+ + o(|x - x_0|) \ .$$

Contradiction.

Analogously one can check that (ii)(a) and (ii)*(a) are equivalent.

Remark. As we have seen, viscosity solutions can be characterized in different ways. The definition is clearly closed under uniform limits.

The disadvantage is that it could produce undesirable solutions like

$$u(x) = \alpha_1 x_1^+ + \alpha_2 x_1^-$$

with any α_1, α_2 such that

$$G(\alpha_1, 0) \leq 0, \quad G(\alpha_2, 0) \leq 0 \ .$$

Extra care will be necessary to construct solutions with the desired geometric measure-theoretic properties.

2.3. Comparison principles

Strictly speaking, the definition of viscosity subsolution (supersolution) involves a condition at those points of the free boundary that are regular from the left (right). These conditions turn out to be not enough for comparison purposes.

Suppose we intend to prove that a viscosity subsolution v cannot touch a solution u from below at a point $x_0 \in F(u) \cap F(v)$. Then it is natural to require the existence at x_0 of a touching ball from *the right* (not from the left) and a proper asymptotic behavior for v near x_0 that could force a contradiction. Therefore it is useful to introduce another kind of "subsolution" with these characteristics. It turns out that a natural way to construct such a type of functions is to start from a subsolution or a solution (in the viscosity sense) and to build parallel surfaces. Here is the simplest example.

Let u be a solution in D and take

(2.16) $$v_t(x) = \sup_{B_t(x)} u, \qquad t > 0.$$

Let us examine the properties of v_t. Since v_t is the supremum of a family of translations of u, it is subharmonic both in $\Omega^+(v_t)$ and $\Omega^-(v_t)$. Now let

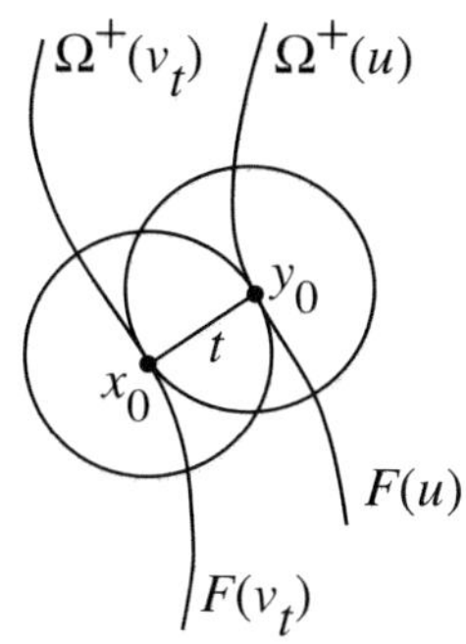

Figure 2.1

$x_0 \in F(v_t)$. This means that $B_t(x_0)$ is touching $F(u)$ from $\Omega^-(u)$ at a point y_0 (see Figure 2.1). Therefore we have the following.

(a) x_0 is regular from the right since $B_t(y_0) \subset \Omega^+(v_t)$ and $B_t(y_0)$ touches $F(v_t)$ at x_0.

(b) y_0 is a regular point from the left for $F(u)$, thus, near y_0,

$$u^-(x) = \beta\langle x - y_0, \nu\rangle^- + o(|x - x_0|) \qquad (\beta > 0)$$

in $B_t(x_0)$, while

$$u^+(x) = \alpha\langle x - y_0, \nu\rangle^+ + o(|x - x_0|) \qquad (\alpha \geq 0)$$

in $\mathcal{C}B_t(x_0)$, with $G(\alpha, \beta) \geq 0$.
Hence, since $v_t(x) \geq u(x + y_0 - x_0)$, near x_0,

$$v_t^+(x) \geq \alpha\langle x - x_0, \nu\rangle^+ + o(|x - x_0|)$$

in $B_t(y_0)$, while

$$v_t^-(x) \leq \beta\langle x - x_0, \nu\rangle^- + o(|x - x_0|)$$

in $\mathcal{C}B_t(y_0)$.

Let us summarize the properties of v_t:

(i) $\Delta v \geq 0$ both in $\Omega^+(v)$ and $\Omega^-(v)$.

(ii) Whenever $x_0 \in F(v)$ has a touching ball $B \subset \Omega^+(v)$, then near x_0, in B

$$v^+(x) \geq \bar{\alpha}\langle x - x_0, \nu\rangle^+ + o(|x - x_0|) \tag{2.17}$$

and in $\mathcal{C}B$

$$v^-(x) \leq \bar{\beta}\langle x - x_0, \nu\rangle^- + o(|x - x_0|) \tag{2.18}$$

with $\bar{\alpha}, \bar{\beta} \geq 0$, and

$$G(\bar{\alpha}, \bar{\beta}) \geq 0\ .$$

We call a function v with properties (i), (ii) an *R-subsolution.* We can now prove the following comparison result:

Lemma 2.1. *Let u, v be a (viscosity) solution and an R-subsolution in D, respectively. Then if $u \geq v$, $u > v$ in $\Omega^+(v)$ and $x_0 \in F(u) \cap F(v)$, x_0 cannot be a regular point from the right.*

Proof. If x_0 is a regular point from the right, we have, near x_0, in a touching ball $B \subset \Omega^+(v)$, nontangentially,

$$u^+(x) = \alpha\langle x - x_0, \nu\rangle^+ + o(|x - x_0|) \ , \quad v^+(x) \geq \bar{\alpha}\langle x - x_0, \nu\rangle^- + o(|x - x_0|)$$

and in $\mathcal{C}B$, nontangentially,

$$u^-(x) = \beta\langle x - x_0, \nu\rangle^- + o(|x - x_0|) \ , \quad v^-(x) \geq \bar{\beta}\langle x - x_0, \nu\rangle^+ + o(|x - x_0|)$$

with

$$G(\alpha, \beta) \leq 0 \quad \text{and} \quad G(\bar{\alpha}, \bar{\beta}) \geq 0 \ . \tag{2.19}$$

Since $u \geq v$, we have $\alpha \geq \bar{\alpha}$ and $\beta \leq \bar{\beta}$. The strict monotonicity of G and (2.19) imply $\alpha = \bar{\alpha}$, $\beta = \bar{\beta}$. But $u - v$ is a positive superharmonic function in $\Omega^+(v)$. By the Hopf principle, since x_0 is a regular point, we have $(u - v)(x) \geq |x - x_0|$ with $\varepsilon > 0$, radially into $\Omega^+(v)$ along ν from x_0, a contradiction. □

A more refined version, of a "continuous deformation" nature, is the following theorem that we will use later.

Theorem 2.2. *Let v_t, $0 \leq t \leq 1$, be a family of R-subsolutions, continuous in $\bar{\Omega} \times [0, 1]$. Let u be a solution continuous in $\bar{\Omega}$. Assume that*

(i) *$v_0 \leq u$ in Ω,*

(ii) *$v_t \leq u$ on $\partial\Omega$ and $v_t < u$ in $\overline{\Omega^+(v_t)} \cap \partial\Omega$ for $0 \leq t \leq 1$,*

(iii) *every point $x_0 \in F(v_t)$ is a regular point from the right,*

(iv) *the family $\Omega^+(v_t)$ is continuous, that is, for every $\varepsilon > 0$,*

$$\Omega^+(v_{t_1}) \subset \mathcal{N}_\varepsilon(\Omega^+(v_{t_2}))$$

whenever $|t_1 - t_2| \leq \delta(\varepsilon)$.

Then

$$v_t \leq u \quad in \ \ \Omega$$

for every $t \in [0, 1]$.

Proof. Let $E = \{t \in [0, 1] : v_t \leq u \text{ in } \bar{\Omega}\}$. E is obviously closed. Let us show that it is open. If $v_{t_0} \leq u$, from (ii) and the strong maximum principle it follows that $v_{t_0} < u$ in $\Omega^+(v_{t_0})$. Since every point of $F(v_{t_0})$ is regular from the right, Lemma 2.1 and (ii) imply that $\overline{\Omega^+(v_{t_0})}$ is compactly contained in $\Omega^+(w)$, up to $\partial\Omega$.

From assumption (iv), the openness of E follows. □

Remark. Comparisons of this nature are necessary when a maximum principle (or uniqueness) is not available. For instance, the classical reflection method of Alexandrov, of Serrin and of Gidas, Ni, and Nirenberg is of this nature.

The family v_t constructed in (2.16) is an admissible family for the comparison in the previous theorem. It can be used for a comparison principle that says: *if u_1, u_2 are solutions such that $u_1 \le u_2$ and near $\partial\Omega$, $\sup_{B_t(x)} u_1 \le u_2(x)$, then also in the interior of Ω, $\sup_{B_t(x)} u_1 \le u_2(x)$,* keeping in particular $F(u_2)$ t-away from $F(u_1)$.

We shall see the usefulness of this kind of principle in proving strong regularity results for the free boundary.

Chapter 3

The Regularity of the Free Boundary

3.1. Weak results

We now go back to our general free boundary problem of Section 1.1 and examine the regularity of the free boundary.

Philosophically speaking, we can divide the free boundary smoothness results into two categories: "measure-theoretic or weak" and "higher order or strong" results.

The measure-theoretic regularity of the free boundary amounts to saying that $F(u)$ has finite locally $(n-1)$-Hausdorff measure and, for each $x \in F(u)$,

$$H^{n-1}(F(u) \cap B_r(x)) \sim r^{n-1} \sim H^{n-1}(F^*(u) \cap B_r(x))$$

where $F^*(u)$ denotes the reduced part of $F(u)$. In particular

$$H^{n-1}\lfloor_{F(u)}\,, \quad H^{n-1}\lfloor_{F^*(u)} \text{ and } \Delta u^+$$

are all positive measures, with bounded (from above and below) density with respect to each other.

As we have seen in Sections 1.2 and 1.3 for the one-phase singular perturbation problem, these weak regularity properties of the free boundary follow if one can prove two essential features of the solution:

(a) u is locally Lipschitz across the free boundary,

(b) u^+ has linear growth and is nondegenerated away from $F(u)$, i.e.,

$$u^+(x) \geq c\, d(x, F(u))\,, \qquad x \in \Omega^+(u)\,,$$

and

$$\sup_{B_r(x_0)} u^+ \geq Cr\ , \qquad x_0 \in F(u)\ .$$

Let us see what the minimal assumptions are for achieving this kind of result. A parallel with viscosity solutions of elliptic equations $\mathcal{L}v = \mathrm{Tr}(A(x)D^2v) = 0$ is in order.

The linear growth and the nondegeneracy for u^+ correspond to having a Harnack inequality (controlled growth) for v and this is true, for instance, if $\mathcal{L}$ is strictly elliptic ($A(x) \geq \lambda I$) and A is bounded measurable.

For our free boundary problem, the parallel requirement is that $0 < c < u_\nu^+ \leq C$ in the viscosity sense. More precisely if at $x_0 \in F(u)$ there is a touching ball $B \subset \Omega^+(u)$, no matter how small, then u has linear behavior

$$u(x) = \alpha\langle x - x_0, \nu\rangle^+ - \beta\langle x - x_0, \nu\rangle^- + o(|x - x_0|)$$

with, for instance

$$\alpha = G(\beta)\ .$$

Strict ellipticity corresponds to the strict monotonicity of G, with $G(0) > 0$. Ellipticity from above ($u_\nu^+ \leq C$) is then assured by the monotonicity formula (see Theorem 12.7).

In Chapter 6 we shall construct solutions of our free boundary problem, precisely satisfying properties (a) and (b). Therefore their free boundaries will enjoy the above measure-theoretic regularity.

3.2. Weak results for one-phase problems

The heuristic discussion in the previous section leads naturally to the following list of weak results for one-phase problems. Suppose u is a nonnegative continuous function in B_1, harmonic in its positivity set $\Omega^+(u)$. As usual let $F(u) = \partial\Omega^+(u) \cap B_1$ be the free boundary. At every $x_0 \in F(u)$ at which there is a touching ball $B \subset \Omega^+(u)$, we know u has the linear nontangential behavior

$$u(x) = \alpha\langle x - x_0, \nu\rangle^+ + o(|x - x_0|) \tag{3.1}$$

with $\alpha > 0$, $x \in B$.

Lemma 3.1. *If $\alpha \leq C$, then u is Lipschitz in $B_{1/2}$.*

Proof. (See the beginning of Section 12.1.) □

So, a uniform bound from above for u_ν^+ in the viscosity sense gives Lipschitz continuity.

The next result we would like to have is the equivalent of Lemma 1.5:

Pseudolemma 3.2. *If $\alpha \geq c > 0$, then in $B_{1/2}$*

$$u^+(x) \geq c\, d(x, F(u))\ .$$

Therefore, a uniform bound from below for u_ν^+ would give us linear growth.

This is not true in general. The function $(\log|x|)^+$ in two dimensions provides a counterexample. We will in fact construct solutions with this property in Chapter 6 using Perron's method.

Lemma 3.3. *If $\alpha \leq C$ and if we assume the conclusions of Pseudolemma 3.2, then for every $x_0 \in F(u) \cap B_{1/2}$,*

$$\sup_{B_\rho(x_0)} u \geq c\rho, \tag{3.2}$$

$$|\Omega^+(u) \cap B_\rho(x_0)| \geq c\rho^n, \tag{3.3}$$

$$\left(\rho < \frac{1}{3}\right).$$

Proof. First, one proves that if $x_0 \in B_{1/2}$ and $d = d(x_0, F(u))$, there exists $\gamma > 0$ such that

$$\sup_{B_d(x_0)} u \geq (1+\gamma)u(x_0)\ .$$

This follows as in Lemma 1.10.

Then (3.2) follows, constructing a polygonal along which u grows linearly, exactly as in the proof of Theorem 1.9. The density estimate (3.3) follows as in Corollary 1.11. □

Thus, from Lipschitz continuity and linear growth, we deduce nondegeneracy and uniform density of $\Omega^+(u)$ along $F(u)$. Notice that the conclusions of the above lemmas hold in every connected component of $\Omega^+(u)$.

Under the hypothesis of Lemma 3.3 we deduce now the measure-theoretic properties of $F(u)$. We recall that $\mathcal{N}_\delta(F)$ denotes a δ-neighborhood of F.

Theorem 3.4. *Let u be Lipschitz and nondegenerate in B_1. Let $x_0 \in F(u) \cap B_{1/2}$ and $0 < \varepsilon < R < \frac{1}{2}$. Then the following quantities are comparable with R^{n-1}:*

(a) $\frac{1}{\varepsilon}|\{0 < u < \varepsilon\} \cap B_R|$,

(b) $\frac{1}{\varepsilon}|\mathcal{N}_\varepsilon(F(u)) \cap B_R|$,

(c) $N\varepsilon^{n-1}$, *where N is the number of any family of balls of radius ε, with finite overlapping, covering $F(u) \cap B_R$,*

(d) $H^{n-1}(F(u) \cap B_R)$. *In particular, $\Omega^+(u)$ is a set of finite perimeter.*

Proof. (Compare with Theorem 1.12 and Lemmas 1.13–1.14.) The key points are to prove that if $x_0 \in F(u) \cap B_{1/2}$, then

$$\int_{\{0<u<\varepsilon\}\cap B_R(x_0)} |\nabla u|^2 \sim \varepsilon R^{n-1} \tag{3.4}$$

and

$$\int_{B_\varepsilon(x_0)} |\nabla u|^2 \sim \varepsilon^n. \tag{3.5}$$

The inequalities in (3.5) follow from Lipschitz continuity and linear growth as in Lemma 1.15. To prove (3.4), set $u_{\varepsilon,s} = (\min(u,\varepsilon) - s)^+$, $0 < s < \varepsilon$, $u_\varepsilon = u_{\varepsilon,0}$. Then ($B_R = B_R(x_0)$)

$$0 = \int_{B_R} u_{\varepsilon,s} \Delta u = \int_{\partial B_R} u_{\varepsilon,s} u_\nu \, d\sigma - \int_{B_R \cap \{s<u<\varepsilon\}} |\nabla u|^2.$$

Therefore, letting $s \to 0$, since u is Lipschitz and $u_{\varepsilon,s} \leq \varepsilon$,

$$\int_{B_R \cap \{0<u<\varepsilon\}} |\nabla u|^2 \leq c\varepsilon R^{n-1} .$$

To get the opposite inequality, let g be the Green's function for B_R and let

$$w(x) = -\frac{1}{(\sigma R)^n} \int_{B_{\sigma R}} g(x,y)\, dy .$$

The function w satisfies $\Delta w = (\sigma R)^{-n} \chi_{B_{\sigma R}}$ in B_R, $w_{|\partial B_R} = 0$. Then

$$w \leq c(\sigma R)^{2-n} \quad \text{in } \; B_R \setminus B_{\sigma R}$$

and

$$-w_\nu \sim R^{1-n} \quad \text{on } \; \partial B_R .$$

We write

$$\frac{1}{\varepsilon} \int_{B_R} u\, u_\varepsilon \Delta w = \frac{1}{\varepsilon} \int_{\partial B_R} u\, u_\varepsilon w_\nu \, d\sigma + \frac{1}{\varepsilon} \int_{B_R} w\, \Delta(u\, u_\varepsilon) .$$

Now, by the linear growth of u,

$$-\frac{1}{\varepsilon} \int_{\partial B_R} u\, u_\varepsilon w_\nu \, d\sigma \geq c_0 R$$

and, for σ small,

$$\left| \frac{1}{\varepsilon} \int_{B_R} u\, u_\varepsilon \Delta w \right| \leq \fint_{B_{\sigma R}} u^+ \leq \bar{c} \sigma R < \frac{c_0}{2} R .$$

Hence

$$c(\sigma R)^{2-n} \frac{1}{\varepsilon} \int_{B_R \cap \{0<u<\varepsilon\}} |\nabla u|^2 \geq \frac{1}{\varepsilon} \int_{B_R} w \Delta(u u_\varepsilon) \geq \frac{c_0}{2} R,$$

which is the desired inequality.

With (3.4) and (3.5) at hand, consider a finite overlapping family of balls $B_\varepsilon(x_j)$ for $F(u) \cap B_R$. Then

$$\begin{aligned}\int_{UB_\varepsilon(x_j)} |\nabla u|^2 &\sim \sum \int_{B_\varepsilon(x_j)} |\nabla u|^2 \\ &\sim \sum |B_\varepsilon(x_j)| \sim N\varepsilon^n \sim \sum |B_{2\varepsilon}(x_j)| \\ &\geq |\mathcal{N}_\varepsilon(F(u)) \cap B_R| \geq C \int_{UB_\varepsilon(x_j)} |\nabla u|^2 .\end{aligned}$$

Therefore the quantities (b) and (c) are comparable.

Moreover, since u is Lipschitz and nondegenerate we have

$$|\{0 < u < \varepsilon\} \cap B_R| \sim |\mathcal{N}_\varepsilon(F(u)) \cap B_R|$$

since for proper choices of c we can make

$$\mathcal{N}_{c\varepsilon}(F(u)) \cap B_R \subset \{0 < u < \varepsilon\} \cap B_R$$

or vice versa.

It follows that the quantities (a), (b) and (c) are all comparable to R^{n-1}.

Finally, let $\{B_{r_j}(x_j)\}$, $x_j \in F(u)$, be a finite covering of $F(u) \cap B_R$ by balls of radius $r_j < \varepsilon$ that approximates $H^{n-1}(F(u) \cap B_R)$. Let $r < \min r_j$ and let $\{B_r(x_j^k)\}$ be a finite overlapping covering for $F(u) \cap B_{r_j}(x_j)$. Then on one hand

$$\sum_{k,j} |\partial B_r(x_j^k)| \leq cR^{n-1}$$

by the argument above with $\varepsilon = r$. On the other hand

$$\sum_{k} |\partial B_r(x_j^k)| \geq cr_j^{n-1}$$

again by the above discussion with $R = r_j$. This implies

$$H^{n-1}(F(u) \cap B_{r_j}(x_j)) \geq cr_j^{n-1} \geq cr^{n-1}$$

and the last equivalence follows easily. □

An immediate corollary is the representation theorem (see Theorems 1.19–1.20)

$$\Delta u = gH^{n-1}_{\lfloor F(u)}$$

on $B_{1/2}$ with g H^{n-1}-measurable on $F(u) \cap B_{1/2}$ and $0 \leq c \leq g \leq C$. Clearly $g = u_\nu^+$ in the sense of measures:

$$-\int_{B_{1/2}} \nabla u \cdot \nabla \varphi = \int_{F(u) \cap B_{1/2}} \varphi g \, dH^{n-1} \quad \forall\, \varphi \in C_0^\infty(B_{1/2}) .$$

By combining Lipschitz continuity and nondegeneracy with the monotonicity formula, it is possible to prove other useful properties of a topological nature. For instance, if along the boundary of a connected component A of $\Omega^+(u)$ the zero set has uniform positive density, then A is a *nontangentially accessible domain*, an important property of A that allows one to use all the Harnack principles in Chapter 11. We will discuss this result in Section 12.3.

3.3. Strong results

"Strong" regularity results correspond to higher order regularity for solutions of elliptic equations. Asking for more than just uniform ellipticity (for instance, coefficients in C^α) will produce more regular solutions (in $C^{2,\alpha}$). Analogously, requiring more than just boundedness and strict positivity of u_ν^+ will imply higher order regularity of $F(u)$.

In order to clarify the natural starting requirements to u_ν^+, let us try to (heuristically, but see Sections 6.7–6.9) classify global solutions, somewhat as in minimal surface theory.

We know that $F^*(u)$ is, in particular, a set of finite perimeter and therefore almost every point, with respect to $H^{n-1}\lfloor F^*(u)$, is a differentiability point. That is, if x is one of those points, at x there is a well-defined normal vector $\nu = \nu(x)$ such that if we set

$$\begin{aligned} \Omega_r^+ &= \{y : r(y-x) \in \Omega^+(u)\}, \\ P^+ = P^+(x,\nu) &= \{y : \langle y-x, \nu\rangle > 0\}, \\ \pi = \pi(x,\nu) &= \{y : \langle y-x, \nu\rangle = 0\} \end{aligned}$$

and if we let $B = B(x)$ a (small) ball centered at x, then $\operatorname{Per}(\Omega_r^+ \cap B)$ converges in the sense of vector measures to $\operatorname{Per}(P^+ \cap B)$; that is, for any continuous vector field $\vec{\varphi}$,

$$\int_{\partial\Omega_r^+\cap B} \langle\vec{\varphi},\nu\rangle\, d\operatorname{Per} \xrightarrow{r\to 0} \int_\pi \langle\vec{\varphi},\nu\rangle\, dH^{n-1}\,.$$

In particular, $\partial\Omega_r^+ \cap B$ converges uniformly to $\pi \cap B$ and $\Omega_r^+ \cap B$ converges uniformly to $P^+ \cap B$.

In fact, suppose not; then there exists $\varepsilon > 0$ such that for any r_j there is $x_j \in \partial\Omega_{r_j}^+$ with $d(x_j,\pi) > \varepsilon$. Therefore $\operatorname{Per}(\partial\Omega_{r_j}^+ \cap B_{\varepsilon/2}(x_j)) \ge c\varepsilon^{n-1}$, a contradiction for r_j small.

It follows that, for any sequence $\{r_j\}$, there is a subsequence $\{r_{j_k}\}$ such that

$$u_k(y) = \frac{1}{r_{j_k}} u(r_{j_k} y)$$

converges uniformly in any compact subset of $\mathbb{R}^n$ (by the Lipschitz continuity of u). The limit u_∞ will also be a Lipschitz function, harmonic and positive in $P^+(x,\nu)$ and harmonic and negative in $P^-(x,\nu)$. Therefore it must be a two-plane solution,

$$u_\infty(y) = \alpha\langle y - x, \nu\rangle^+ - \beta\langle y - x, \nu\rangle^-$$

for some $\alpha > 0$ (by the nondegeneracy of u^+) and $\beta \geq 0$, that, in principle could depend on the particular subsequence. Furthermore it must be

$$\alpha = G(\beta)$$

so that u_∞ is a global solution of our free boundary problem.

But now, the monotonicity formula (Lemma 12.5) implies that

$$\alpha \cdot \beta = J \tag{3.6}$$

with J independent on the subsequence. Indeed, suppose that α_1, β_1 and α_2, β_2 correspond to two different converging subsequences $\{u_k^1\}$ and $\{u_k^2\}$. If $\alpha_1 > \alpha_2$, the free boundary condition implies $\beta_1 > \beta_2$, contradicting (3.6).

Thus, all candidates α and β must be the same and the limit u_∞ is unique.

This means that the sequence u_k has a nice asymptotic configuration and that the original $F(u)$ is flat around x. A flatness hypothesis is thus a natural starting point for getting higher order results. As in minimal surface theory we split the main result into two steps:

(1) to prove that in a neighborhood of a "flat" point the free boundary is a Lipschitz graph,

(2) Lipschitz free boundaries are $C^{1,\gamma}$.

We start with step (2) in the next chapter.

Chapter 4

Lipschitz Free Boundaries Are $C^{1,\gamma}$

4.1. The main theorem. Heuristic considerations and strategy

We start now to study the regularity of the free boundary in the following problem (f.b.p. in the sequel): to find a function u such that, in the cylinder $C_1 = B_1'(0) \times (-1,1)$, $B_1'(0) \subset \mathbb{R}^{n-1}$,

$$\begin{aligned} &\Delta u = 0 \text{ in } \Omega^+(u) = \{u > 0\} \text{ and } \Omega^-(u) = \{u \le 0\}^0, \\ &u_\nu^+ = G(u_\nu^-) \text{ on } F(u) = \partial\Omega^+(u) . \end{aligned} \tag{4.1}$$

We assume that $F(u)$ is given by the graph of a Lipschitz function $x_n = f(x')$, $x' \in B_1'(0)$, with Lipschitz constant L and $f(0) = 0$. We want to prove that in $B_{1/2}'(0)$, f is a $C^{1,\gamma}$-function. Precisely, the main result is the following ([C1]):

Theorem 4.1. *Let u be a viscosity solution of f.b.p. in C_1. Suppose $0 \in F(u)$ and*

(i) *$\Omega^+(u) = \{(x', x_n) : x_n > f(x')\}$ where f is a Lipschitz function with Lipschitz constant L,*

(ii) *G is continuous, strictly increasing and there exists $N > 1$ such that*

$$s^{-N}G(s)$$

is decreasing.

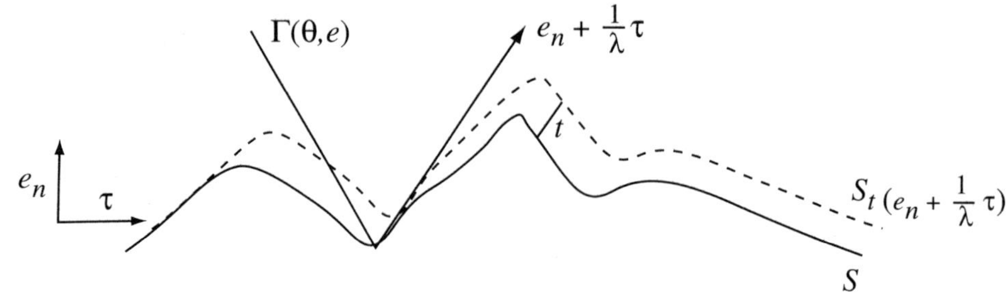

Figure 4.1

Then in $B'_{1/2}(0)$, f *is a* $C^{1,\gamma}$*-function, for some* $0 < \gamma \leq 1$, $u \in C^{1,\gamma}(\bar{\Omega}^{\pm}(u))$ *and* (4.1) *holds in a classical sense.*

Let us discuss some of the ideas that lead to the proof, showing first a connection with the theory of minimal surfaces and in particular with the following result by DeGiorgi: *Let* S *be a minimal surface in* B_1. *Assume* S *is the graph of a Lipschitz function* $x_n = w(x')$, *with Lipschitz constant* λ. *Then* S *is* $C^{1,\alpha}$ *in* $B_{1/2}$, *for some* $0 < \alpha \leq 1$.

The proof of DeGiorgi's theorem is as follows. Since

$$D_i\left(\frac{D_i w}{(1+|\nabla w|^2)^{1/2}}\right) = 0,$$

any directional derivative $v = D_\tau w$ satisfies the equation

$$D_i\left(\frac{\delta_{ij}D_j v}{(1+|\nabla w|^2)^{1/2}} - \frac{D_i w D_j w D_j v}{(1+|\nabla w|^2)^{3/2}}\right) = 0,$$

that is,

$$D_i(a_{ij}D_j v) = 0$$

where

$$a_{ij} = \frac{(1+|\nabla w|^2)\delta_{ij} - D_i w D_j w}{(1+|\nabla w|^2)^{3/2}},$$

a measurable, uniformly elliptic matrix.

Applying DeGiorgi's Hölder continuity theorem, it follows that v is Hölder continuous, achieving in this way the result. Let us stress the main lemma in the DeGiorgi proof. Let v be a solution of $D_i(a_{ij}D_j v) = 0$ in B_1 and let $|v| \leq 1$. If $|\{v < 0\}| \geq \frac{1}{2}|B_1|$, then $v \leq \lambda < 1$ in $B_{1/2}$. The possibility of rescaling and iterating this lemma gives a geometric decay of the Lipschitz constant in dyadic balls and therefore the Hölder continuity of v.

Let us rephrase the previous argument in more geometric terms. First of all, the Lipschitz continuity of w can be expressed in the following way.

Let $\Gamma(\theta, e_n)$ be the cone with axis e_n and opening θ given by

$$\lambda = \operatorname{cotan} \theta.$$

For any vector τ, denote by S_τ the surface obtained when translating S by τ. Then S is Lipschitz with constant λ if for any $\tau \in \Gamma(\theta, e_n)$, "$S_\tau$ stays above S". In other words, if $|\tau| = 1$, $\langle \tau, e_n \rangle = 0$ and $\gamma \leq 1/\lambda$, the family of surfaces $S_{t(e_n+\gamma\tau)}$, $t > 0$, stays above S. If we choose $\gamma = 1/\lambda$, it may happen that $S_{t(e_n+\gamma\tau)}$ becomes tangent to S at some point.

In principle, the comparison theorem precludes this from happening but it does not give any quantitative information. It is Harnack's inequality, applied to the first derivatives of w, that supplies what is needed. In this context Harnack's inequality says: if the distance between the surfaces S and $S_{t(e_n+\gamma\tau)}$ is of order t at one point, then it is of order t in a neighborhood of that point. Now, depending on the direction of the normal to the surface, if $\gamma = 1/\lambda$, one of the two surfaces $S_{t(e_n\pm\gamma\tau)}$ separates from S by a distance of order t, in at least half of the points. If we suppose that the "good" surface is $S_{t(e_n+\gamma\tau)}$, then by Harnack's inequality, $S_{t(e_n+\gamma\tau)}$ stays ct-away from S in $B_{1/2}$. But this is exactly what DeGiorgi's lemma says. Indeed, the tangential direction $e_n + \frac{1}{\lambda}\tau$ is no longer a critical direction in $B_{1/2}$, since it is now possible to translate S along the direction $(1-c)e_n + \frac{1}{\lambda}\tau$ staying above S.

Thus, if we adjust our system of coordinates, the Lipschitz norm of w can be improved in $B_{1/2}$ by a factor $(1-c')$. It is another instance of the fact that "ellipticity has the virtue of propagating instantaneously a perturbation all over the domain of definition of S".

Going back to the proof of Theorem 4.1, the idea is to use similar considerations and arguments to improve geometrically the Lipschitz constant of the free boundary in dyadic cylinders

$$\mathcal{C}_{2^{-k}} = B'_{2^{-k}} \times (-2^{-k}, 2^{-k}) \ .$$

We know from Section 11.4 that if in the cylinder $\mathcal{C}_1$ the free boundary is given by a Lipschitz graph $x_n = f(x')$, then, in a smaller cylinder, the solution u is increasing along every τ in a cone $\Gamma(\theta, e_n)$. The opening θ of that cone detects how flat the level sets of u are and therefore, improving the Lipschitz constant of $F(u)$ amounts to an increase in the opening θ. Our starting hypothesis (see Corollary 11.13) will be the existence of the cone of monotonicity $\Gamma(\theta, e_n)$ and we shall show that in correspondence to the dyadic cylinders $C_{2^{-k}}$ there exists a sequence of monotonicity cones $\Gamma(\theta_k, \nu_k)$ with the following properties:

(i) $\Gamma(\theta_{k+1}, \nu_{k+1}) \subset \Gamma(\theta_k, \nu_k) \qquad (\theta_0 = \theta)$.

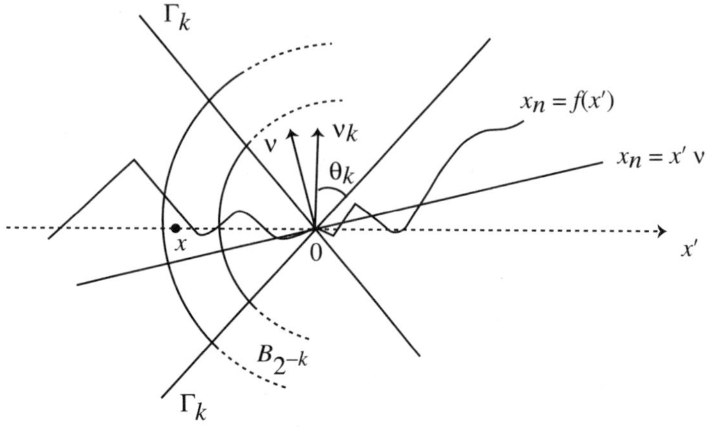

Figure 4.2. $|f(x') - x' \cdot \nu| \le c\delta_k |x|$

(ii) If $\delta_k = \frac{\pi}{2} - \theta_k$, then

$$\delta_k \le c\, b^k, \qquad c = c(\eta, \theta), \quad b = b(\eta, \theta), \quad 0 < b < 1,$$
$$|\nu_{k+1} - \nu_k| \le \delta_k - \delta_{k+1}\,.$$

This clearly implies that $\nu_k \to \nu$ and that $F(u)$ is $C^{1,\gamma}$ at 0, with normal ν. In fact, if $|x'| \approx 2^{-k}$, we have

$$|f(x') - x' \cdot \nu| \le c\delta_k |x'| \tag{4.2}$$

and the speed of δ_k gives for the modulus of continuity of ∇f,

$$\omega(r) = \sup_{|x'-y'|<r} |\nabla f(x') - \nabla f(y')| \le c\, r^\gamma$$

with $\gamma = -\log_2 b$.

The same procedure can be applied in a neighborhood of any free boundary point x_0. Since the bounds in (ii) are uniform w.r.t. $x_0 \in F(u) \cap B_{1/2}$ (say), then we conclude that $F(u) \cap B_{1/2}$ is a $C^{1,\gamma}$-graph.

Here is the main strategy of the proof:

Step 1: To improve the Lipschitz constant away from the free boundary, say, in a neighborhood of $x_0 = (\frac{3}{4}e_n)$.

Step 2: To carry the information in Step 1 to the free boundary, in $B_{1/2}$, giving up a little bit of the interior improvement.

Step 3: To rescale and repeat Steps 1 and 2, observing the invariance of the problem under elliptic dilations.

4.2. Interior improvement of the Lipschitz constant

Since we are dealing on both sides of the free boundary with two positive harmonic functions, i.e., u^+ and u^-, and since we are assuming that $F(u)$ is a Lipschitz graph w.r.t. the direction x_n, we can apply Corollary 11.13 to conclude that, in a neighborhood of the free boundary, $D_\tau u \geq 0$ along every $\tau \in \Gamma(\theta, e_n)$ with $\theta \leq \frac{1}{2}$ arcotan L or $\theta \geq \frac{\pi}{4} - \frac{1}{2}$ artan $L \equiv \theta_0$.

We call $\Gamma(\theta, e_n)$ the *monotonicity cone*. The existence of such a cone means that near $F(u)$ all the level sets of u are uniformly Lipschitz surfaces w.r.t. the same direction e_n.

We may suppose that this happens in the whole cylinder $\mathcal{C}_1$ by using, if necessary, the invariance by elliptic dilations of the problem.

We will call $\delta_0 = \frac{\pi}{2} - \theta$ the *defect angle* because it measures how far the level sets of u are from being *flat*. Notice that if $\nu = \nabla u/|\nabla u|$ and if $\alpha(\sigma, \tau)$ denotes the angle between the vectors σ, τ, we have $\alpha(\nu, e_n) \leq \delta_0$ so that

$$|\nabla u| \geq D_{e_n} u = |\nabla u| \cdot \cos\alpha(\nu, e_n) \geq |\nabla u| \cos\delta_0,$$

i.e., $D_{e_n} u$ and $|\nabla u|$ are equivalent.

To improve the Lipschitz constant means to increase the opening θ of the cone of monotonicity. This amounts to showing that there exists a monotonicity cone $\Gamma(\theta_1, \nu_1)$ containing $\Gamma(\theta, e_n)$, with $\delta_0 \leq \lambda\delta_1$, $\lambda = \lambda(n, \theta_0) < 1$.

In this section we show how it is possible to get this improvement in a neighborhood of an interior point, say $x_0 = \frac{3}{4} e_n$. The key point is to observe that the information for starting the procedure described in Section 4.1 is stored in the direction of $\nabla u(x_0)$. Indeed, let $\nu = \nu(x_0) = \nabla u(x_0)/|\nabla u(x_0)|$ and let $H(\nu)$ be the hyperplane orthogonal to ν. If $\sigma \in \Gamma(\theta, e_n)$, $|\sigma| = 1$, then $D_\sigma u(x_0) \geq 0$ and, if $\sigma \in \partial\Gamma(\theta, e_n)$, in principle, $D_\sigma u(x_0)$ may be zero. This happens if the cone is tangent to $H(\nu)$ along a generatrix in the σ direction. However, as soon as $\operatorname{dist}(H(\nu), \sigma) > 0$, then $D_\sigma u(x_0) > 0$, leaving room for an enlargement of the cone of monotonicity.

Precisely, we have

$$\frac{D_\sigma u(x_0)}{D_{e_n} u(x_0)} = \frac{\langle \sigma, \nu\rangle}{\langle e_n, \nu\rangle} \geq \langle \sigma, \nu\rangle = \operatorname{dist}(\sigma, H(\nu)). \tag{4.3}$$

From Harnack's inequality, in $C_{1/8}(x_0) = B'_{1/8}(x_0) \times (-\frac{1}{8}, \frac{1}{8})$,

$$D_\sigma u \geq c_0 \langle \sigma, \nu\rangle D_{e_n} u; \tag{4.4}$$

that is, if $\tau(\sigma)$ is the unit vector in the direction $\sigma - c_0\langle \sigma, \nu\rangle e_n$,

$$D_{\tau(\sigma)} u \geq 0 \ .$$

We show that the family $\{\tau(\sigma); \sigma \in \Gamma(\theta, e_n)\}$ contains a new cone of directions $\Gamma(\theta_1, \nu_1)$, strictly larger than $\Gamma(\theta, e_n)$. In fact, formula (4.4) implies

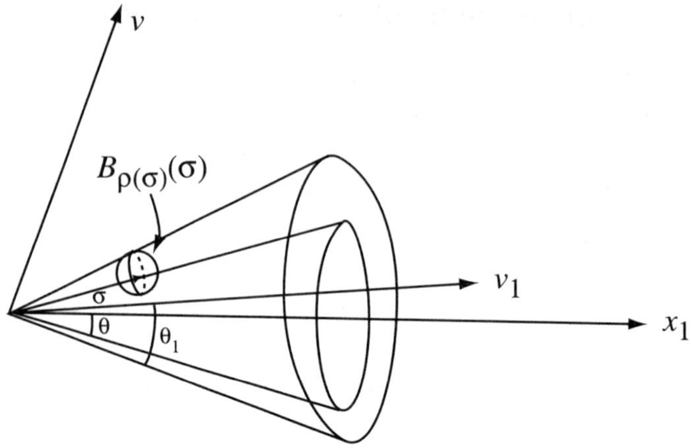

Figure 4.3. The enlarged cone

that the gain in the opening is measured by the quantity $\mathcal{E}(\sigma) = c_0\langle \sigma, \nu\rangle$, $|\sigma| = 1$, $\sigma \in \Gamma(\theta, e_n)$.

This implies that for a small $\mu > 0$, for any vector $\sigma \in \partial\Gamma(\theta, e_n)$ there exist a ball $B_{\rho(\sigma)}(\sigma)$ where

$$\rho(\sigma) = |\sigma|\mu\langle\sigma,\nu\rangle = |\sigma|\mu \sin(E(\sigma)),$$
$$E(\sigma) = \frac{\pi}{2} - \alpha(\sigma,\nu)\ ,$$

such that the directional derivative of u is nonnegative along any vector in $B_{\rho(\sigma)}(\sigma)$. The envelope of the balls $B_{\rho(\sigma)}(\sigma)$ contains a cone $\Gamma(\theta_1, \nu_1)$ that contains $\Gamma(\theta, e_n)$ and with an opening $\theta_1 > \theta$.

This is precisely stated and proved in the following theorem.

Theorem 4.2 (Intermediate cone). *Let $0 < \theta_0 < \theta < \frac{\pi}{2}$ and, for a unit vector ν, let $H(\nu)$ be the hyperplane perpendicular to ν. Assume that the cone $\Gamma(\theta, e) \subset H(\nu)$ and for any $\sigma \in \bar{\Gamma}(\theta, e)$ put*

$$E(\sigma) = \frac{\pi}{2} - \alpha(\sigma,\nu)\ .$$

Moreover, for a small positive μ put

$$\rho(\sigma) = |\sigma|\mu\sin(E(\sigma))\ , \qquad S_\mu = \bigcup_{\sigma\in\bar{\Gamma}(\theta,e)} B_{\rho(\sigma)}(\sigma)\ .$$

Then there exist $\bar{\theta}$ and $\lambda = \lambda(\mu, \theta_0) < 1$ such that

$$\Gamma(\theta, e) \subset \Gamma(\bar{\theta}, \bar{e}) \subset S_\mu$$

and

$$\frac{\pi}{2} - \bar{\theta} \leq \lambda\Big(\frac{\pi}{2} - \theta\Big)\ .$$

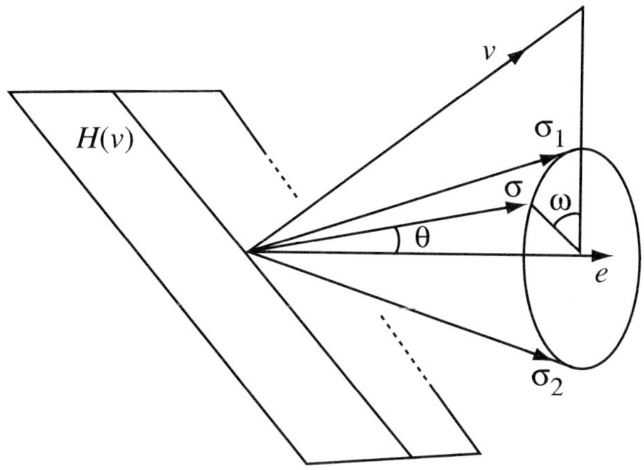

Figure 4.4

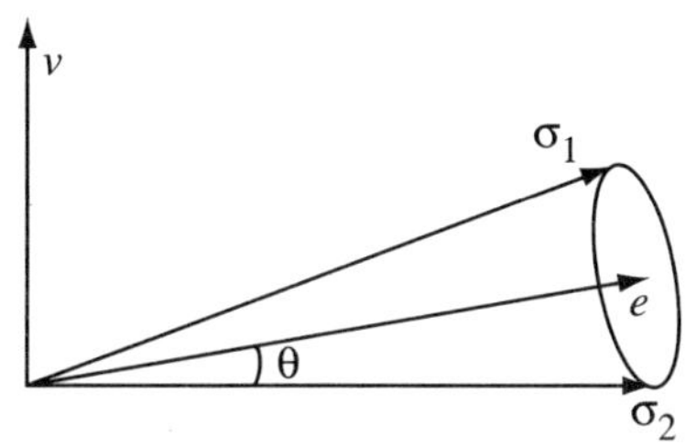

Figure 4.5. The worst situation: $\langle\sigma_2, \nu\rangle = 0$

Proof. Put $\delta = \frac{\pi}{2} - \theta$ and let σ_1, σ_2 (unit vectors) be the two generatrices of $\Gamma(\theta, e)$ belonging to $\operatorname{span}\{\nu, e\}$. Suppose that σ_1 is the nearest to ν of the two. Thus,

$$\alpha(\sigma_1, \nu) \le \frac{\pi}{2} - 2\theta\,, \quad \alpha(\sigma_2, \nu) \le \frac{\pi}{2}\,. \tag{4.5}$$

These two directions give the maximum and the minimum gain in the opening of the cone $\Gamma(\theta, e)$, respectively. By replacing, if necessary, ν by $\bar\nu$ such that $\bar\nu \in \operatorname{span}\{e, \nu\}$, $|\,\bar\nu\,| = 1$, $\langle\bar\nu, \sigma_2\rangle = 0$, we reduce ourselves to the equality case in (4.5). This case, indeed, is the worst possible since $\langle\sigma, \bar\nu\rangle \le \langle\sigma, \nu\rangle$ for any $\sigma \in \Gamma(\theta, e)$, diminishing the opening gain in each direction. Assume therefore that equality holds in (4.5). In this case $\langle\sigma_2, \nu\rangle = 0$ (no gain), while

$$\langle\sigma_1, \nu\rangle = 2\sin\delta\sin\theta \ge 2\sin\theta_0\sin\delta \qquad \text{(maximum gain)}$$

so that

$$\rho(\sigma_1) \ge 2\mu\sin\theta_0\sin\delta$$

and in the plane $\operatorname{span}\{\nu, e\}$ we have an increase in angle estimated from below by $C_0(\mu, \theta_0)\delta$.

Now consider a generatrix $\sigma \in \partial\Gamma(\theta, e)$, $|\sigma| = 1$, and let ω be the solid angle between the planes span$\{e, \sigma\}$ and span$\{e, \nu\}$. Then from the cosine law of spherical trigonometry we have

$$\begin{aligned}\langle\sigma, \nu\rangle &= \cos\alpha(e,\nu)\cdot\cos\theta + \sin\alpha(e,\nu)\sin\theta\cdot\cos\omega \\ &= \sin\delta\cos\delta(1+\cos\omega) \geq \sin\theta_0(1+\cos\omega)\sin\delta\ .\end{aligned}$$

Therefore, if $\omega \leq \frac{99}{100}\pi$ (say), we can say that the increase in angle is estimated from below by $C_1(\mu, \theta_0)\delta$.

Put $\bar{e} = \gamma\delta e^1 + e$, where $e^1 \in$ span$\{e, \nu\}$, $|e^1| = 1$, $\langle e^1, e\rangle = 0$, $\gamma \leq \frac{1}{3}C_1(\mu, \theta_0)$, and let

$$S'_\mu = \{\bar{\sigma} : \bar{\sigma} = \sigma + \rho(\sigma)\sigma^1,\ \sigma \in \partial\Gamma(\theta, e)\}$$

where $|\sigma^1| = 1$, $\langle\sigma^1, \sigma\rangle = 0$, $\sigma^1 \in$ span$\{e, \sigma\}$. Then, if $\gamma = \gamma(\mu, \theta_0)$ is small enough, for every $\bar{\sigma} \in \partial S'_\mu$,

$$\alpha(\,\bar{\sigma}, \bar{e}\,) \geq \theta + \gamma\delta \equiv \bar{\theta}\ .$$

Thus $S'_\mu \subset S_\mu$ and contains the cone $\Gamma(\,\bar{\theta}, \bar{e})$ with

$$\frac{\pi}{2} - \bar{\theta} \leq (1-\gamma)\Big(\frac{\pi}{2} - \theta\Big)\ . \qquad \square$$

Remark 4.3. Theorem 4.2 holds also if we fix any θ', $\frac{\theta}{2} \leq \theta' < \theta$, and put, for every $\sigma \in \bar{\Gamma}(\theta', e)$,

$$\begin{aligned}E(\sigma) &= \frac{\pi}{2} - \alpha(\sigma,\nu) - (\theta - \theta'), \\ \rho(\sigma) &= |\sigma|\sin\big(\theta - \theta' + \mu E(\sigma)\big), \\ S_\mu &= \bigcup_{\sigma\in\bar{\Gamma}(\theta', e)} B_{\rho(\sigma)}(\sigma).\end{aligned}$$

The constant λ still depends only on μ and θ_0.

This remark allows a better control of the opening gain and it will be useful in the sequel.

Applying Theorem 4.2 to our situation, we get

Lemma 4.4 (Interior gain). *There exists a cone $\Gamma(\bar{\theta}_1, \bar{\nu}_1) \supset \Gamma(\theta, e_n)$ with*

$$\bar{\delta}_1 \leq \bar{\lambda}\delta_0 \qquad \Big(\bar{\delta}_1 = \frac{\pi}{2} - \bar{\theta}_1\Big)$$

where $\bar{\lambda} = \bar{\lambda}(\theta_0, n) < 1$, such that, in $C_{1/8}(x_0)$

$$D_\sigma u(x) \geq 0$$

for every $\sigma \in \Gamma(\bar{\theta}_1, \bar{\nu}_1)$.

The question is now how to propagate the information contained in Lemma 4.4 to the free boundary.

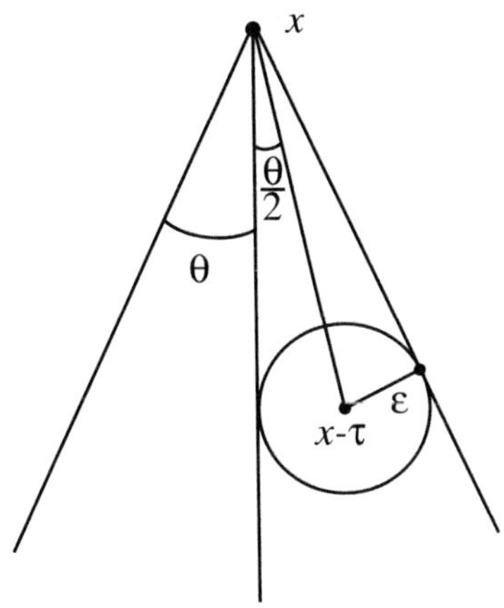

Figure 4.6

4.3. A Harnack principle. Improved interior gain

The monotonicity properties of u can be reformulated in a more flexible form by introducing a suitable function that measures the cone opening. Indeed, let $\tau \in \Gamma(\frac{\theta}{2}, e_n)$ be a vector with small norm and put

$$\varepsilon = |\tau| \sin \frac{\theta}{2} \ .$$

Then the monotonicity of u along the directions of $\Gamma(\theta, e_n)$ amounts to asking that

$$v_\varepsilon(x) \equiv \sup_{B_\varepsilon(x)} u(y - \tau) \leq u(x)$$

for $x \in C_{1-\varepsilon}$ and every (small) $\tau \in \Gamma(\frac{\theta}{2}, e_n)$.

In terms of v_ε one can refine Lemma 4.4, thanks to the following Harnack-type lemma.

Lemma 4.5 (Harnack principle). *Let $0 \leq u_1 \leq u_2$ be harmonic functions in $B_R = B_R(0)$. Let $\varepsilon \leq R/8$ and assume that in $B_{R-\varepsilon}$*

$$v_\varepsilon(x) = \sup_{B_\varepsilon(x)} u_1(y) \leq u_2(x) \tag{4.6}$$

and furthermore

$$v_\varepsilon(0) \leq (1 - b\varepsilon) u_2(0) \qquad (b > 0). \tag{4.7}$$

Then for some $\bar{c} = \bar{c}(R)$, $\mu = \mu(R, n)$, we have in $B_{\frac{3}{4}R}$

$$v_{(1+\mu b)\varepsilon}(x) \leq u_2(x) - \bar{c} b \varepsilon u_2(0). \tag{4.8}$$

Proof. For $|\sigma| < 1$

$$w(x) = u_2(x) - u_1(x + \varepsilon\sigma)$$

is harmonic and positive (by (4.6)) in $B_{R-\varepsilon}$. By Harnack's inequality and (4.6), in $B_{\frac{3}{4}R}$,

$$w(x) \geq cw(0) \geq cb\varepsilon u_2(0) \ .$$

Shauder estimates and Harnack's inequality again give

$$|\nabla u_1(x)| \leq \frac{c}{R}u_1(0) \leq \frac{c}{R}u_2(0)$$

in $B_{\frac{3}{4}R}$. It follows that

$$\begin{aligned} u_2(x) - u_1(x + (1 + \mu b)\varepsilon\sigma) &= w(x) + u_1(x + \varepsilon\sigma) - u_1(x + (1 + \mu b)\varepsilon\sigma) \\ &\geq cb\varepsilon u_2(0) - \frac{c\mu b}{R}\varepsilon u_2(0) \\ &\geq \bar{c}b\varepsilon u_2(0) \end{aligned}$$

if $\mu = \mu(R, n)$ is chosen small. □

We apply Lemma 4.5 in $B_{1/6}(x_0)$ to

$$u_1(x) = u(x - \tau) \quad \text{and} \quad u_2(x) = u(x) \ .$$

The only nontrivial hypothesis to check is (4.7).

Let $y \in B_\varepsilon(x)$ and notice that if $\tau \in \Gamma(\frac{\theta}{2}, e_n)$ and

$$\bar{\tau} = \tau - (y - x),$$

then $\alpha(\tau, \bar{\tau}) \leq \frac{\theta}{2}$, since $|\bar{\tau} - \tau| = |x - y| \leq |\tau| \sin\frac{\theta}{2}$. Also

$$|\bar{\tau}| \geq |\tau| - |\tau| \sin\frac{\theta}{2} \geq \frac{1}{2}|\tau|$$

since $\frac{\theta}{2} < \frac{\pi}{4}$. Therefore $D_{\bar{\tau}}u \geq 0$ and using Harnack's inequality for both $D_{\bar{\tau}}u$ and u, together with Theorem 11.10, we deduce that

$$\begin{aligned} \inf_{B_{1/8}(x_0)} D_{\bar{\tau}}u \geq c_0\langle\nu, \bar{\tau}\rangle|\nabla u(x_0)| &\geq c\langle\nu, \bar{\tau}\rangle u(x_0) \\ &\geq c_1|\bar{\tau}| \cos\alpha(\nu, \bar{\tau})\Big(\sup_{B_{1/8}(x_0)} u\Big) \\ &\geq b\varepsilon \sup_{B_{1/8}(x_0)} u \end{aligned}$$

where $b = b(\tau) = C\cos(\frac{\theta}{2} + \alpha(\nu, \tau))$.

It follows that, for every $x \in B_{1/8}(x_0)$,

$$u(x - \bar{\tau}) \leq u(x) - D_{\bar{\tau}}u(\tilde{x}) \leq (1 - b\varepsilon)u(x) \ ,$$

which gives, in particular, for $\varepsilon < \frac{1}{100}$ (say),

$$\sup_{B_\varepsilon(x_0)} u(y - \tau) \leq (1 - b\varepsilon)u(x_0)$$

and the hypotheses of Lemma 4.5 are satisfied.

We conclude that

Lemma 4.6. *There exist positive constants $\bar{c}$ and μ, depending only on θ_0, n, such that, for each small vector $\tau \in \Gamma(\frac{\theta}{2}, e_n)$ and every $x \in B_{1/8}(x_0)$*

$$\sup_{B_{(1+\mu b)\varepsilon}(x)} u(y-\tau) \le u(x) - \bar{c} b \varepsilon u(x_0). \tag{4.9}$$

Notice that (4.9) gives a quantitative estimate of the ε-shift between the level sets of u and those of its τ-translation and implies Lemma 4.4, perhaps with a slightly different enlarged cone, which we still denote by $\Gamma(\bar{\theta}_1, \bar{\nu}_1)$.

To see this, observe that, for $\theta_0 \le \theta < \frac{\pi}{2}$, in the notation of Theorem 4.2 and Remark 4.3, with $\theta' = \theta/2$,

$$\begin{aligned}
(1+b\mu)\varepsilon &= |\tau|\Big(\sin\frac{\theta}{2}\Big)\Big[1 + c\mu\cos\Big(\frac{\theta}{2} + \alpha(\nu,\tau)\Big)\Big] \\
&= |\tau|\Big(\sin\frac{\theta}{2}\Big)\left[1 + c\mu \sin E(\tau)\right] \\
&\ge |\tau| \sin\Big(\frac{\theta}{2} + \bar{\mu} E(\tau)\Big) \equiv \rho(\tau)
\end{aligned}$$

with $\bar{\mu} = \mu c \frac{\theta_0}{2}$.

This means that if

$$S_{\bar{\mu}} = \bigcup_{\tau \in \Gamma(\frac{\theta}{2}, e_n)} B_{\rho(\tau)}(\tau)$$

and $\tau \in S_{\bar{\mu}}$, then

$$D_\tau u \ge 0$$

and, in particular, $\Delta v \ge 0$ also in the intermediate cone $\Gamma(\bar{\theta}_1, \bar{\nu}_1)$. This completes Step 1.

4.4. A continuous family of R-subsolutions

At this point the situation is as follows:

- in $B_{1-\varepsilon}$

$$v_\varepsilon(x) \le u(x) \tag{4.10}$$

  which amounts to the monotonicity of u along the direction of the original cone $\Gamma(\theta, e_n)$,

- in $B_{1/8}(x_0)$, $x_0 = \frac{3}{4} e_n$,

$$v_{(1+b\mu)\varepsilon}(x) \le u(x) - \bar{c}\varepsilon b u_2(x_0) \tag{4.11}$$

  with $b = b(\tau) = c\cos(\frac{\theta}{2} + \alpha(\nu,\tau))$, $\tau \in \Gamma(\frac{\theta}{2}, e_n)$, which implies the monotonicity of u in the larger cone $\Gamma(\bar{\theta}_1, \bar{\nu}_1)$.

The purpose is now to carry this information to the free boundary by finding for instance that, for some intermediate $\bar{\mu}$, an inequality of the type

$$v_{(1+b\bar{\mu})\varepsilon}(x) \leq u(x)$$

holds in $B_{1/2}$.

The idea is to use a continuous deformation method based on the comparison Theorem 2.2 to transfer the improvement in $B_{1/8}(x_0)$ to $B_{1/2}(0)$. The key point is the construction of a delicate family of subsolutions of the type considered in (2.16) but with the radius of the ball $B_t(x)$ dependent on x itself, i.e., $t = \varphi(x)$. In fact, the family

$$v_t(x) = \sup_{B_t(x)} u$$

with t constant can only detect a uniform enlargement of the monotonicity cone, and, as such, one cannot exploit the interior gain.

For this purpose we ask the question: what are the conditions on a variable radius $t(x)$ so that for any harmonic function u, v_t will always be subharmonic.

Here is the fundamental lemma.

Lemma 4.7. *Let φ be a C^2-positive function satisfying in B_1*

$$\Delta\varphi \geq \frac{C|\nabla\varphi|^2}{\varphi} \tag{4.12}$$

for $C = C(n)$ large enough. Let u be continuous, defined in a domain Ω so large that the function

$$w(x) = \sup_{|\sigma|=1} u(x + \varphi(x)\nu)$$

is well defined in B_1.

Then if u is harmonic in $\{u > 0\}$, w is subharmonic in $\{w > 0\}$.

Proof. Let us normalize the situation assuming that $w(0) > 0$, $\varphi(0) = 1$ and $w(0) = \sup_{B_1} u$ is attained at $x = e_n$. We will show that

$$\liminf_{r\to 0} \frac{1}{r^2}\left[\fint_{B_r} w(x)\, dx - w(0)\right] \geq 0\,.$$

Choose the system of coordinates so that

$$\nabla\varphi(0) = \alpha e_1 + \beta e_n\,.$$

We estimate $w(x)$ from below for x near the origin by

$$w(x) \geq u(x + \varphi(x)\sigma)$$

with an appropriate choice of $\sigma = \frac{\sigma^*}{|\sigma^*|}$, given by

$$\sigma^* = \sigma^*(x) = e_n + (\beta x_1 - \alpha x_n)e_1 + \gamma \sum_{i=2}^{n-1} x_i e_i$$

where γ is to be chosen later.

Notice that

$$|\sigma^*|^2 = 1 + (\beta x_1 - \alpha x_n)^2 + \gamma^2 \sum_{i=1}^{n-1} x_i^2 .$$

Put $y(x) = x + \varphi(x)\sigma(x)$. We have $y(0) = e_n$ and

$$\begin{aligned} y(x) = x + &\left\{1 + \langle \nabla\varphi(0), x\rangle + \frac{1}{2} D_{ij}\varphi(0)x_i x_j + o(|x|^2)\right\} \\ &\cdot \left\{e_n + (\beta x_1 - \alpha x_n)e_1 + \gamma \sum_{i=2}^{n-1} x_i e_i\right\} \\ &\cdot \left\{1 - \frac{1}{2}(\beta x_1 - \alpha x_n)^2 - \frac{1}{2}\gamma^2 \sum_{i=2}^{n-1} x_i^2 + o(|x|^4)\right\} . \end{aligned}$$

We can write the above expression as

$$y(x) = e_n + \{\text{first order terms}\} + \{\text{quadratic terms}\} + o(|x|^2) .$$

The first order term is

$$y_1(x) = x + (\alpha x_1 + \beta x_n)e_n + (\beta x_1 - \alpha x_n)e_1 + \gamma \sum_{i=2}^{n-1} x_i e_i ,$$

which can be written in the form

$$y_1(x) = Mx$$

where

$$M = \begin{pmatrix} 1+\beta & & & & -\alpha \\ 0 & 1+\gamma & & & 0 \\ \vdots & & \ddots & & \vdots \\ 0 & & & 1+\gamma & 0 \\ \alpha & 0 & \cdots & 0 & 1+\beta \end{pmatrix} .$$

Since $\det M = (1+\gamma)^{n-2}[(1+\beta)^2 + \alpha^2]$, if we choose γ such that

$$(1+\gamma)^2 = (1+\beta)^2 + \alpha^2 ,$$

the transformation

$$x \to y_1(x)$$

can be thought of as a rotation (given by the matrix $M/(1+\gamma)$) followed by a $(1+\gamma)$-dilation.

Put

$$y^*(x) = e_n + y_1(x)\ .$$

Then the quadratic term is given by

$$y(x) - y^*(x) = \frac{1}{2}\left[D_{ij}\varphi(0)x_ix_j - (\beta x_1 - \alpha x_n)^2 - \gamma^2\sum_{i=2}^{n-1} x_i^2\right] e_n + O(|\nabla\varphi(0)|^2|x|^2)e_0$$

where $e_0 \perp e_n$ and $|e_0| = 1$. Then

$$\begin{aligned}\fint_{B_r} w(x)\,dx - w(0) &\geq \fint_{B_r} u(y(x))\,dx - u(y(0))\\ &= \fint_{B_r} [u(y(x)) - u(y^*(x))]\,dx\\ &\quad + \fint_{B_r} [u(y^*(x)) - u(y(0))]\,dx\\ &= \fint_{B_r} [u(y(x)) - u(y^*(x))]\,dx\end{aligned}$$

since $u(y^*(x))$ is harmonic.

Evaluate now $u(y) - u(y^*)$. Observe first that since $w(0) = u(e_n) = u(y(0))$, $\nabla u(y(0))$ must point in the direction of e_n. Then

$$\begin{aligned}u(y) - u(y^*) &= \nabla u(y^*)\cdot(y - y^*) + O(|y - y^*|^2)\\ &= \nabla u(e_n)\cdot(y \quad y^*) + O(|y - y^*|^2)\\ &= \frac{1}{2}|\nabla u(e_n)|\cdot\left\{D_{ij}\varphi(0)x_ix_j - (\beta x_1 - \alpha x_n)^2 - \gamma^2\sum_{i=2}^{n-2} x_i^2\right\}\\ &\quad + O(|x|^4)\end{aligned}$$

and hence

$$\frac{1}{r^2}\fint_{B_r} [u(y) - u(y^*)]\,dx = \frac{1}{2n}|\nabla u(e_n)|\{\Delta\varphi(0) - [\beta^2 + \alpha^2 + (n-2)\gamma^2]\} \geq 0$$

if

$$\Delta\varphi(0) \geq |\nabla\varphi(0)|^2 + (n-2)\gamma^2\ .$$

Since $\gamma^2 \leq |\nabla\varphi(0)|^2$, the last inequality is satisfied if

$$\Delta\varphi(0) \geq c|\nabla\varphi(0)|^2$$

with $c \geq n - 1$. □

Remark 4.8. We point out that for a C^2-positive function φ,

$$\varphi\Delta\varphi = C|\nabla\varphi|^2$$

if and only if

$$\varphi^{1-c} \text{ is harmonic.}$$

In particular, if φ^{1-c} takes values a_1 on $\partial B_1(0)$ and a_2 on $\partial B_{1/8}(x_0)$, it will take intermediate values "in between", that is, strictly inside $B_1(0) \setminus B_{1/8}(x_0)$, and so will φ.

Given a solution u of our free boundary problem and a function φ satisfying the properties of Lemma 4.7, we consider the function v_φ defined by

$$v_\varphi(x) = \sup_{B_{\varphi(x)}(x)} u(y) \ .$$

v_φ is continuous and we know that v_φ is subharmonic both in $\{v_\varphi < 0\}$ and $\Omega^+(v_\varphi) = \{v_\varphi > 0\}$. To complete the comparison, we need to examine which kind of condition v_φ satisfies on $F(v_\varphi) = \partial\Omega^+(v_\varphi)$. We start with the asymptotic behavior of v_φ at $F(v_\varphi)$.

Lemma 4.9. *Let u be a continuous function and let*

$$v_\varphi(x) = \sup_{B_{\varphi(x)}(x)} u$$

where φ is a positive C^2-function with $|\nabla\varphi| < 1$. Assume that

$$x_1 \in \partial\Omega^+(v_\varphi) \ , \quad y_1 \in \partial\Omega^+(u)$$

and that (Figure 4.6)

$$y_1 \in \partial B_{\varphi(x_1)}(x_1) \ .$$

Then

(a) *x_1 is a regular point from the right for $F(v_\varphi)$,*

(b) *If $\nu = \frac{y_1 - x_1}{|y_1 - x_1|}$ and near y_1, nontangentially,*

$$u^+(y) = \alpha\langle y - y_1, \nu\rangle^+ + o(|y - y_1|) \tag{4.13}$$

or

$$u^-(y) = \beta\langle y - y_1, \nu\rangle^- + o(|y - y_1|), \tag{4.14}$$

then near x_1, nontangentially,

$$v^+(x) \geq \alpha\langle x - x_1, \nu + \nabla\varphi(x_1)\rangle^+ + o(|x - x_1|) \tag{4.15}$$

or

$$v^-(x) \leq \beta\langle x - x_1, \nu + \nabla\varphi(x_1)\rangle^- + o(|x - x_1|) \ . \tag{4.16}$$

(c) *If $F(u)$ is a Lipschitz graph, with Lipschitz constant λ, and $|\nabla\varphi|$ is small enough (i.e., $|\nabla\varphi| \leq c(\lambda) \ll 1$), then $F(v_\varphi)$ is a Lipschitz graph with Lipschitz constant*

$$\lambda' \leq \lambda + c_1 \sup|\nabla\varphi| \ .$$

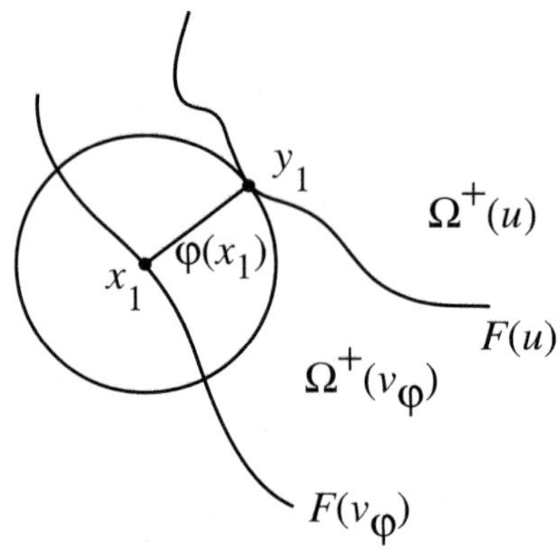

Figure 4.7

Proof. (a) Notice that $\Omega^+(v_\varphi)$ contains the set

$$K = \{|x - y_1|^2 < \varphi(x)^2\}$$

since for $|x - y_1| < \varphi(x)$, we have $v_{\varphi(x)}(x) > u(y_1) > 0$. The boundary of K is a C^2-surface, since along ∂K

$$\nabla(|x - y_1|^2 - \varphi(x)^2) = 2(x - y_1 - \varphi(x)\nabla\varphi(x)) \neq 0$$

because $|\nabla\varphi| < 1$. Now, $x_1 \in \partial K$ so that (a) is proven.

(b) Near x_1

$$\varphi(x) = \varphi(x_1) + \langle x - x_1, \nabla\varphi(x_1)\rangle + o(|x - x_1|) \ .$$

Hence, if

$$y = x + \varphi(x)\nu$$

and (4.13) holds, we have, since $y_1 = x_1 + \varphi(x_1)\nu$,

$$\begin{aligned} v_\varphi(x) \geq u(y) &= \alpha\langle x + \varphi(x)\nu - y_1, \nu\rangle^+ + o(|y - y_1|) \\ &= \alpha\langle x - x_1 + [\varphi(x) - \varphi(x_1)]\nu, \nu\rangle^+ + o(|x - x_1|) \\ &= \alpha\langle x - x_1, \nu + \nabla\varphi(x_1)\rangle^+ + o(|x - x_1|) \ . \end{aligned}$$

In the same way, (4.14) implies (4.16).

(c) $\Omega^+(u)$ is the union of convex cones with vertices on $F(u)$ and therefore we can suppose that $\Omega^+(u)$ is above the graph of a smooth convex cone $x_n = f(x')$, $x' \in \mathbb{R}^{n-1}$.

Then $\nu = \frac{y_1 - x_1}{|y_1 - x_1|}$ is the inner unit normal to a supporting plane π to $F(u)$ at y_1 and it must lie in a cone with axis e_n and opening artan λ.

On the other hand, the surfaces $S_1 = \partial K$ and

$$S_2 = \{\text{dist}(x, \pi)^2 = \varphi(x)^2\}$$

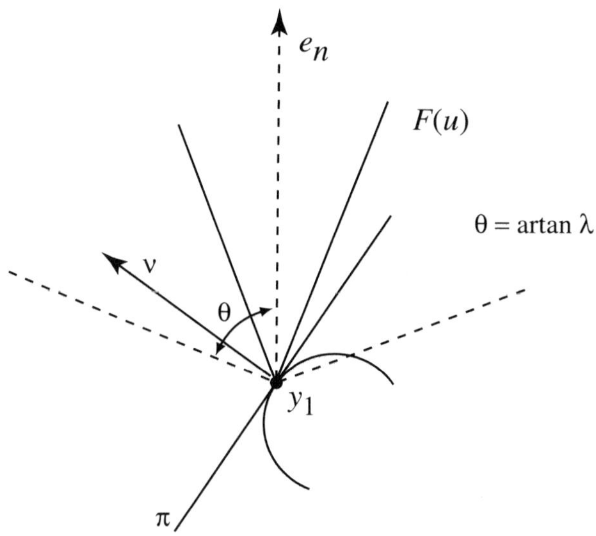

Figure 4.8

are tangent to $F(v_\varphi)$ at x_1 from above and from below, respectively. Indeed, if $x \in F(v_\varphi)$, $\operatorname{dist}(x, F(u)) = \varphi(x)$ so that $\operatorname{dist}(x, \pi) \leq \varphi(x)$.

Both surfaces are smooth with unit normal vector at x_1 parallel to

$$\bar{\nu} = \nu + \nabla\varphi(x_1) .$$

Therefore, if $a(\tau_1, \tau_2)$ denotes the angle between the vectors τ_1, τ_2,

$$\begin{aligned} a(\bar{\nu}, e_n) &\leq a(\nu, e_n) + a(\nu, \bar{\nu}\,) \\ &\leq \operatorname{artan}\lambda + \operatorname{arsin}|\nabla\varphi(x_1)| \\ &\leq \operatorname{artan}\lambda + c_0|\nabla\varphi(x_1)|. \end{aligned}$$

Now, since $|\nabla\varphi| < 1$,

$$\tan(\operatorname{artan}\lambda + c_0|\nabla\varphi(x_1)|) \leq \frac{\lambda + c_1|\nabla\varphi(x_1)|}{1 - \lambda c_1|\nabla\varphi(x_1)|}.$$

If $|\nabla\varphi| \leq \frac{1}{c_1(1+\lambda)}$, then

$$\tan a(\bar{\nu}, e_n) \leq \lambda + c_1|\nabla\varphi(x_1)|;$$

that is, $F(v_\varphi)$ is Lipschitz with Lipschitz constant

$$\lambda' \leq \lambda + c_1 \sup|\nabla\varphi| . \qquad \square$$

An important corollary is

Lemma 4.10. *Let u be a viscosity solution of our free boundary problem. If φ satisfies the hypotheses of Lemmas 4.9 and 4.7, then the following hold.*

(a) *v_φ is subharmonic in both $\Omega^+(v_\varphi)$ and $\Omega^-(v_\varphi)$.*

(b) *Every point of $F(v_\varphi)$ is regular from the right.*

(c) *At every point $x_1 \in F(v_\varphi)$, v_φ satisfies the asymptotic inequality*

$$v_\varphi(x) \ge \bar\alpha\langle x - x_1, \bar\nu\rangle^+ - \bar\beta\langle x - x_1, \bar\nu\rangle^- + o(|x - x_1|) \tag{4.17}$$

with

$$\frac{\bar\alpha}{1 - |\nabla\varphi(x_1)|} \ge G\left(\frac{\bar\beta}{1 + |\nabla\varphi(x_1)|}\right). \tag{4.18}$$

Proof. Let $v_\varphi(x_1) = u(y_1)$. Then $y_1 \in F(u)$ and (4.13), (4.14) hold with $\alpha \ge G(\beta)$. Put

$$\bar\nu = \frac{\nu + \nabla\varphi(x_1)}{|\nu + \nabla\varphi(x_1)|}, \qquad \bar\alpha = \alpha|\nu + \nabla\varphi(x_1)|\ , \ \bar\beta = \beta|\nu + \nabla\varphi(x_1)|\ .$$

Then from (4.15), (4.16)

$$v_\varphi(x) \ge \bar\alpha\langle x - x_1, \bar\nu\rangle^+ - \bar\beta\langle x - x_1, \nu\rangle^- + o(|x - x_1|)$$

and $\alpha \ge G(\beta)$ gives (4.18). □

Inequality (4.18) says that v_φ is "almost" a subsolution, due to the fact that $\nabla\varphi$ is not identically zero. We shall later perturb v_φ to make it a subsolution. For the moment we will construct a family of functions φ_t, satisfying the hypotheses of Lemmas 4.9 and 4.7, such that $v_{\varepsilon\varphi_t}$ carries the monotonicity gain from $B_{1/8}(x_0)$ to the free boundary as t goes from 0 to 1.

This means that we want $\varphi_t = 1$ along, say, ∂B_1, $\varphi_t \approx 1 + ctb$ on $\partial B_{1/8}(x_0)$ and $\varphi_t \approx 1 + \bar\mu tb$ in $B_{1/2}$.

Lemma 4.11. *Let $0 < r \le \frac{1}{8}$. Then there exist positive $\lambda = \lambda(r)$, $h = h(r)$ and a C^2 family of functions φ_t, $0 \le t \le 1$, defined in $\bar B_1 \setminus B_{r/2}(\frac{3}{4}e_n)$ such that*

(i) $1 \le \varphi_t \le 1 + th$,

(ii) $\varphi_t\Delta\varphi_t \ge C|\nabla\varphi_t|^2$,

(iii) $\varphi_t \equiv 1$ *outside* $B_{7/8}$,

(iv) $\varphi_{t|B_{1/2}} \ge 1 + \lambda th$,

(v) $|\nabla\varphi_t| \le Cth$.

Proof. Recalling Remark 4.8, let ψ_0 be a smooth superharmonic function in $\bar B_1 \setminus B_{r/2}(\frac{3}{4}e_n)$ with, say, $\psi_0 = 1$ on $\partial B_{r/2}(\frac{3}{4}e_n)$, $\psi_0 \equiv 2$ outside $B_{7/8}$, $1 \le \psi_0 \le 2$ in $\bar B_1$ and $\psi_0 \le 2 - \gamma$ $(0 < \gamma < 2)$ on $\bar B_{1/2}$. Choose $c > 1$ and put

$$\varphi_0 = \psi_0^{1/1-c}\ .$$

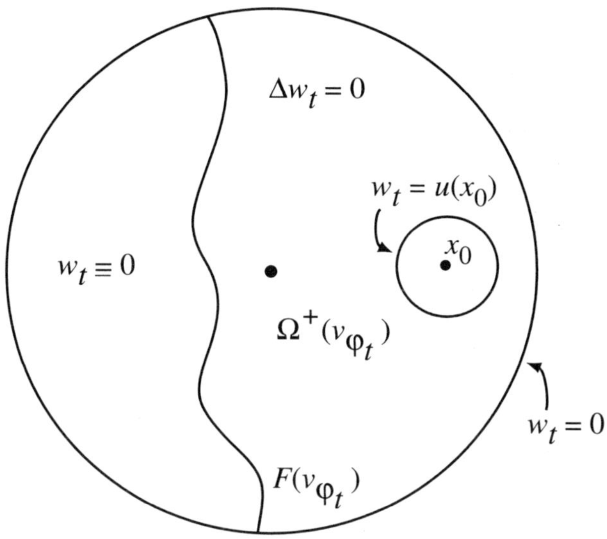

Figure 4.9

Then $\varphi_0 \Delta\varphi_0 \geq C|\nabla\varphi_0|^2$, $2^{1/1-c} \leq \varphi_0 \leq 1$ in $\bar{B}_1$, $\varphi_0 \equiv 2^{1/1-c}$ outside $B_{7/8}$ and $\varphi_0 - 2^{1/1-c} \geq C(\gamma) > 0$ in $B_{1/2}$. It is now easy to check that the family

$$\varphi_t = 1 + th\frac{\varphi_0 - 2^{1/1-c}}{1 - 2^{1/1-c}} \qquad (0 \leq t \leq 1)$$

satisfies (i)–(v), provided h is small enough. □

We now go back to the solution u of our free boundary problem and construct the family v_{φ_t} with φ_t as in Lemma 4.10. We will perturb v_{φ_t} by adding a correction term that makes it a family of subsolutions for which the comparison Theorem 2.2 applies. This correction term is a multiple of the harmonic measure of $\partial B_{1/8}(x_0)$ with respect to the domain $\{v_{\varphi_t} > 0\}$, extended by zero in the complement.

Lemma 4.12. *Let u be a solution of our free boundary problem and let φ_t be the family constructed in Lemma 4.11 with $r = 1/8$. Let w_t be a continuous function in $\Omega = \bar{B}_{9/10} \setminus B_{1/8}(x_0)$ defined by (Figure 4.9)*

$$\begin{cases} \Delta w_t = 0 \quad in \quad \Omega^+(v_{\varphi_t}) \cap \Omega \equiv \Omega_t, \\ w_t \equiv 0 \quad in \quad \bar{\Omega} \setminus \Omega_t, \\ w_t = 0 \quad on \quad \partial B_{9/10}, \\ w_t = u(x_0) \quad on \quad \partial B_{1/8}(x_0). \end{cases}$$

Then for a small constant c, h *and any* $\varepsilon > 0$ *small enough,*

$$V_t = v_{\varepsilon\varphi_t} + c\varepsilon w_t \qquad (0 \le t \le 1)$$

is a family of subsolution.

Proof. The subharmonicity of V_t in $\Omega^+(V_t)$ and $\Omega^-(V_t)$ follows from Lemma 4.7. We have to check that V_t has the correct asymptotic behavior. Notice that $F(V_t) = F(v_{\varepsilon\varphi_t})$. From Lemma 4.10, $v_{\varepsilon\varphi_t}$ satisfies the inequality (4.17) with

$$\frac{\bar{\alpha}}{1 - \varepsilon|\nabla\varphi_t|} \ge G\left(\frac{\bar{\beta}}{1 + \varepsilon|\nabla\varphi_t|}\right).$$

Since $|\nabla\varphi_t| \equiv 0$ outside $B_{7/8}$, the right inequality is satisfied by $v_{\varepsilon\varphi_t}$ and hence by V_t since w_t is positive. Inside $B_{7/8} \cap \Omega^+(V_t)$ we use the comparison Theorem 11.5. Let $x_1 \in F(V_t) \cap B_{7/8}$. Choosing ε (and therefore $\varepsilon|\nabla\varphi_t|$) small, from Lemma 4.9(c) we have that $F(V_t)$ are uniformly Lipschitz domains. Therefore, in a neighborhood of x_1,

$$\frac{v_{\varepsilon\varphi_t}}{w_t} \le c$$

since at an interior point the values of $v_{\varepsilon\varphi_t}$ and w_t are comparable. Therefore, from the asymptotic development of Lemma 4.11 we deduce that

$$V_t^+(x) = (v_{\varepsilon\varphi_t} + c\varepsilon w_t)^+(x) \ge \alpha^*\langle x - x_1, \nu\rangle^+ + o(|x - x_1|)$$

with $\alpha^* \ge (1 + c\varepsilon)\bar{\alpha}$. To complete the proof of the lemma, we must prove that if we choose h small enough in the definition of φ_t, then

$$\alpha^* \ge G(\bar{\beta}) \ .$$

From the properties of G, $s^{-N}G(s)$ is decreasing. Hence

$$\bar{\beta}^{-N}G(\bar{\beta}) \le \left[\frac{\bar{\beta}}{1 + \varepsilon|\nabla\varphi_t|}\right]^{-N} G\left(\frac{\bar{\beta}}{1 + \varepsilon|\nabla\varphi_t|}\right)$$

or

$$G(\bar{\beta}) \le (1 + \varepsilon|\nabla\varphi_t|)^N G\left(\frac{\bar{\beta}}{1 + \varepsilon|\nabla\varphi_t|}\right) \le \frac{(1 + \varepsilon|\nabla\varphi_t|)^N}{1 - \varepsilon|\nabla\varphi_t|}\ \frac{\alpha^*}{1 + c\varepsilon} \ .$$

Since $|\nabla\varphi_t| \le cht$, the proof is complete if h is small enough. □

Now we are ready for Steps 2 and 3.

4.5. Free boundary improvement. Basic iteration

We now use the family of subsolutions constructed in Lemma 4.11 along with Theorem 2.2 to get an improvement in the opening of the monotonicity cone up to the free boundary.

Lemma 4.13. *Let $u_1 \le u_2$ be two solutions of our free boundary problem in B_1, with $F(u_2)$ Lipschitz and $0 \in F(u_2)$. Assume that in $B_{1-\varepsilon}$*

$$v_\varepsilon(x) = \sup_{B_\varepsilon(x)} u_1 \le u_2(x), \tag{4.19}$$

that for $b > 0$, small,

$$v_\varepsilon(x_0) \le (1 - b\varepsilon) u_2(x_0) \qquad \left(x_0 = \tfrac{3}{4} e_n\right)$$

and that

$$B_{1/8}(x_0) \subset \Omega^+(u_1) \ .$$

Then, for ε small enough, there exists $\bar\mu$ (depending only on n, λ in Lemma 4.11 and the Lipschitz constant of $F(u_2)$) such that in $B_{1/2}$

$$v_{(1+\bar\mu b)\varepsilon}(x) \le u_2(x) \ .$$

Proof. Put, for $0 \le t \le 1$,

$$\bar v_t(x) = \sup_{B_{\varepsilon\varphi_{bt}}(x)} u_1 + Cb\varepsilon w_{bt}$$

where w_t is as in Lemma 4.12. Then $\bar v_t$ is a family of subsolutions. Let us check that $\bar v_t$ satisfies the hypotheses of Theorem 2.2 in $\Omega = B_{9/10} \setminus B_{1/8}(x_0)$ with respect to u_2.

(i) $\bar v_0 \le u_2$ is clear from (4.19) in $B_{9/10} \setminus \Omega^+(\bar v_0)$. In $\Omega^+(\bar v_0)$ it follows from Lemma 4.5 and the maximum principle, since $w_0 = u(x_0)$ on $\partial B_{1/8}(x_0)$.

(ii) Follows again from Lemma 4.5 and the maximum principle, provided h in Lemma 4.11 is kept small enough; to ensure strict inequality along $\partial B_1 \cap \overline{\Omega^+(\bar v_t)}$, we may replace ε with any smaller ε'.

(iii) Follows from Lemma 4.9(a).

(iv) Follows from the definition of $\bar v_t$.

We conclude that $\bar v_t \le u_2$ for each $t \in [0, 1]$. In particular

$$\bar v_1 \le u_2$$

which means that

$$\sup_{B_{(1+\bar\mu b)\varepsilon}(x)} u_1 \le u_2(x)$$

in $B_{1/2}$ since $\varphi_{b|_{B_{1/2}}} \ge 1 + \lambda b h \equiv 1 + \bar\mu b$. □

We apply Lemma 4.12 to

$$u_1(x) = u(x - \tau) \ \text{ and } \ u_2(x) = u(x)$$

with $\tau \in \Gamma(\frac{\theta}{2}, e_n)$.

Thanks to Lemma 4.6 all the hypotheses are satisfied and therefore we conclude that, in $B_{1/2}$, for every small vector $\tau \in \Gamma(\frac{\theta}{2}, e_n)$

$$\sup_{B_{(1+\bar{\mu}b)\varepsilon}} u(y-\tau) \le u(x).$$

The immediate consequence is

Lemma 4.14. *Let u be a solution of our free boundary problem in $\mathcal{C}^1$. Assume that, for some $0 < \bar{\theta} < \theta \le \pi/2$, u is monotonically increasing along any direction $\tau \in \Gamma(\theta, e_n)$. Then there exist $\lambda = \lambda(\bar{\theta}, n) < 1$ and a cone $\Gamma(\theta_1, \nu_2) \supset \Gamma(\theta, e_n)$ such that*

$$\delta_1 \le \lambda\delta_0 \qquad \left(\delta_1 = \frac{\pi}{2} - \theta_1\right)$$

and in $\mathcal{C}_{1/2}$, $D_\sigma u \ge 0$ for every $\sigma \in \Gamma(\theta_1, \nu_1)$.

We are now ready for Step 3.

Proof of Theorem 4.1. We repeat Lemma 4.14 inductively, observing that if u is a solution of our free boundary problem, then $u_h(x) = \frac{u(hx)}{h}$ is also a solution in the corresponding domain. We get that, in $\mathcal{C}_{2^{-k}}$, u is monotone increasing along a cone of directions

$$\Gamma(\theta_k, \nu_k) \qquad (k \ge 0)$$

with $\theta_0 = \theta$, $\nu_0 = e_n$ and

$$\Gamma(\theta_{k+1}, \nu_{k+1}) \supset \Gamma(\theta_k, \nu_k),$$

$$\delta_{k+1} \le \lambda\delta_k \qquad \left(\delta_k = \frac{\pi}{2} - \theta_k\right).$$

It follows that $\delta_k \le \delta_0\lambda^k$ and hence that the free boundary is $C^{1,\alpha}$ at the origin for some $\alpha = \alpha(\lambda) > 0$. $\square$

Chapter 5

Flat Free Boundaries Are Lipschitz

5.1. Heuristic considerations

The second step in the proof of strong regularity results for the free boundary consists basically in showing that if the free boundary is uniformly close to a nice asymptotic configuration, then it is actually Lipschitz. This is the case, for instance, when the dilations

$$u_\lambda(x - x_0) = \frac{u(\lambda(x - x_0))}{\lambda}$$

around a differentiability point x_0 of $F(u)$ converges to a two-plane solution

$$\alpha x_1^+ - \beta x_1^-$$

in a suitable system of coordinates.

It is convenient for us to replace "closeness" to a nice limit configuration by a "flatness" condition, expressed by ε-monotonicity, along a large cone of directions $\Gamma(\theta_0, \varepsilon)$, a notion treated extensively in Section 11.5

In particular, our basic hypothesis will be: there are a unit vector e and an angle θ_0 with $\theta_0 > \frac{\pi}{4}$ (say) and $\varepsilon > 0$ (small) such that, for every $\varepsilon' \geq \varepsilon$,

$$\sup_{B_{\varepsilon' \sin\theta_0}(x)} u(y - \varepsilon' e) \leq u(x).$$

The simplest theorem that can be proved is the following ([C2]).

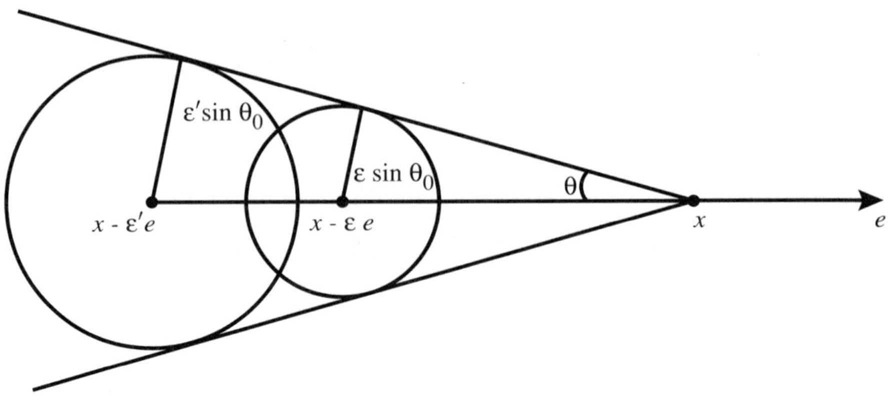

Figure 5.1. Cone of ε-monotonicity

Theorem 5.1. *Let $\frac{\pi}{4} < \theta_0 < \frac{\pi}{2}$ and let u be a viscosity solution in $\mathcal{C}_1 = B_1' \times (-1,1)$ of the free boundary problem*

$$\Delta u = 0 \quad in \quad \Omega^+(u) \; and \; \Omega^-(u),$$
$$u_\nu^+ = G(u_\nu^-)$$

where $G : \mathbb{R}^+ \to \mathbb{R}^+$ is strictly monotone and $s^{-N}G(s)$ is decreasing for some large N. Assume $u(0) = 0$.

Then there exists $\varepsilon = \varepsilon(\theta_0, G)$ such that if u is ε-monotone in $\mathcal{C}_{1-\varepsilon} = B_{1-\varepsilon}' \times (-1+\varepsilon, 1-\varepsilon)$ along any direction τ in the cone $\Gamma(\theta_0, e)$, then u is fully monotone in $C_{1/2} = B_{1/2}' \times (-\frac{1}{2}, \frac{1}{2})$ along any direction $\tau \in \Gamma(\theta_1, e)$, with $\theta_1 = \theta_1(\theta_0, \varepsilon)$.

In particular, this implies that the free boundary $F(u)$ is Lipschitz and therefore by Theorem 4.1 is also $C^{1,\gamma}$.

The problem with the hypotheses of Theorem 5.1 is that if β is zero but $u^- \not\equiv 0$, u^- could be very degenerate, that is, very close to zero, and not ε-monotone for any ε. We will return later to this situation proving a more powerful theorem by making u^- negligible in that case.

The strategy of the proof of Theorem 5.1 is based on the following inductive argument.

Supposing ε small and θ_0 close to $\pi/2$, we show that u is ε_k-monotone along any direction of a cone $\Gamma(\theta_k, e_1)$ in the cylinders

$$\tilde{\mathcal{C}}_k = B_{\rho_k} \times (-1, 1)$$

where

(a) $\varepsilon_k = \lambda^k \varepsilon, \qquad 0 < \lambda < 1,$
(b) $\theta_k - \theta_{k-1} \leq c\varepsilon_k^p, \qquad 0 < p < 1,$

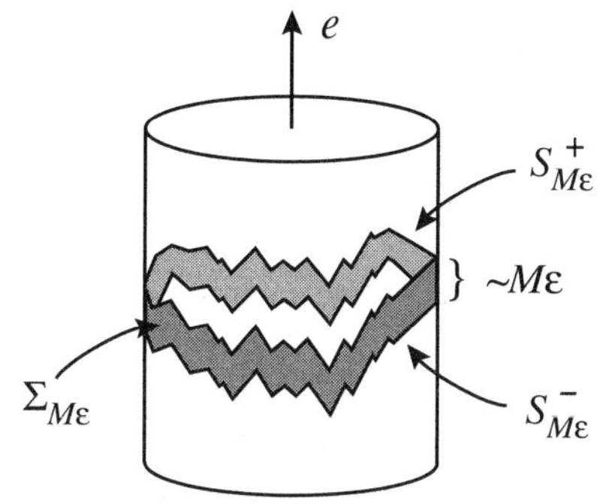

Figure 5.2. The intersection of $S_{M\varepsilon}$ with the boundary of the cylinder: the "bad influence" region

(c) $\rho_k - \rho_{k-1} \leq c\varepsilon_k^{p'}, \qquad 0 < p' < 1,$

with c, λ, p, p' depending only on θ_0, u, and $p' < p$.

Then we will get that in $\mathcal{C}_{1/2}$, u is fully monotone in a little smaller cone of the type

$$\Gamma(\theta_0 - c\varepsilon^{k_0}, e)$$

for some positive $k_0 = k_0(\theta_0)$. Therefore $F(u)$ is Lipschitz. Let us examine why it is possible to implement the above strategy. The main reasons are the following.

(1) From Lemma 11.12, we know that, for M large, outside an $M\varepsilon$-strip around the free boundary, u becomes automatically fully monotone along $\tau \in \Gamma(\theta, e)$. Thus, the improvement of ε-monotonicity is needed only in an $M\varepsilon$-strip $S_{M\varepsilon}$ between two Lipschitz graphs (in the direction e).

In particular, the only bad influence on the achievement of full monotonicity comes from the intersection $\Sigma_{M\varepsilon}$ of the strip with the boundary of the cylinder.

(2) Let us try to transfer the "good information" (full monotonicity), available on the top and the bottom of the strip, to its interior. The idea is to use the same type of continuous deformation argument as in Theorem 4.1, constructing a suitable family of continuous perturbation, as the maximum of u on balls of variable radius $\varphi = \varphi(x)$. What should we require of φ?

Certainly, we should require $\sup_{B_{\varphi(x)}(x)} u = v(x)$ to be a subharmonic function in its positive and negative sets. Therefore we ask

$$\varphi\Delta\varphi \geq C|\nabla\varphi|^2 \qquad (C \gg 1)\ .$$

As in Section 4.4, a suitable correction has to be added to v to make it an R-subsolution; this requires a control of $|\nabla\varphi|$.

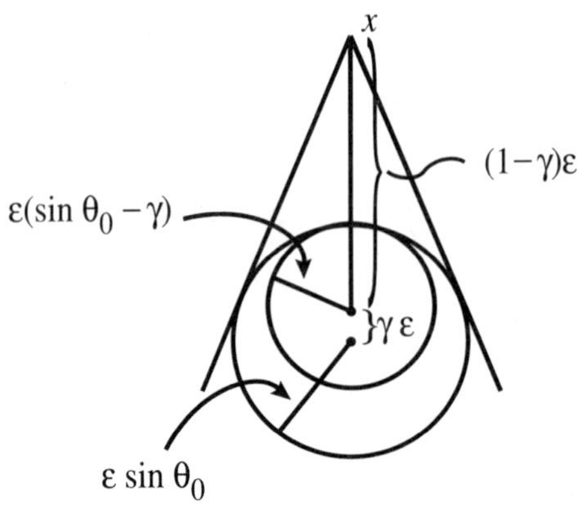

Figure 5.3. A loss in ε-monotonicity

We now have to carefully balance the size of φ in the strip. We know that, everywhere,

$$\sup_{B_{\varepsilon \sin \theta_0}} u(y - \varepsilon e) \leq u(x) .$$

Since we are willing to give up a small part of the original cone, we may translate a little less, say $(1-\gamma)\varepsilon$, and pass from $\sin \theta_0$ to $\sin \theta_0 - \gamma$. Thus, we certainly have everywhere

$$\sup_{B_{\varepsilon(\sin \theta_0 - \gamma)}(x)} u(y - (1-\gamma)e) \leq u(x)$$

since

$$B_{\varepsilon(\sin \theta_0 - \gamma)}(x - (1-\gamma)\varepsilon e) \subset B_{\varepsilon \sin \theta_0}(x - \varepsilon e) .$$

On the other hand, on the top $S^+_{M\varepsilon}$ and the bottom $S^-_{M\varepsilon}$ of the $M\varepsilon$-strip, we know that u is fully monotone so that there we can write

$$\sup_{B_{(1-\gamma)\varepsilon \sin \theta_0}(x)} u(y - (1-\gamma)\varepsilon e) \leq u(x) .$$

This means that

(a) we can take $\varphi(x) \approx (1-\gamma)\varepsilon \sin \theta_0$ on $S^+_{M\varepsilon}$ and $S^-_{M\varepsilon}$, with a gain of $\gamma\varepsilon$ in monotonicity with *no loss* in the opening θ_0 of the cone;

(b) we are forced to take $\varphi(x) \approx \varepsilon(\sin \theta_0 - \gamma)$ near $\Sigma_{M\varepsilon}$, the intersection of the strip with ∂C_1, with a loss of order $\gamma(1 - \sin \theta_0)$ in the opening of the cone.

(3) This last loss is clearly very bad for our iteration argument, unless it decreases very fast with the distance from $\partial \mathcal{C}_1$. Indeed, we can control the effect of $\Sigma_{M\varepsilon}$ on the variable radius, away from $\partial \mathcal{C}_1$, by means of an estimate

for the harmonic measure $\omega^x(\Sigma_{M\varepsilon})$ of that strip. According to that estimate (see Lemma 5.3),

$$\omega^x(\Sigma_{M\varepsilon}) \sim M\varepsilon \frac{d(x, \partial\mathcal{C}_1)}{d(x, \Sigma_{M\varepsilon})^2}$$

so that, if we are inside $S_{M\varepsilon}$ at a distance at least $C\varepsilon$ from $\partial\mathcal{C}_1$ since $d(x, \partial\mathcal{C}_1)$ and $d(x, \Sigma_{M\varepsilon})$ are comparable, we have

$$\omega^x(\Sigma_{M\varepsilon}) \sim \varepsilon\, d(x, \partial\mathcal{C}_1)^{-1}$$

and

$$|\nabla\omega^x(\Sigma_{M\varepsilon})| \leq C\,\varepsilon\, d(x, \partial\mathcal{C}_1)^{-2} .$$

As a consequence, the effect of $\Sigma_{M\varepsilon}$ on the variable radius φ will dissipate ε-away from $\partial\mathcal{C}_1$ as

$$\varepsilon\, d(x, \partial\mathcal{C}_1)^{-1}\operatorname{osc}\varphi$$

and $|\nabla\varphi|$ will decay as

$$\varepsilon\, d(x, \partial\mathcal{C}_1)^{-2}\operatorname{osc}\varphi .$$

This means that if we stay at a distance of order ε^a from $\partial\mathcal{C}_1$, $0 < a < 1$, φ will decrease from its optimal value (corresponding to no loss in monotonicity) of a quantity of order ε^{1-2a} with a sacrifice of $C\varepsilon^{1-2a}$ in the opening of the cone. At the same time, the perturbation in the free boundary condition requires a correction of order $|\nabla\varphi| \sim \varepsilon^{1-2a}$.

(4) Writing the family of variable radii in the form $\varepsilon\varphi_t$, we are lead to the following requirements for φ_t:

(a) $\varphi_t\Delta\varphi_t \geq C|\nabla\varphi_t|^2$.

(b) $\varphi_0 \leq \sin\theta_0 - \gamma$.

(c) $\varphi_t \leq \sin\theta_0 - \gamma$ near the edges of $S_{M\varepsilon}$ and $\varphi_{\bar{t}} = (1-\gamma)\sin\theta_0 - \varepsilon^\sigma$, with suitable $0 < \sigma < 1$ (the ε^σ is necessary for the correction term) for at least one $\bar{t} > 0$, inside the strip at a distance ε^a from $\partial\mathcal{C}_1$.

(d) The various parameters have to be chosen such that $a < \sigma$ and, for θ_0 close to $\pi/2$, $\gamma \leq \sin\theta_0 - \frac{1}{2}$ (say).

In this way, letting $\lambda = 1 - \gamma$, one has

$$\sup_{B_{\varphi_{\bar{t}}(x)}(x)} u(y - \lambda\varepsilon) \leq u(x)$$

giving $\lambda\varepsilon$-monotonicity in a cone $\Gamma(\theta_1, e)$, with $\theta_0 - \theta_1 \leq C\varepsilon^\sigma$, ε^a-away from ∂C_1.

That is, we gain a geometric increment of monotonicity, sacrificing a geometrically decaying amount of angle and radius.

It is clear that a careful adjustment would produce full monotonicity in $\mathcal{C}_{1/2}$ at least in a cone $\Gamma(\theta_0/2, e)$.

5.2. An auxiliary family of functions

In this section we construct a perturbation family of functions adapted to a Lipschitz δ-strip.

Lemma 5.2. *Let A be the graph of a Lipschitz function $x_n = f(x')$ in $B_1' \subset \mathbb{R}^{n-1}$ with $f(0) = 0$ and Lipschitz norm L. Let $\mathcal{C} = \bar{B}_1' \times [-2L, 2L]$. Then given $\delta > 0$ (small), there exists a family of C^2-functions φ_t, $0 \le t \le 1$, such that*

(a) $1 \le \varphi_t \le 1 + t$,

(b) $\varphi_t \Delta \varphi_t \ge C|\nabla \varphi_t|^2$,

(c) $\varphi_t \approx 1$ *on*
$$A_\delta = \{x : d(x, A \cap \partial\mathcal{C}) < \delta\},$$

(d) *in the set* $\{x : d(x, \partial\mathcal{C}) > \delta\}$
$$\varphi_t \ge 1 + t\left[1 - \frac{C\delta}{d(x, \partial\mathcal{C})^2}\right],$$

(e) $|\nabla \varphi_t| \le \frac{Ct}{\delta}$.

Proof. Let ψ_0 be the harmonic function in $\mathcal{C}$ with boundary values given by a δ-smoothing of $\chi_{A_{2\delta}}$. In particular
$$\psi_0 \equiv 1 \quad \text{on} \quad A_\delta \cap \partial\mathcal{C},$$
$$\psi_0 \equiv 0 \quad \text{outside} \quad A_{3\delta} \cap \partial\mathcal{C} \ .$$

Claim. *We have the following.*

(i) *At a distance at least δ from $\partial\mathcal{C}$,*
$$\psi_0 \le c\delta/d(x, \partial\mathcal{C})^2 \quad \text{and} \quad |\nabla\psi_0| \le c/\delta.$$

(ii) *In A_δ,*
$$\psi_0 \ge c > 0 \ .$$

Assume the claim. We expand, truncate and mollify of order δ defining
$$\psi_1 = \min\{c\psi_0((1-\delta)x), 1\} * \xi_\delta \ .$$

If c is chosen large enough, ψ_1 has the following properties:

(i) ψ_1 is defined and superharmonic in $\mathcal{C}$.

(ii) $|\nabla\psi_1| \le c/\delta$.

(iii) $\psi_1 \equiv 1$ on $A_\delta \cap \mathcal{C}$.

(iv) In the set $\{x : d(x, \partial\mathcal{C}) > \delta\}$,
$$\psi_1 \le c\delta/d(x, \partial\mathcal{C})^2.$$

We next define

$$\psi_2 = \left(\frac{1+\psi_1}{2}\right)^{1/(1-2c)}$$

with $c > 1$, the constant appearing in Lemma 5.2. Then

$$0 \geq \Delta(\psi_2^{1-2c}) = (1-2c)\psi_2^{-2c}\Delta\psi_2 + (1-2c)(-2c)\psi_2^{-2c-1}|\nabla\psi_2|^2$$

or

$$\psi_2\Delta\psi_2 \geq 2c|\nabla\psi_2|^2 \ .$$

Moreover, $\psi_{2|A_\delta} \equiv 1$, $1 \leq \psi_2 \leq 2^{1/(2c-1)}$ and, if $d(x, \partial\mathcal{C}) > \delta$,

$$\psi_2(x) \geq 2^{1/(2c-1)} - c\delta\, d(x, \partial\mathcal{C})^{-2} \ .$$

Finally, the function

$$\varphi_t = 1 + t\frac{(\psi_2 - 1)}{2^{1/(2c-1)} - 1}, \qquad 0 \leq t \leq 1,$$

has all the properties (a)–(e) listed above. □

The claim follows from the following estimate for the harmonic measure of a thin Lipschitz strip.

Lemma 5.3. *Let $\mathcal{C}$, f, A, A_δ be as in Lemma 5.2 and let $S_\delta = A_\delta \cap \partial\mathcal{C}$. Let $w(x) = \omega^x(S_\delta)$, the harmonic measure of S_δ in $\mathcal{C}$. Then the following hold.*

(a) *If $d(x, \partial\mathcal{C}) \geq \delta$,*

$$w(x) \leq c\frac{\delta d(x, \partial\mathcal{C})}{d(x, S_\delta)^2} \ . \tag{5.1}$$

(b) *If $d(x, A \cap \partial\mathcal{C}) \leq \bar{c}d(x, \partial\mathcal{C}) \leq \bar{c}\delta$,*

$$w(x) \geq c > 0$$

where c depends only on $\bar{c}$.

Proof. We are interested in the region near S_δ. There, the boundary of $\mathcal{C}$ is smooth so that the Poisson kernel for the cylinder $P(x, y)$ behaves like that of a half plane. That is, for $y \in \partial\mathcal{C}$,

$$P(x, y) \sim \frac{d(x, \partial\mathcal{C})}{|x-y|^n} \ . \tag{5.2}$$

Moreover,

$$w(x) = \int_{\partial\mathcal{C}} P(x, y)\chi_{S_\delta}(y)\, dy \ .$$

If we make the Lipschitz transformation

$$z' = x' \ , \qquad z_n = x_n - f(x')$$

that straightens the strip S_δ, the estimate (5.2) remains true and therefore, in the points of interest,

$$w(x) \sim h(z) \qquad (z = z(x))$$

with h the harmonic function in the infinite cylinder satisfying

$$h_{|\partial\mathcal{C}} = \chi_{|z_n|<\delta} \ .$$

Thus, h has the asymptotic two-dimensional behavior of h^*, the harmonic function in the half plane $\{(z_1, z_n) : z_1 > 0\}$ with boundary values equal to 1 in $S^*_\delta = \{|z_n| < \delta,\ z_1 = 0\}$. Since

$$h^*(z_1, z_n) = \int_{-\delta}^{\delta} \frac{z_1}{(z_n - s)^2 + z_1^2}\, ds \le c \frac{\delta z_1}{d(z, S^*_\delta)^2},$$

(a) and (b) follow after the inverse Lipschitz transformations. □

5.3. Level surfaces of normal perturbations of ε-monotone functions

We are going to use the perturbation family φ_t of Lemma 5.2 in order to construct our basic family of R-subsolution. First we need to study the level surfaces of

$$v(x) = \sup_{B_{\varphi(x)}(x)} u$$

where φ is a smooth positive function and u is ε-monotone.

We do it through the following lemma.

Lemma 5.4. *Let $\varphi \in C^2(B_1)$ be a positive function and let u be ε-monotone along every $\tau \in \Gamma(\theta, e)$ in a domain D so large that the function*

$$v(x) = \sup_{B_{\varphi(x)}(x)} u$$

is well defined in B_1.

Assume that

$$\sin\bar\theta \le \frac{1}{1 + |\nabla\varphi|}\left(\sin\theta - \frac{\varepsilon}{2\varphi}\cos^2\theta - |\nabla\varphi|\right).$$

Then v is monotone in the cone $\Gamma(\bar\theta, e)$; in particular its level surfaces are Lipschitz graphs, in the direction of e, with Lipschitz constant $\tilde L \le \operatorname{cotg}\bar\theta$.

Proof. Let $v(x_0) = u(y_0)$, with $|x_0 - y_0| = \varphi(x_0)$. We first estimate the maximum angle $\alpha(y_0 - x_0, e)$ between the vectors $y_0 - x_0$ and e. Consider $y_0 + \varepsilon e$. If $B_{\varepsilon\sin\theta}(y_0 + \varepsilon e)$ intersects $B_{\varphi(x_0)}(x_0)$ at a point z, then $u(y_0) < u(z)$, so

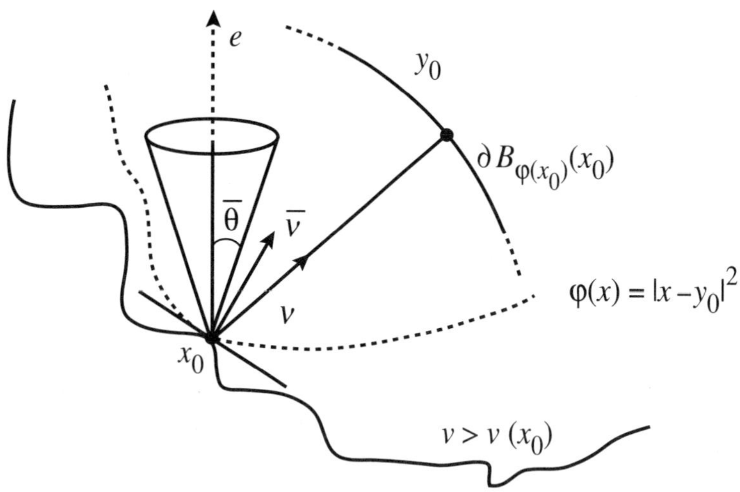

Figure 5.4

that the maximum is achieved when $B_{\varphi(x_0)}(x_0)$ is tangent to $B_{\varepsilon \sin\theta}(y_0+\varepsilon e)$. Hence, suppose that $B_{\varepsilon\sin\theta}(y_0 + \varepsilon e)$ touches $B_{\varphi(x_0)}(x_0)$ at a point z_0. Since

$$\begin{aligned}|\varepsilon \sin\theta + \varphi(x_0)|^2 &= |y_0 + \varepsilon e - x_0|^2 \\ &= |y_0 - x_0|^2 + \varepsilon^2 - 2\varepsilon\langle y_0 - x_0, e\rangle \\ &= \varepsilon^2 + \varphi(x_0)^2 + 2\varepsilon\varphi(x_0)\cos\alpha,\end{aligned}$$

we have

$$\cos\alpha = \sin\theta - \frac{\varepsilon}{2\varphi(x_0)}\cos^2\theta\ . \tag{5.3}$$

As in Lemma 4.9, the set

$$\{x : v(x) \geq v(x_0)\}$$

contains the domain

$$\{x : |x - y_0|^2 \leq \varphi(x)^2\}$$

that has the unit normal

$$\bar{\nu} = (\nu + \nabla\varphi(x_0))/|\nu + \nabla\varphi(x_0)|$$

at x_0, where $\nu = (y_0 - x_0)/|y_0 - x_0|$.

Hence, if $\bar{\theta} \leq \frac{\pi}{2} - \alpha(\bar{\nu}, e)$, the set $\{x : v(x) \geq v(x_0)\}$ contains $x_0 + \Gamma(\bar{\theta}, e) \cap B_\rho$, for some $\rho > 0$.

The requirement on $\bar{\theta}$ is equivalent to the inequality

$$\sin\bar{\theta} \leq \cos\alpha(\bar{\nu}, e) = \langle\bar{\nu}, e\rangle.$$

Since $|\nu+\nabla\varphi(x_0)|\le 1+|\nabla\varphi(x_0)|$, from (5.3)

$$\begin{aligned}\langle\bar\nu,e\rangle &\ge \frac{1}{1+|\nabla\varphi(x_0)|}\langle\nu+\nabla\varphi(x_0),e\rangle\\ &\ge \frac{1}{1+|\nabla\varphi(x_0)|}\left(\sin\theta-\frac{\varepsilon}{2\varphi(x_0)}\cos^2\theta-|\nabla\varphi(x_0)|\right).\end{aligned}$$

Therefore it is enough to require that

$$\sin\bar\theta\le\frac{1}{1+|\nabla\varphi(x_0)|}\left(\sin\theta-\frac{\varepsilon}{2\varphi(x_0)}\cos^2\theta-|\nabla\varphi(x_0)|\right). \qquad \square$$

We will use Lemma 5.3 with $\theta=\theta_0>\frac{\pi}{4}$ (say) and $\varphi=\sigma\varphi_t$, $\frac12\varepsilon<\sigma<\varepsilon$, and φ_t as in Lemma 5.2. If moreover we keep $|\nabla\varphi|<c\varepsilon$, v will be monotone in a cone $\Gamma(\bar\theta,e)$ with $\bar\theta$ strictly positive.

5.4. A continuous family of R-subsolutions

At this point we have all the ingredients for constructing a family of R-subsolutions, adapted for a comparison theorem.

Let u be a solution of our free boundary problem in the cylinder $\mathcal{C}_1=B_1'\times(-2,2)$. In the sequel we set $\mathcal{C}_\rho=B_\rho'\times(-2\rho,2\rho)$. If u is ε-monotone along $\tau\in\Gamma(\theta_0,e)$, $\theta_0>\pi/4$, then (see Proposition 11.14) $F(u)$ is contained in a ε-neighborhood $\mathcal{N}_\varepsilon(A)$ of the graph A of a Lipschitz continuous function f with Lipschitz constant $L<1$. Moreover, for a large M (see Lemma 11.12) u is fully monotone outside $\mathcal{N}_{M\varepsilon}(A)$. Now let φ_t be the family of functions constructed in Lemma 5.2:

$$v_t(x)=\sup_{B_{\sigma\varphi_t(x)}(x)}u \qquad (0\le t<1)$$

with $\frac12\varepsilon<\sigma<2\varepsilon$. Then v_t is well defined in $\mathcal{C}_{1-4\varepsilon}$. We set $\Omega^+(v_t)=\mathcal{C}_{1-4\varepsilon}\cap\{v_t>0\}$. Also let w_t be the harmonic function in $\Omega^+(v_t)\cap\mathcal{N}_{CM\varepsilon}(A)$ with boundary values

$$w_t=\begin{cases}u & \text{on } \partial\mathcal{N}_{CM\varepsilon}(A)\cap\Omega^+(v_t),\\ 0 & \text{otherwise.}\end{cases}$$

We extend w_t to zero outside $\Omega^+(v_t)$.

Lemma 5.5. *For $0\le t\le 1$, $\eta>0$, define in $\mathcal{C}_{1-4\varepsilon}$,*

$$\bar v_t=v_t+\eta w_t\ .$$

There exist positive constants c_1,c_2, depending only on N in Theorem 5.1 such that if

(i) $\theta_0\ge\pi/4$,

(ii) $\eta\ge c_1\sigma/\delta$ *(δ from the construction of φ_t, $\delta\gg\varepsilon$),*

(iii) $\sigma/\delta \le c_2$,

then $\bar{v}_t$ is an R-subsolution of our free boundary problem in $\mathcal{C}_{1-4\varepsilon}$.

Proof. We first make sure that $F(v_t) = \partial\Omega^+(v_t) \cap \mathcal{C}_{1-4\varepsilon}$ is uniformly Lipschitz. According to Lemma 5.4,

$$\frac{1}{1+\sigma|\nabla\varphi_t|}\left(\sin\theta_0 - \frac{\varepsilon}{2\sigma\varphi_t}\cos^2\theta_0 - \sigma|\nabla\varphi_t|\right)$$

must be kept strictly positive, that is, since $|\nabla\varphi_t| \le c/\delta$,

$$\frac{1}{1+c\sigma/\delta}\left(\sin\theta_0 - \frac{\varepsilon}{2\sigma}\cos^2\theta_0 - c\frac{\sigma}{\delta}\right)$$

must be strictly positive. If, say, $\theta_0 > \pi/4$, $\frac{1}{2}\varepsilon < \sigma < 2\varepsilon$ and σ/δ is small, this is ensured.

We now want $\bar{v}_t$ to satisfy the appropriate free boundary conditions. Since $\varphi_t \equiv 1$ is a δ-neighborhood of A_δ (see Lemma 5.2) and $\delta \gg \varepsilon$, it is enough to consider points, say, 6ε-away from $\mathcal{C}_{1-4\varepsilon}$. According to the calculations in Lemma 4.9, any point $x_0 \in F(v_t)$ is regular from the right and near x_0,

$$v_t(x) \ge \alpha\langle x-x_0, \nu+\nabla\varphi(x_0)\rangle^+ - \beta\langle x-x_0, \nu+\nabla\varphi(x_0)\rangle^- + o(|x-x_0|)$$

with $\alpha = G(\beta)$ and where ν is the unit inward normal to the touching ball $B \subset \Omega^+(v_t)$ at x_0. Concerning the correction term w_t, if we stay at a distance greater than (say) 6ε from $\partial\mathcal{C}_{1-4\varepsilon}$, then from the comparison Theorem 11.6, near x_0,

$$w_t \ge cv_t$$

with c depending on θ_0, n.

Hence, 10ε-away from $\partial\mathcal{C}_1$, we have

$$\bar{v}_t \ge \tilde{\alpha}\langle x-x_0, \tilde{\nu}\rangle^+ + \beta\langle x-x_0, \tilde{\nu}\rangle^- + o(|x-x_0|)$$

where

$$\tilde{\alpha} \ge (1+c\eta)(1-c\sigma|\nabla\varphi_t|)\alpha\ , \quad \tilde{\beta} \le (1+c\sigma|\nabla\varphi_t|)\beta$$

and $\tilde{\nu} = (\nu+\nabla\varphi(x_0))/|\nu+\nabla\varphi(x_0)|$.

For $\bar{v}_t$ to be an R-subsolution, we want

$$\tilde{\alpha} \ge G(\tilde{\beta});$$

that is, since $\alpha = G(\beta)$,

$$(1+c\eta)(1-c\sigma|\nabla\varphi_t|)G(\beta) \ge G\Big((1+c\sigma|\nabla\varphi_t|)\beta\Big)\ . \tag{5.4}$$

Since $|\nabla\varphi_t| \le ct/\delta$, $s^{-N}G(s)$ is decreasing and G is increasing, we have

$$1-c\sigma|\nabla\varphi_t| \ge 1-c\frac{\sigma t}{\delta}\ , \quad 1+c\sigma|\nabla\varphi_t| \le 1+c\frac{\sigma t}{\delta}$$

and

$$G(\beta) \geq \left(1 + c\frac{\sigma t}{\delta}\right)^{-N} G\left(\left(1 + c\frac{\sigma t}{\delta}\right)\beta\right) .$$

Therefore it is enough to show that

$$(1 + c\eta)\left(1 - c\frac{\sigma t}{\delta}\right) \geq \left(1 + c\frac{\sigma t}{\delta}\right)^{N} .$$

This is possible if $\sigma t/\delta \gg 1$ and $\eta \gg \sigma/\delta$, both depending on N. □

5.5. Proof of Theorem 5.1

Before going to the proof of Theorem 5.1, we need to show that for an ε-monotone solution u of our free boundary problem, at least $M\varepsilon$-away from the free boundary, $|\nabla u(x)|$ behaves like $u(x)/\operatorname{dist}(x, F(u))$ since u becomes fully monotone and its level surfaces become Lipschitz graphs.

This is the content of the following lemma.

Lemma 5.6. *Let $u \in C(\mathcal{C}_1)$, $u \geq 0$, $u(0) = 0$ be ε-monotone along $\Gamma(\theta, e_n)$, harmonic in $\Omega^{\pm}(u)$.*

There exist $\varepsilon_0 = \varepsilon_0(n)$, $M = M(n)$, $C = C(n, \theta)$ such that if $\varepsilon \leq \varepsilon_0$ and $x \in \mathcal{C}_{1/2}$, $\operatorname{dist}(x, F(u)) > CM\varepsilon$, then

$$|\nabla u(x)| \sim \frac{u(x)}{\operatorname{dist}(x, F(u))} . \tag{5.5}$$

Proof. Let $x \in \Omega^+(u)$, $\operatorname{dist}(x, F(u)) = d_x$. The inequality

$$|\nabla u(x)| \leq cu(x)/d_x$$

comes from standard interior estimates and Harnack's inequality.

Thus, let us prove that if $d_x \geq CM\varepsilon$,

$$|\nabla u(x)| \geq cu(x)/d_x .$$

$F(u) = \partial\Omega^+(u)$ is contained in a $(1 - \sin\theta)\varepsilon$-strip bounded by two Lipschitz functions with Lipschitz constant $L = \operatorname{cotg}\theta$. Moreover, from Lemma 11.15, u is fully monotone outside an $M\varepsilon$ neighborhood of $F(u)$, with $M = M(n)$, large enough.

Let $x_0 \in \Omega^+(u)$, $d_{x_0} = 10M\varepsilon$ and set $u(x_0) = a$. The level surface $\{u = a\}$ is a Lipschitz surface in the ball $B_{4\eta\varepsilon}(x_0)$ at distance greater than $8M\varepsilon$ from $F(u)$, if $\eta = \eta(L)$ is chosen properly small.

Consider the cylinder

$$T_\varepsilon[-20M\varepsilon, 20M\varepsilon] \times B'_{\eta\varepsilon}(x_0) ,$$

and denote by ω the harmonic measure in T_ε.

Since $|\partial T_\varepsilon \cap \{u = 0\}| \geq \gamma|\partial T_\varepsilon|$ with $0 < \gamma < 1$, $\gamma = \gamma(n, \theta)$, from Lemma 11.19 we also have

$$\omega^x(\partial T_\varepsilon \cap \{u = 0\}) \geq \gamma'|\partial T_\varepsilon|$$

with $0 < \gamma' < 1$, $\gamma' = \gamma'(n, \theta)$.

Then from the subharmonicity of u in T_ε,

$$a \leq \int_{\partial T_\varepsilon} u \, d\omega^x \leq (1 - \gamma') \max_{\partial T_\varepsilon} u \ .$$

Therefore, there exists $y_0 \in \partial T_\varepsilon$ such that

$$u(y_0) \geq \frac{1}{(1 - \gamma')} a \equiv k_0 a > a$$

and the level surface $\{u = k_0 a\} \cap B_{4\eta\varepsilon}(y_0)$ has distance from $F(u)$ less than $CM\varepsilon$. Notice that $\{u > a\} \supset \{u > k_0 a\}$. Therefore, if $x \in \{u > k_0 a\}$, we have, since $(u - a)^+$ is harmonic in $T_\varepsilon \cap \{u > a\}$, from Theorem 11.10,

$$(u(x) - a)^+ \leq c|\nabla u(x)| d(x, \partial\{u > a\}) \leq c|\nabla u(x)| \, d_x \ .$$

On the other hand,

$$u(x) - a = \frac{1}{k_0} u(x) - a + \left(1 - \frac{1}{k_0}\right) u(x) \geq \left(1 - \frac{1}{k_0}\right) u(x)$$

and (5.5) follows if $d_x \geq CM\varepsilon$ and $|x' - x_0'| < \eta\varepsilon$. Since $x_0 \in \Omega^+(u)$ is an arbitrary point at distance $10M\varepsilon$ from $F(u)$, (5.5) holds if $x \in \mathcal{C}_{1/2} \cap \Omega^+(u)$ with $d_x \geq CM\varepsilon$. □

We are now ready for the proof of Theorem 5.1. The basic inductive lemma follows.

Lemma 5.7. *Let u be a solution of our free boundary problem in $\mathcal{C}_1$, ε-monotone along the cone of directions $\Gamma(\theta, e_n)$ for some $\varepsilon_0 > \varepsilon$, with $\frac{\pi}{4} < \theta_0 \leq \theta \leq \frac{\pi}{2}$. Assume $u(0) = 0$. Then there exist positive $c_0 = c_0(\theta_0)$, $\varepsilon_0(\theta_0)$ and $\lambda = \lambda(\theta_0)$, $\lambda < 1$, such that u is $\lambda\varepsilon$-monotone along the cone of directions $\Gamma(\theta - c_0\varepsilon^{1/4}, e_n)$ in $\mathcal{C}_{1/2}$.*

Proof. Let $\lambda < 1$, to be chosen later, and let

$$u_1(x) = u(x - \lambda\varepsilon e_n) \ .$$

From the ε-monotonicity of u (for $1 - \lambda < \sqrt{2}/2$), we have

$$\sup_{B_{\varepsilon(\sin\theta - (1-\lambda))}(x)} u_1 \leq u(x) \equiv u_2(x) \tag{5.6}$$

since

$$B_{\varepsilon(\sin\theta - (1-\lambda))}(x - \lambda\varepsilon e_n) \subset B_{\varepsilon \sin\theta}(x - \varepsilon e_n) \ .$$

Notice that, choosing a slightly smaller radius $\varepsilon(\sin\theta - (1 - \lambda'))$, $\lambda' < \lambda$, we have strict inequality in (5.6).

On the other hand, u is fully monotone outside $\mathcal{N}_{M\varepsilon}$, an $M\varepsilon$-neighborhood of the graph of a Lipschitz function $x_n = f(x')$. Therefore

$$\sup_{B_{\lambda\varepsilon\sin\theta}(x)} u_1 \le u_2(x)$$

for any $x \notin \mathcal{N}_{M\varepsilon}$.

Our purpose is to obtain, by means of the family φ_t, an intermediate radius $\sigma\varphi_t$, such that

$$\sup_{B_{\sigma\varphi_t}(x)} u_1 \le u_2(x)$$

for $x \in \mathcal{C}_{1-c\varepsilon^{1/8}} \cap \mathcal{N}_{CM\varepsilon}$.

We now fix the various parameters according to the requirements in Lemmas 5.2–5.5. We choose

$$\sigma = \varepsilon(\sin\theta - (1-\lambda) - \varepsilon^2)\,, \quad \lambda \ge \frac{3}{2} - \frac{\sqrt{2}}{2}\,, \quad \eta = \varepsilon^{1/4}\,, \quad \delta = \varepsilon^{1/2}\,.$$

To give room for the control of the correction term ηw_t, of order $\varepsilon^{1/4}$, we limit t to make sure that

$$\sigma\varphi_t \le \varepsilon(\lambda\sin\theta - \bar{c}\varepsilon^{1/4}) \tag{5.7}$$

for some $\bar{c}$ to be chosen later, that is,

$$[\sin\theta - (1-\lambda)](1+t) \le \lambda\sin\theta - \bar{c}\varepsilon^{1/4}$$

(with a slightly smaller $\bar{c}$) or

$$1 + t \le \frac{\lambda\sin\theta - \bar{c}\varepsilon^{1/4}}{\sin\theta - (1-\lambda)} \le \frac{\lambda\sin\frac{\pi}{4}}{\sin\frac{\pi}{4} - (1-\lambda)}\,.$$

To have equality in (5.7) for some $\bar{t}$, $0 < \bar{t} \le 1$, it is then enough to choose λ so close to 1 to have

$$\frac{\lambda\sin\frac{\pi}{4}}{\sin\frac{\pi}{4} - (1-\lambda)} \le 2\,.$$

We want to show that the family $\bar{v}_t$ so chosen satisfies

$$\bar{v}_t \le u_2$$

in $\mathcal{C}_{1-c\varepsilon^{1/8}} \cap \mathcal{N}_{CM\varepsilon}$, for $0 \le t \le \bar{t}$, if c is large enough. Since $\bar{v}_t$ is an R-subsolution, we have only to check that $\bar{v}_t < u_2$ on the boundary.

On the positive side, along

$$\partial\mathcal{N}_{CM\varepsilon} \cap \mathcal{C}_{1-4\varepsilon}$$

we have, for $R_1 \le R_2 \le \lambda\varepsilon\sin\theta$,

$$\sup_{B_{R_1}(x)} u_1 \le \sup_{B_{R_2}(x)} u_1 - (R_2 - R_1)|\nabla u_1(x)| \le \left[1 - \frac{R_2 - R_2}{CM\varepsilon}\right] u_2(x)$$

since, by Lemma 5.6, $u(x) \sim d(x, F(u))|\nabla u(x)|$.

With $R_1 = \sigma\varphi_t$ and $R_2 = \lambda\varepsilon\sin\theta$, we get, for $0 \le t \le \bar{t}$, $R_2 - R_1 \ge \bar{c}\varepsilon^{1+1/4}$, so that, adjusting the constant $\bar{c}$,

$$v_t \le (1 - \tilde{c}^{1/4})u_2 < u_2 - \varepsilon^{1/4} w_t$$

since $w_t \le cu_2$ along $\partial\mathcal{N}_{CM\varepsilon} \cap \mathcal{C}_{1-4\varepsilon}$, by Harnack's inequality. Thus,

$$\bar{v}_t < u_2 \ .$$

On $\partial\mathcal{C}_{1-4\varepsilon} \cap \mathcal{N}_{CM\varepsilon}$, $w_t = 0$ and since $\varepsilon^{1/2} \gg \varepsilon$, $\varphi_t \equiv 1$ in an ε-neighborhood of this set, so that

$$\bar{v}_t = v_t < u_2, \qquad 0 \le t \le \bar{t} \ .$$

Consider now the set

$$E = \{t : 0 \le t \le \bar{t},\, \bar{v}_t \le u_2 \ \text{ in } \mathcal{C}_{1-4\varepsilon} \cap \mathcal{N}_{CM\varepsilon}\}.$$

$E \ne \emptyset$ since $0 \in E$, via the maximum principle argument. E is obviously closed. To see that it is open, we show that if $v_t \le u_2$ in $\mathcal{C}_{1-4\varepsilon} \cap \mathcal{N}_{CM\varepsilon}$, then

$$\Omega^+(v_t) \subset\subset \Omega^+(u) \cap \mathcal{N}_{CM\varepsilon} \cap \mathcal{C}_{1-4\varepsilon} \ .$$

If not, $F(v_t)$ and $F(u_2)$ have to touch at some point x_0, which is therefore a regular point from the right for both free boundaries.

Moreover, since $\varphi_t \equiv 1$ in an ε-neighborhood of $\partial\mathcal{C}_{1-4\varepsilon} \cap \mathcal{N}_{CM\varepsilon}$, it must be $d(x_0, \partial\mathcal{C}_{1-4\varepsilon}) \ge c\varepsilon$ and near x_0, by the comparison Theorem 11.6, we have

$$w_t \ge cv_t \ .$$

We can now reach a contradiction as in the proof of Lemma 4.12.

We conclude that $E = [0, \bar{t}]$ or

$$\bar{v}_t \le u_2 \ \text{ in } \ \mathcal{C}_{1-4\varepsilon} \cap \mathcal{N}_{CM\varepsilon}$$

for every $0 \le t \le \bar{t}$. If we stay more inside, that is, in $\mathcal{C}_{1-c\varepsilon^{1/8}} \cap \mathcal{N}_{CM\varepsilon}$, from (d) of Lemma 5.2 we deduce that

$$\varphi_t \ge 1 + t\left[1 - \frac{c\varepsilon^{1/2}}{d(x, \partial\mathcal{C}_{1-4\varepsilon})^2}\right] \ge 1 + t(1 - c\varepsilon^{1/4}) \ .$$

For the maximum t, $t = \bar{t}$, we have equality in (5.7), so that, in $\mathcal{C}_{1-c\varepsilon^{1/8}} \cap \mathcal{N}_{CM\varepsilon}$,

$$\sigma\varphi_{\bar{t}} \ge \varepsilon\lambda\sin\theta - \bar{c}\varepsilon^{1+1/4} - ct\varepsilon^{1+1/4} \ge \varepsilon(\lambda\sin\theta - \tilde{c}\varepsilon^{1/4})$$

which implies

$$\sup_{B_{\varepsilon(\lambda\sin\theta - \tilde{c}\varepsilon^{1/4})}(x)} u_1 \le u_2(x) \ .$$

Since

$$\lambda\sin\theta - \tilde{c}\varepsilon^{1/4} \ge \lambda\sin(\theta - c_0\varepsilon^{1/4})$$

for a suitable positive c_0, the proof is complete. □

Proof of Theorem 5.1. Iterating Lemma 5.7 gives a sequence of domains

$$C_k = B_{R_k} \times (-4R_k L, 4R_k L)$$

where $R_k = 1 - \sum_{p=1}^{k} (c\lambda^p \varepsilon)^{1/8}$, in which u is $\lambda^k\varepsilon$-monotone along any direction τ of a cone $\Gamma(\theta_k, e_n)$ where

$$\theta_k = \theta_0 - \sum_{p=1}^{k} (c\lambda^p \varepsilon)^{1/4} .$$

To complete the proof, it is enough to choose $\varepsilon_0 = \varepsilon(\theta_0)$ so small that

$$R_k > R_\infty \geq \frac{1}{2} \text{ and } \theta_k > \theta_\infty \geq \frac{1}{2}\left(\theta_0 + \frac{\pi}{4}\right) . \qquad \square$$

5.6. A degenerate case

Theorem 5.1 holds also for one-phase problems where $u^- \equiv 0$, ending up, using Theorem 4.1, with a $C^{1,\gamma}$ free boundary, which in particular implies $u_\nu^+ \geq c > 0$ on $F(u)$.

As we will see clearly in the parabolic case, in the hypothesis of ε-monotonicity (for both u^+, u^-) in a flat cone there is hidden, in a rather subtle way, a nondegenerate free boundary condition of the positive part u^+: $u_\nu^+ \geq c > 0$ in the viscosity sense.

The next theorem deals with a case in which u^- could be degenerate, that is, very close to zero without being identically zero. For instance if u^+ is ε-monotone along $\tau \in \Gamma(\theta, e_n)$ and $\partial\Omega^+(u)$ is contained in an ε-neighborhood of the graph A of a Lipschitz function $x_n = f(x')$, it could be that $u^- \equiv 0$ below A and $u^- > 0$ somewhere between A and $\Omega^+(u)$. In this case u^- is not ε-monotone for any ε.

The following theorem covers this situation ([C2]).

Theorem 5.8. *Let u be a solution of our free boundary problem in C_1 with $0 \in F(u)$. Suppose that the following hold.*

(i) $\alpha_0 \leq u^+(x)/d(x, F(u)) \leq \alpha_1$, *i.e., u^+ is Lipschitz and has linear growth.*

(ii) *G is increasing, $s^{-N}G(s)$ is decreasing and moreover G is Lipschitz continuous with $G(0) > 0$.*

Then there exist

(a) *a $\theta_0 < \pi/2$,*

(b) *an ε_0*

such that if u^+ is ε-monotone along every $\tau \in \Gamma(\theta_0, e_n)$ for some $\varepsilon \leq \varepsilon_0$, then u^+ is fully monotone in $C_{1/2}$ along any $\tau \in \Gamma(\theta_1, e_n)$, with $\theta_1 = \theta_1(\theta_0, \varepsilon_0)$.

In particular, $F(u)$ is the graph of a Lipschitz function and therefore, by Theorem 4.1, it is a graph of a $C^{1,\gamma}$-function.

The strategy of the proof is to balance the situation in which u^+ highly predominates over u^- and the case in which u^- is not too small with respect to u^+.

The next basic lemma deals with this dichotomy; A always denotes the graph of the Lipschitz function f. It says that as long as u^- is very small, we can improve the ε-monotonicity of u^+ and when u^- goes through a threshold, the whole of u becomes ε-monotone.

Lemma 5.9. *Let u be as in Theorem 5.8 and let*

$$m = \sup_{\mathcal{C}_{7/8}} |u| \ .$$

Then there exist θ_0 and ε_0 such that if $\theta \geq \theta_0 > \theta_1 > \pi/4$ and $\varepsilon \leq \varepsilon_0$, we have the following alternative: there is a large constant K such that

(a) *if $u^-(-\frac{1}{2}e_n) < Km\varepsilon$, then u^+ is $\lambda\varepsilon$-monotone in $\mathcal{C}_{1-\varepsilon^{1/8}}$ along any $\tau \in \Gamma(\theta - \varepsilon^{1/4}, e_n)$ with $\lambda = \lambda(\theta_0) < 1$,*

(b) *if $u^-(-\frac{1}{2}e_n) \geq Km\varepsilon$, then u is $c\varepsilon^{1/8}$-monotone along any $\tau \in \Gamma(\theta_1, e_n)$ in a δ_0-neighborhood of A in $\mathcal{C}_{1/2}$, $\delta_0 = \delta_0(\theta_0)$.*

Proof. Let $\rho \gg \varepsilon$ and let $x_0 \in F(u) \cap \mathcal{C}_{1-2\rho}$ be a point where $F(u)$ has a touching ball contained in $\{u \leq 0\}$. Then near x_0,

$$u^-(x) = \alpha\langle x - x_0, \nu\rangle^+ - \beta\langle x - x_0, \nu\rangle^- + o(|x - x_0|) \ .$$

From the monotonicity formula and (i) in Theorem 5.8, we have

$$0 \leq \beta \leq c\frac{\alpha_1}{\alpha_0}\frac{m}{\rho} \ .$$

In particular this implies that u^- is Lipschitz continuous in $\mathcal{N}_\rho(F(u)) \cap \mathcal{C}_{1-2\rho}$ with Lipschitz constant

$$\mathrm{Lip}(u^-) \leq \bar{c}\, m/\rho \ .$$

Therefore, if $x \in \mathcal{N}_\varepsilon(A) \cap \mathcal{C}_{1-2\rho}$, then

$$u^-(x) \leq \bar{c}\, m\varepsilon/\rho \ . \tag{5.8}$$

Now assume alternative (a), that is, $u^-(-\frac{1}{2}e_n) < Km\varepsilon$.

We want to control u^- below the graph A in $\mathcal{C}_{1-2\rho}$. Let v be the harmonic function in $\{x_n < f(x')\} \cap \mathcal{C}_{1-2\rho}$ with boundary values

$$v = \begin{cases} 0 & \text{on } A, \\ u^- & \text{otherwise.} \end{cases}$$

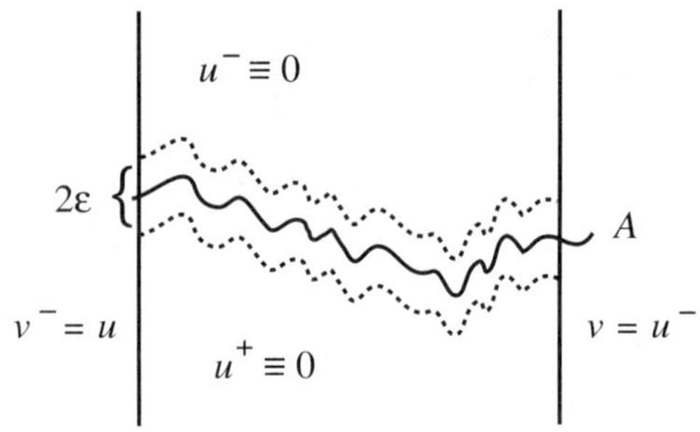

Figure 5.5

Then by the maximum principle, in $\{x_n \leq f(x')\} \cap \mathcal{C}_{1-2\rho}$

$$v \leq u^- \leq v + \bar{C}m\varepsilon/\rho \ . \tag{5.9}$$

On the other hand, from the boundary and interior Harnack inequalities, there is a large $k = k(\theta_0)$ such that, in $\mathcal{C}_{1-2\rho} \cap \{x_n < f(x')\}$,

$$v(x) \leq c\frac{v(-\frac{1}{2}e_n)}{\rho^k} \leq c\frac{u^-(-\frac{1}{2}e_n)}{\rho^k} \ . \tag{5.10}$$

The monotonicity formula gives now, at a regular point $x_0 \in F(u)$,

$$\beta^2 \leq c\frac{1}{\rho^2}\int_{B_\rho(x_0)} |\nabla u^-|^2 |x - x_0|^{2-n}\, dx \leq \frac{c}{\rho^2}(\sup_{B_\rho(x_0)} u^-)^2$$

or

$$\beta \leq \frac{c}{\rho}\left\{\frac{m\varepsilon}{\rho} + \frac{u^-(-\frac{1}{2}e_n)}{\rho^k}\right\} \ . \tag{5.11}$$

Choose ρ so that $\rho^{k+1} = \varepsilon^{1/2}$. Then from (5.11)

$$\beta \leq Cm\left\{\varepsilon^{1-\frac{1}{k+1}} + \varepsilon^{1/2}\right\} \leq Cm\varepsilon^{1/2} \ .$$

We can now proceed as in Lemma 5.7 with

$$u_1(x) = u^+(x - \lambda\varepsilon e_n)$$

and

$$u_2(x) = u^+(x) \ .$$

Construct the family $\bar{v}_t$ corresponding to u^+ instead of u. To make $\bar{v}_t$ an R-subsolution, at a point $x_0 \in F(\bar{v}_t)$, regular from the right, we must require (see formula (5.4)):

$$(1 + c\eta)\left(1 - c\frac{\sigma t}{\delta}\right) G(0) \geq G(Cm\varepsilon^{1/2}) \ .$$

Since $G(Cm\varepsilon^{1/2}) \le G(0) + Cm\varepsilon^{1/2}$, it is enough that

$$\eta \ge c\left\{\frac{\sigma}{\delta} + \varepsilon^{1/2}\right\} \tag{5.12}$$

which is satisfied with the usual choices $\sigma \sim \varepsilon$, $\delta = \varepsilon^{1/2}$, $\eta = \varepsilon^{1/4}$.

Then the proof goes as in Lemma 5.7 and we get that u^+ is $\lambda\varepsilon$-monotone along every $\tau \in \Gamma(\theta - C\varepsilon^{1/4}, e_n)$ in $\mathcal{C}_{1-\varepsilon^{1/8}}$, as before.

If alternative (b) holds, choose $\rho = 1/8$. Then (5.11) gives

$$\beta \le cu^-\left(-\frac{1}{2}e_n\right)$$

so that in $\mathcal{C}_{7/8}$

$$u^-(x) \le cu^-\left(-\frac{1}{2}e_n\right) d(x, F(u)) . \tag{5.13}$$

For the auxiliary function v, we use the results in Section 11.4 to deduce that, in $\mathcal{N}_{\delta_0}(A) \cap \{x_n < f(x')\}$, with $\delta_0 = \delta_0(\theta_0)$,

(1) v is (fully) monotone along every direction τ in the cone $\Gamma(\theta_1, -e_n)$, if $\theta_1 \le \theta_0 - \frac{\pi}{8}$ (say);

(2) if $x_1, x_2 \in \mathcal{N}_{\delta_0}(A) \cap \{x_n < f(x')\}$ and

$$\tau = \frac{x_2 - x_1}{|x_2 - x_1|} \in \Gamma(\theta_1, -e_n),$$

then for $\frac{1}{2} \le \lambda \le 1$ and $x = x_1 + \lambda(x_2 - x_1)$,

$$D_\tau v(x) \ge c\frac{v(x_2)}{\delta_0} . \tag{5.14}$$

Furthermore, let $h = h(z)$ be the auxiliary harmonic function of the form

$$h(z) = r^a g(\sigma) \qquad (z = r\sigma,\ r = |z|,\ \sigma \in \partial B_1)$$

with g the first eigenfunction of the Laplace-Beltrami operator in the polar cap

$$\langle \sigma, -e_n \rangle \le \theta_0 .$$

Then for $x_0 \in A \cap \mathcal{C}_{7/8}$ and $x - x_0 \in \Gamma(\theta_1, -e_n)$, using Harnack's inequality and the maximum principle,

$$v(x) \ge cv\left(-\frac{1}{2}e_n\right) h(x - x_0) \ge c\left\{u^-\left(-\frac{1}{2}e_n\right) - C\varepsilon m\right\} h(x - x_0).$$

Thus, if the constant K in (b) is large enough and $x - x_0 \in \Gamma(\theta_1, -e_n)$, we can write

$$v(x) \ge c\,u^-\left(-\frac{1}{2}e_n\right) |x - x_0|^a . \tag{5.15}$$

We are now ready to prove the $c\varepsilon^{1/8}$-monotonicity of u^- along every $\tau \in \Gamma(\theta_1, -e_n)$, in $\mathcal{N}_{\delta_0/2}(A) \cap \{x_n < f(x')\}$. Let x_1, x_2 satisfy

$$c_2\varepsilon^{1/8} \geq |x_1 - x_2| \geq c_1\varepsilon^{1/8}$$

and let $x_1 - x_2 \in \Gamma(\theta_1, -e_n)$. We want to show that

$$u^-(x_2) \geq u^-(x_1) \; . \tag{5.16}$$

If $x_1 \in \Omega^+(u)$, the inequality is trivial, so it is enough to consider x_1 below the graph $\{x_n = f(x') + \varepsilon\} = A_\varepsilon$.

If x_1, x_2 are both below A, then from (5.4)

$$v(x_2) - v(x_1) \geq c\frac{v(x_2)}{\delta_0}\varepsilon^{1/8}$$

and since (5.15) gives

$$v(x_2) \geq cu^-\left(-\frac{1}{2}e_n\right)\varepsilon^{a/8}, \tag{5.17}$$

we obtain

$$v(x_2) - v(x_1) \geq \frac{c}{\delta_0}u^-\left(-\frac{1}{2}e_n\right)\varepsilon^{(a+1)/8} \; .$$

On the other hand, from (5.9), we have

$$v \leq u^- \leq v + cu^-\left(-\frac{1}{2}e_n\right)\varepsilon$$

in $\mathcal{C}_{7/8}$. Thus

$$\begin{aligned} u^-(x_2) - u^-(x_1) &\geq v(x_2) - v(x_1) - cu^-\left(-\frac{1}{2}e_n\right)\varepsilon \\ &\geq \frac{c}{\delta_0}u^-\left(-\frac{1}{2}e_n\right)\left\{\varepsilon^{(a+1)/8} - c\varepsilon\right\} \; . \end{aligned}$$

If we choose θ_0 close enough to $\pi/2$, then a is close to 1. Hence we may assume $a < 7$ and get

$$u^-(x_2) - u^-(x_1) \geq 0 \; .$$

If x_2 is below A and x_1 is between A and A',

$$u^-(x_2) - u^-(x_1) \geq v(x_2) - cu^-\left(-\frac{1}{2}e_n\right)\varepsilon \geq \frac{c}{\delta_0}u^-\left(-\frac{1}{2}e_n\right)\left\{\varepsilon^{a/8} - c\varepsilon\right\} \geq 0$$

and alternative (b) is proved. □

Proof of Theorem 5.8. Either we are already in alternative (b) or we iterate alternative (a) until we reach (if ever) alternative (b). □

A variant of Theorem 5.8 follows.

Theorem 5.10. *In the hypotheses of Theorem 5.8 replace the ε-monotonicity of u^+ with the following condition: there exist $\bar{\theta} < \frac{\pi}{2}$ and $\bar{\varepsilon}$ such that if for $\varepsilon < \bar{\varepsilon}$, $F(u)$ is contained in an ε-neighborhood of the graph of a Lipschitz function $x_n = f(x')$, with Lipschitz norm*

$$\mathrm{Lip}(f) \leq \tan\left(\frac{\pi}{2} - \bar{\theta}\right) .$$

Then the same conclusions hold.

Proof. We show that u^+ is $C\varepsilon$-monotone in $\Gamma(\theta, e_n)$ for suitable θ and ε, thus reducing Theorem 5.10 to Theorem 5.8.

Let θ_0 be as in Theorem 5.8 and choose $\bar{\theta} > \theta_0$. Consider the harmonic function v in $\mathcal{C}_{7/8}$, vanishing on $A_\varepsilon = \{x_n = f(x') - c_0\varepsilon\}$ and $v = u^+$ on $\partial\mathcal{C}_{7/8} \cap \{x_n > f(x') - c_0\varepsilon\}$. Choose c_0 in such a way that $A_\varepsilon \subset \mathcal{C}\Omega^+(u)$.

Then by the maximum principle, Harnack's inequality, and the linear growth of u^+,

$$u^+(x) + 2c_0\varepsilon \geq v(x) \geq u^+(x) \geq c_1 d(x, A_\varepsilon) - c_2\varepsilon$$

in $\Omega^+(u)$, if c_0 is large enough.

Moreover, for any $\theta_0 < \theta < \bar{\theta}$, v is fully monotone along any $\tau \in \Gamma(\theta, e_n)$ in a δ_0-neighborhood of A_ε in $\mathcal{C}_{3/4}$, with $\delta_0 = \delta_0(\theta_0, \bar{\theta})$. In particular, if

$$\delta_0 \geq d(x, A_\varepsilon) > c_3\varepsilon,$$

we get (Lemma 5.5)

$$D_\tau v(x) \sim \frac{v(x)}{d(x, A_\varepsilon)} \geq c_1 - c_2\frac{\varepsilon}{d(x, A_\varepsilon)} \geq \frac{1}{2}c_1 > 0$$

if $\varepsilon \leq c_1\delta_0/2c_2$. Therefore, in $\Omega^+(u)$, if $\tau \in \Gamma(\theta, e_n)$,

$$\begin{aligned} u^+(x + c_4\varepsilon\tau) - u^+(x) &\geq v(x + c_4\varepsilon\tau) - v(x) - 2c_0\varepsilon \\ &\geq c(c_4 - c_3)\varepsilon - 2c_0\varepsilon > 0 \end{aligned}$$

if c_4 is large enough. □

Chapter 6

Existence Theory

6.1. Introduction

In this section we construct a viscosity solution of our free boundary problem in a smooth (Lipschitz) bounded domain $\Omega \subset \mathbb{R}^n$.

We use a variant of the Perron method, by taking the infimum of suitable (admissible) supersolutions and having in mind producing solutions with all the measure-theoretic properties considered in Section 3.2, eventually falling under the hypotheses of one of the regularity theorems of Section 5.6 after an appropriate rescaling.

As a warning to the fact that this is not obvious, consider the singular perturbation problem

$$\Delta u_\varepsilon = f_\varepsilon(u_\varepsilon)$$

with f_ε as in Section 1.2. When $\varepsilon \to 0$, we expect a solution to converge to a function u, harmonic when positive or negative, and satisfying on $\partial\Omega^+(u)$ the free boundary condition

$$(u_\nu^+)^2 - (u_\nu^-)^2 = 1 \ .$$

A one-dimensional solution $u_\varepsilon = u_\varepsilon(x_n)$, with slope $\alpha < 1$ at $+\infty$, will "bounce" in the layer $0 < u_\varepsilon < \varepsilon$ and it comes out with opposite slope $-\alpha$.

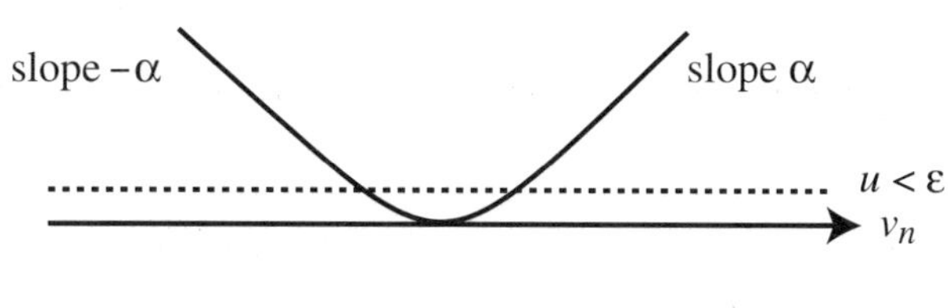

Figure 6.1

Formally, in the limit we get a two-plane function

$$u(x) = \alpha|x_n|$$

which is a viscosity solution of the free boundary problem. Clearly, the free boundary does not have the usual measure-theoretic properties; for instance its reduced part is empty.

To avoid this anomalous solution, one must carefully select the class of admissible supersolutions.

There are two main requirements:

- They must be suitable for comparison principles, like the R-subsolutions defined in Section 4.3.
- If we want the infimum of the selected supersolutions to be a solution in the viscosity sense, it is reasonable to ask them also to be supersolutions in the viscosity sense. This requires the appropriate asymptotic behavior at a regular point from the *right*.

We achieve these two goals by requiring that every point of their free boundary be regular from the *left*, with the appropriate "concave" asymptotic behavior.

Precisely, let Ω be a bounded Lipschitz domain in $\mathbb{R}^n$.

Definition 6.1. A function w continuous in $\bar{\Omega}$ is an *admissible supersolution* if the following hold.

(a) $\Delta w \leq 0$ in $\Omega^+(w) = \{w > 0\}$ and $\Omega^-(w) = \{w \leq 0\}^0$.

(b) Every point $x_0 \in F(w) = \partial\Omega^+(w) \cap \Omega$ is regular from the *left* and near x_0

$$w^+(x) \leq \alpha\langle x - x_0, \nu\rangle^+ + o(|x - x_0|),$$
$$w^-(x) \geq \beta\langle x - x_0, \nu\rangle^- + o(|x - x_0|)$$

with

$$\alpha < G(\beta) ,$$

where ν is the normal unit vector to the ball at the touching point x_0, inward to $\Omega^+(w)$.

Remark. No uniformity whatsoever is required on the size of the tangent ball.

We denote by $\mathcal{F}$ the family of *admissible supersolutions*. The Perron method also requires a *subsolution minorant*.

Definition 6.2. A continuous function $\underline{u}$ in $\bar{\Omega}$ is a *strict minorant* if the following hold.

(a) $\underline{u}$ is locally Lipschitz.

(b) $\Delta \underline{u} \geq 0$ in $\Omega^+(\underline{u})$ and $\Omega^-(\underline{u})$.

(c) Every point $x_0 \in F(\underline{u})$ is regular from the right and near x_0,

$$\underline{u}^-(x) \leq \beta\langle x - x_0, \nu\rangle^- + o(|x - x_0|),$$
$$\underline{u}^+(x) \geq \alpha\langle x - x_0, \nu\rangle^+ + o(|x - x_0|)$$

with

$$\alpha > G(\beta) \ .$$

Thus, $\underline{u}$ is an R-subsolution. Observe that if $w \in \mathcal{F}$ and $w_{|\partial\Omega} \geq \underline{u}_{|\partial\Omega}$, then $w > \underline{u}$ in Ω.

The main theorem follows ([C3]).

Theorem 6.1. *Let Ω be a Lipschitz domain and let g be a continuous function on $\partial\Omega$. Suppose $\mathcal{F}$ is nonempty and that there exists a strict minorant $\underline{u}$ with $\underline{u} = g$ on $\partial\Omega$. Then if G is strictly increasing and $G(0) > 0$, the function*

$$u = \inf\{w : w \in \mathcal{F},\ w \geq \underline{u} \text{ in } \bar{\Omega}\}$$

is a viscosity solution of our free boundary Dirichlet problem.

Example. Assume that Ω is the annulus $B_2 \setminus \bar{B}_1$, $g > 0$ on ∂B_2, $g < 0$ on ∂B_1. If we consider $u_0 = 0$ on $\partial B_{2-\delta}$ and harmonic otherwise, we obtain a subsolution for δ small. On the other hand if $u_1 = 0$ on $\partial B_{1+\delta}$ and if it is harmonic otherwise, we obtain a supersolution. We will construct a solution u between u_0 and u_1.

We will call u a *minimal viscosity solution*. As a byproduct of the proof, u turns out to be locally Lipschitz and nondegenerate. This and the fact that u is constructed as an infimum of "nice" R-supersolutions gives that the perimeter of the reduced part $F^*(u)$ of the free boundary is equivalent to the surface measure (Section 6.7).

Then after an appropriate rescaling of u around a point of $F^*(u)$, the free boundary becomes flat, in the sense of Theorem 5.10, and therefore it is a $C^{1,\gamma}$-surface.

The strategy of the proof of Theorem 6.1 consists in the following steps.

(1) u^+ is locally Lipschitz.

(2) u^- is locally Lipschitz (therefore u is locally Lipschitz).

(3) u^+ is nondegenerate.

(4) u is a viscosity supersolution.

(5) u is a viscosity subsolution.

6.2. u^+ is locally Lipschitz

We first show that given w in $\mathcal{F}$, its harmonic replacement $\tilde{w}$ in $\Omega^+(w)$ still belongs to the family $\mathcal{F}$.

Lemma 6.2. *Let $w \in \mathcal{F}$. There is $\tilde{w} \in \mathcal{F}$ such that*

(a) $\Delta\tilde{w} = 0$ *in* $\Omega^+(\tilde{w})$,

(b) $\tilde{w}^+ \le w$, $\tilde{w}^- = w^-$, $\Omega^+(\tilde{w}) \subseteq \Omega^+(w)$ *and* $F(\tilde{w}) \subseteq F(w)$,

(c) $\tilde{w} \ge \underline{u}$.

Proof. We solve the Dirichlet problem $\Delta\tilde{w} = 0$ in $\Omega^+(w)$, $\tilde{w} = w$ on $\partial\Omega^+(w)$.

Since Ω is Lipschitz and $F(w) = \partial\Omega^+(w) \cap \Omega$ has a tangent ball from outside at every point, $\tilde{w}$ is uniquely defined and continuous in $\bar{\Omega}^+(w)$. Clearly $\Omega^+(\tilde{w}) \subseteq \Omega^+(w)$ and since $\tilde{w}$ may become identically zero in some connected component of $\Omega^+(w)$, it could be that $\Omega^+(\tilde{w}) \subset \Omega^+(w)$. Nevertheless $F(\tilde{w}) \subseteq F(w)$. Indeed if $x_0 \in F(\tilde{w}) \setminus F(w)$, then x_0 is an interior point of a domain where $\Delta\tilde{w} = 0$, $\tilde{w} \ge 0$, $\tilde{w} \not\equiv 0$ in any neighborhood of x_0. Since $\tilde{w}(x_0) = 0$, we have a contradiction.

Therefore any point of $F(\tilde{w})$ is regular from the left. If w^+ has the asymptotic development

$$w^+(x) \le \alpha\langle x - x_0, \nu\rangle^+ + o(|x - x_0|),$$

then since $\tilde{w}^+ \le w^+$,

$$\tilde{w}^+(x) \le \alpha\langle x - x_0, \nu\rangle^+ + o(|x - x_0|)$$

and therefore

$$\tilde{w}^-(x) = w^-(x) \ge \beta\langle x - x_0, \nu\rangle^- + o(|x - x_0|)$$

with

$$\alpha < G(\beta) \ .$$

Thus, $\tilde{w} \in \mathcal{F}$ and clearly $\tilde{w} \ge \underline{u}$. □

The Lipschitz continuity of u^+ is a consequence of the following lemma.

Lemma 6.3. *Let $w \in \mathcal{F}$, with w^+ harmonic in $\Omega^+(w)$. Assume that near $x_0 \in F(w)$,*

$$w^+(x) = \alpha\langle x - x_0, \nu\rangle^+ + o(|x - x_0|) \ .$$

Then the following hold.

(a) *If $h = d(x_0, \partial\Omega)$,*

$$\alpha G^{-1}(\alpha) \le \frac{c}{h^2}\|w\|^2_{L^\infty(\Omega)}. \tag{6.1}$$

(b) *For any* $D \subset\subset \Omega$, w^+ *is locally Lipschitz in* D *with Lipschitz constant* L_D^+ *satisfying*

$$L_D^+ G^{-1}(L_D^+) \le c\left(\frac{\|w\|_{L^\infty(D)}}{d(D,\partial\Omega)}\right)^2 \tag{6.2}$$

where $c = c(\operatorname{diam} D)$.

(c) *In particular,* $\Omega^+(u)$ *is open.*

Proof. (a) From Definition 6.1, w^- has the asymptotic behavior

$$w^-(x) \ge \beta\langle x - x_0, \nu\rangle^- + o(|x - x_0|)$$

with $\alpha < G(\beta)$.

Thus, if $G^{-1}(\alpha) = 0$, (6.1) is trivial and α is bounded.

If $G^{-1}(\alpha) > 0$, then since $\Delta w^- \ge 0$ in $\Omega^-(w)$, we can use the monotonicity formula, Theorem 12.3, and Corollary 12.6, in particular Lemma 6.4, to get

$$\alpha^2\beta^2 \le c\frac{\|w\|^4_{L^\infty(\Omega)}}{h^4}$$

and therefore also (6.1).

To prove (b), let $x_0 \in \Omega^+(w) \cap D$ and let $y_0 \in F(w)$ be such that

$$|x_0 - y_0| = d(x_0, F(w)) = r < \frac{1}{2}d(D, \partial\Omega)\ .$$

Then

$$|\nabla w(x_0)| \le c\frac{w(x_0)}{r}\ .$$

By the usual comparison with a radial harmonic function in the ring $B_r(x_0)\setminus \bar{B}_{r/2}(x_0)$, we get, near y_0,

$$w^+(y) \ge c\frac{w(x_0)}{r}\langle y - y_0, \nu\rangle^+\ .$$

(b) now follows from (a). □

6.3. *u* is Lipschitz

We now prove that u^- also is Lipschitz. Notice first that

Lemma 6.4. *If* $w_1, w_2 \in \mathcal{F}$, $w^* = \min\{w_1, w_2\}$ *also belongs to* $\mathcal{F}$.

Proof. Since

$$\Omega^+(w^*) = \Omega^+(w_1) \cap \Omega^+(w_2),$$

any ball B touching $F(w_1)$ or $F(w_2)$ from the left will also touch $F(w^*)$ from the left.

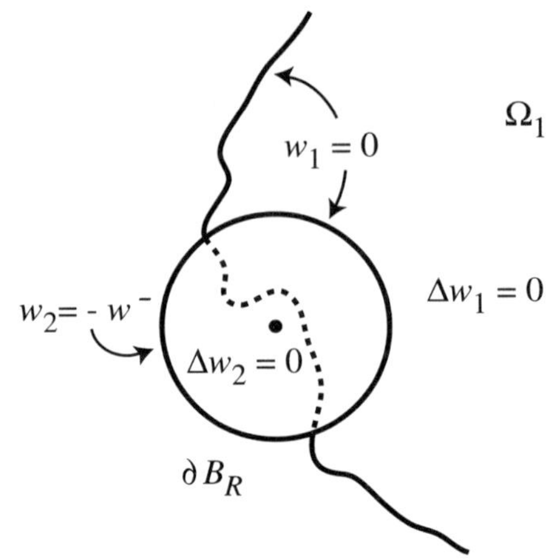

Figure 6.2

The asymptotic developments in (b) of Definition 6.1 are clearly satisfied, since they come from those of w_1 or w_2. □

The proof of the Lipschitz continuity of u^- is based on the following replacement technique.

Let w be a continuous function in $\bar{\Omega}$ and let B_R be a ball, $B_R \subset\subset \Omega$. Let

$$\Omega_1 = \Omega^+(w) \setminus \bar{B}_R.$$

In Ω_1, solve $\Delta w_1 = 0$ with boundary data

$$w_1 = 0 \text{ on } \partial B_R , \quad w_1 = w^+ \text{ on } \partial\Omega_1 \setminus \partial B_R .$$

In B_R, solve $\Delta w_2 = 0$ with boundary data $w_2 = -w^-$. We call the function

$$R(w, B_R) = \begin{cases} w_1 & \text{in } \bar{\Omega}_1, \\ w_2 & \text{in } \bar{B}_R, \\ w & \text{in } \Omega \setminus (\bar{\Omega}_1 \cup \bar{B}_R) \end{cases}$$

the *replacement* of w in B_R. By the maximum principle, $R(w, B_R) \leq w$ in Ω.

We want to show that if $w \in \mathcal{F}$, its replacement in a ball $B_R(x_0)$ of radius $R \ll w^-(x_0)$ is still in $\mathcal{F}$.

Thus, let $w \in \mathcal{F}$ with w^+ harmonic in $\Omega^+(w)$ and $w(x_0) = -h < 0$. Fix $\varepsilon > 0$, small, to be chosen depending only on the local Lipschitz constant of u^+ or, equivalently, on $d_0 = d(x_0, \partial\Omega)$, and let

$$\tilde{w} = R(w, B_{\varepsilon h}(x_0)) .$$

We want to show that $\tilde{w} \in \mathcal{F}$ and $\tilde{w} \geq \underline{u}$.

Note first that the strict minorant $\underline{u}$, being locally Lipschitz, is strictly negative in $B_{\varepsilon h}(x_0)$ for ε small, since $\underline{u}(x_0) < -h$. Hence $\tilde{w} \geq \underline{u}$. Let us see that $\tilde{w} \in \mathcal{F}$.

Clearly, the left ball condition holds for $F(\tilde{w})$. We check the asymptotic behaviors.

If $x_1 \in F(\tilde{w}) \cap F(w)$, since $\tilde{w} \leq w$, $\tilde{w}$ has the correct asymptotic behavior. Let now x_1 be a point on $F(\tilde{w})$ where $w(x_1) > 0$. Hence $x_1 \in \partial B_{\varepsilon h}(x_0) \cap F(\tilde{w})$. Since $w^+(x_0) = 0$ and w^+ is Lipschitz in $B_{d_{0/2}}(x_0)$,

$$\tilde{w} \leq w^+ \leq L_0 \varepsilon h \quad \text{in} \quad B_{2\varepsilon h}(x_0)$$

where L_0 is the local Lipschitz constant of w^+ in $B_{d_{0/2}}(x_0)$. Thus, if v is harmonic in the ring $B_{2\varepsilon h}(x_0) \setminus \bar{B}_{\varepsilon h}(x_0)$, vanishing on $\partial B_{\varepsilon h}(x_0)$, $v = L_0 \varepsilon h$ on $\partial B_{2\varepsilon h}(x_0)$, we have

$$\tilde{w} \leq v$$

in the ring. Moreover, since

$$v(x) \leq C\, d(x, \partial B_{\varepsilon h}(x_0)) \ ,$$

if near x_1 we have

$$\tilde{w}^+(x) = \alpha \langle x - x_1, \nu \rangle^+ + o(|x - x_1|) \ ,$$

it must be that $\nu = \frac{x_1 - x_0}{|x_1 - x_0|}$ and

$$\alpha \leq c_0 \ .$$

On the other hand, in $B_{\varepsilon h}(x_0)$ we have $\tilde{w} \leq 0$, $\Delta \tilde{w} = 0$ and

$$\tilde{w}(x_0) \leq w(x_0) = -h < 0 \ ,$$

so that by Harnack's inequality

$$\tilde{w}(x) \leq -ch \tag{6.3}$$

in $B_{\varepsilon h/2}(x_0)$. Thus, the usual comparison with a radial barrier gives, near x_1, in $B_{\varepsilon h}(x_0)$,

$$\tilde{w}(x) \leq -\frac{ch}{\varepsilon h} \langle x - x_1, \nu \rangle^- \qquad \left(\nu = \frac{x_1 - x_0}{|x_1 - x_0|} \right)$$

which implies

$$\tilde{w}^-(x) \geq \beta \langle x - x_1, \nu \rangle^-$$

with $\beta \geq \frac{c}{\varepsilon}$.

Therefore if we chose ε such that

$$\frac{c}{\varepsilon} > G^{-1}(c_0) \ ,$$

$\tilde{w}$ is an admissible supersolution. Moreover, once ε is fixed, from interior estimates and Harnack's inequality we get,

$$|\nabla \tilde{w}| \leq \frac{c}{\varepsilon h}|\tilde{w}(x_0)| \leq \frac{c}{\varepsilon} \text{ in } B_{\varepsilon h/2}(x_0)$$

which implies the equilipschitzianity of $\tilde{w}$.

We have proved the following lemma.

Lemma 6.5. *Let $w \in \mathcal{F}$, $w(x_0) = -h < 0$, $d_0 = d(x_0, \partial\Omega)$. Then there is $\varepsilon > 0$, depending only on d_0, such that the following hold.*

(a) *The replacement of w in $B_{\varepsilon h}(x_0)$, $\tilde{w} = R(w, B_{\varepsilon h}(x_0))$, belongs to $\mathcal{F}$.*

(b) *$\tilde{w} \leq w$ in Ω, $\tilde{w} \geq \underline{u}$ and*
$$|\nabla \tilde{w}| \leq \frac{c}{\varepsilon}$$
in $B_{\varepsilon h/2}(x_0)$.

Corollary 6.6. *Let x_0 be a point where*

$$u(x_0) = \inf\{w(x_0),\ w \in \mathcal{F},\ w > \underline{u}\} = -h < 0\ .$$

Then there exist a sequence $\{\tilde{w}_k\} \subset \mathcal{F}$, $\tilde{w}_k \geq \underline{u}$, and $\varepsilon > 0$, depending only on d_0, such that the following hold.

(a) *$\tilde{w}_k(x_0) \searrow u(x_0)$.*

(b) *$\Delta \tilde{w}_k = 0$ in $B_{\varepsilon h}(x_0)$.*

(c) *For each k, $\tilde{w}_k$ is Lipschitz in $B_{\varepsilon h}(x_0)$ with Lipschitz constant L_0, depending only on d_0.*

Proof. Let $\{w_k\} \subset \mathcal{F}$, $w_k > \underline{u}$, such that

$$w_k(x_0) \searrow u(x_0)\ .$$

Choose k large so that $w_k(x_0) \leq \frac{u(x_0)}{2} = -\frac{h}{2}$. Now replace w_k with $\tilde{w}_k$ in $B_{\varepsilon h}(x_0)$ and apply Lemma 6.5. □

Corollary 6.7. *u is locally Lipschitz in Ω, continuous in $\bar{\Omega}$ and harmonic in $\Omega^-(u)$ and $\Omega^+(u)$.*

Proof. We have only to show that u^- is Lipschitz and harmonic in $\Omega^-(u)$. Let $u(x_0) = -h < 0$ and let $\{\tilde{w}_k\}$ be as in Corollary 6.6. Then $\tilde{w}_k \searrow u$, uniformly on, say, $B_{\varepsilon h/4}(x_0)$. Indeed suppose $\tilde{w} = \lim_{k\to\infty} \tilde{w}_k$ and $\tilde{w}(x_1) > u(x_1)$, with $x_1 \in B_{\varepsilon h/4}(x_0)$. Consider a new sequence $\tilde{v}_k$ converging to u at x_1 and define $\tilde{u}_k$ the replacement of $\min\{\tilde{v}_k, \tilde{w}_k\}$ in $B_{\varepsilon h/2}(x_0)$. Then $\tilde{u}_k \searrow \tilde{u}$ with $\tilde{u} \leq \tilde{w}$ in $B_{\varepsilon h/2}(x_0)$, $\tilde{u}(x_0) = \tilde{w}(x_0)$, $\tilde{u}(x_1) < \tilde{w}(x_1)$. Since $\tilde{u}$ and $\tilde{w}$ are both harmonic in $B_{\varepsilon h/4}$, this is a contradiction. Thus, u^- is harmonic in $\{u < 0\}$.

Note also that since u^- is subharmonic in Ω, if $u(z) = 0$ for a point $z \in \Omega^-(u)$, then $u \equiv 0$ in the corresponding connected component of $\Omega^-(u)$. The proof is complete. □

Finally

Corollary 6.8. *If $K \subset\subset \Omega$, then there exists $\{w_k\} \subset \mathcal{F}$ such that*

$$w_k \searrow u$$

uniformly in K. If $K \subset\subset \Omega^-(u)$, we may take $w_k \leq 0$ in K.

Proof. The first part follows from the fact that $\{w^+; w \in \mathcal{F}\}$ is an equilipschitz family on $\bar{K}$ and from the previous replacement technique. For the second part let

$$K \subset\subset K_1 \subset\subset \Omega^-(u)$$

and assume K and K_1 are smooth domains. Let v be harmonic in $K_1 \setminus \bar{K}$, with $v = 1$ on ∂K_1 and $v \equiv 0$ in $\bar{K}$. If ε is small and ν is the exterior unit normal to ∂K,

$$\varepsilon v^+_{\nu|\partial K} < G(0) \ . \tag{6.4}$$

If $w_k \searrow u$ uniformly in $\bar{K}_1$, let k be so large that $w_k \leq \frac{\varepsilon}{2}$ in $\bar{K}_1$, and consider

$$\bar{w}_k = \begin{cases} \min\{w_k, \varepsilon v\} & \text{in } \bar{K}_1, \\ w_k & \text{in } \Omega \setminus \bar{K}_1. \end{cases}$$

Then $\bar{w}_k \leq 0$ in K and, thanks to (6.4), $\bar{w}_k \in \mathcal{F}$. □

Remark. u may have lost, of course, the property of being regular from the left ($\Omega^-(u)$). Thus u does not necessarily belong to $\mathcal{F}$ anymore.

6.4. u^+ is nondegenerate

In this section we use the fact that u is the infimum of R-supersolutions to show that u^+ is nondegenerate, with linear growth. As a consequence, $F(w_k) \to F(u)$ locally in Hausdorff distance and $\chi_{\{w_k>0\}} \to \chi_{\{u>0\}}$ in $L^1_{\text{loc}}(\Omega)$.

Lemma 6.9. (a) *If $x \in \Omega^+(u)$, then*

$$u(x) \geq Cd(x, F(u)).$$

(b) *Let $x \in F(u)$ and let A be a connected component of $\Omega^+(u) \cap [B_r(x) \setminus \bar{B}_{r/2}(x)]$ such that*

$$\bar{A} \cap \partial B_{r/2}(x) \neq \emptyset \ , \quad \bar{A} \cap \partial B_r(x) \neq \emptyset \ .$$

Then $\sup_A u \geq Cr$.

(c)

$$\frac{|A \cap B_r(x)|}{|B_r(x)|} \geq C > 0$$

where all the constants C's depend on $d(x, \partial\Omega)$.

Proof. (a) Let $x_0 \in \Omega^+(u)$, set $r = d(x_0, F(u))$ and assume that

$$d(x_0, \Omega^+(\underline{u})) \leq \frac{r}{2} .$$

By the Hopf maximum principle, $\underline{u}$ grows linearly away from $F(\underline{u})$ gives

$$\sup_{B_{3r/4}(x_0)} \underline{u} \geq Cr.$$

Hence

$$\sup_{B_{3r/4}(x_0)} u \geq Cr$$

and by Harnack's inequality

$$u(x_0) \geq Cr .$$

We may therefore assume $\underline{u} \leq 0$ on $B_{r/2}(x_0)$.

We want to show that if $u(x_0) = \sigma \ll r$, we can construct an admissible supersolution smaller than u at some point, giving a contradiction.

Select a sequence $\{w_k\} \subset \mathcal{F}$, uniformly converging to u in $B_R(x_0)$, $r \leq R$. By Harnack's inequality for u, we may choose $w \in \mathcal{F}$, with $w \leq c\sigma$ on $B_{r/2}(x_0)$. Let v be harmonic in the ring $B_{r/2}(x_0) \setminus \bar{B}_{r/4}(x_0)$, $v = 1$ on $\partial B_{r/2}(x_0)$ and $v = 0$ on $\partial B_{r/4}(x_0)$. Let $M > 0$ and define

$$\bar{w} = \begin{cases} 0 & \text{in } \bar{B}_{r/4}(x_0), \\ \min\{w, M\sigma v\} & \text{in } \bar{B}_{r/2}(x_0) \setminus \bar{B}_{r/4}, \\ w & \text{in } \Omega \setminus \bar{B}_{r/2}(x_0) . \end{cases}$$

Then $\bar{w} \geq \underline{u}$. For M large enough, depending only on n, $\bar{w}$ is continuous along $\partial B_{r/2}(x_0)$. On the other hand, along $\partial B_{r/4}(x_0)$, $v_\nu \sim \frac{C}{r}$, so that if $M\sigma \ll r$, $M\sigma v_\nu < G(0)$ and $\bar{w}$ is an admissible supersolution. Since $\bar{w}(x_0) = 0 < u(x_0)$, this is a contradiction. (a) implies that u^+ is a Lipschitz harmonic function in $\Omega^+(u)$ with linear growth. The conclusions in (b) and (c) then follow from Lemma 3.3. □

6.5. u is a viscosity supersolution

In this section we want to prove that u is a viscosity supersolution, that is, u satisfies condition (ii)(a) of Definition 2.4: if $x_0 \in F(u)$ and if at x_0 there is a touching ball $B \subset \Omega^+(u)$, then, near x_0, in B

$$u^+(x) = \alpha\langle x - x_0, \nu\rangle^+ + o(|x - x_0|) \qquad (\alpha > 0)$$

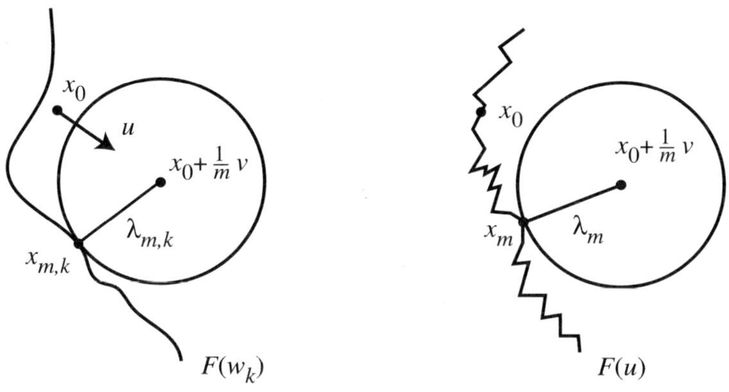

Figure 6.3

while in $\mathcal{C}B$,

$$u^-(x) = \beta\langle x - x_0, \nu\rangle^- + o(|x - x_0|) \qquad (\beta \geq 0)$$

with

$$\alpha \leq G(\beta) \ .$$

This is a consequence of the following slightly more general lemma.

Lemma 6.10. *Assume that $x_0 \in F(u)$ and near x_0*

$$u^+(x) = \alpha\langle x - x_0, \nu\rangle^+ + o(|x - x_0|) \ .$$

Then if

$$u^-(x) = \beta\langle x - x_0, \nu\rangle^- + o(|x - x_0|),$$

we must have

$$\alpha \leq G(\beta) \ .$$

Proof. Let $d_0 = d(x_0, \partial\Omega)$ and let $\{w_k\} \subset \mathcal{F}$, $w_k \searrow u$ uniformly in $B_{d_{0/2}}(x_0)$. Then given a neighborhood $B(x_0)$ of x_0, w_k cannot remain strictly positive in $B(x_0)$, for all k large; otherwise u would be a nonnegative harmonic function, and therefore identically zero.

For each w_k, let

$$B_{m,k} = B_{\lambda_{m,k}}\left(x_0 + \frac{1}{m}\nu\right)$$

be the largest ball centered at $x_0 + \frac{1}{m}\nu$, contained in $\Omega^+(w_k)$, touching $F(w_k)$ at $x_{m,k}$, with unit inward normal $\nu_{m,k}$ at $x_{m,k}$. Then, modulo subsequences, for $k \to +\infty$,

$$\lambda_{m,k} \to \lambda_m, \quad x_{m,k} \to x_m, \quad \nu_{m,k} \to \nu_m$$

with $B_{\lambda_m}(x_0 + \frac{1}{m}\nu)$ touching $F(u)$ at x_m, with unit inward normal ν_m.

Hence from the behavior of u^+,

$$|x_m - x_0| = o\left(\frac{1}{m}\right),$$
$$\frac{1}{m} + o\left(\frac{1}{m}\right) \le \lambda_m \le \frac{1}{m},$$
$$|\nu_m - \nu| = o(1) .$$

Now, since $w_k \in \mathcal{F}$, Definition 6.1 gives, near $x_{m,k}$ in $B_{m,k}$,

$$w_k^+(x) \le \alpha_{m,k}\langle x - x_{m,k}, \nu_{m,k}\rangle^+ + o(|x - x_{m,k}|)$$

and in $\mathcal{C}B_{m,k}$

$$w_k^-(x) \ge \beta_{m,k}\langle x - x_{m,k}, \nu_{m,k}\rangle^- + o(|x - x_{m,k}|)$$

with

$$0 < \alpha_{m,k} \le G(\beta_{m,k}) .$$

Since $w_k^+ \ge u^+$, we deduce

$$\alpha_m = \liminf_{k\to+\infty} \alpha_{m,k} \ge \alpha - \varepsilon_m$$

where $\varepsilon_m \to 0$ as $m \to +\infty$.

The proof will be complete if we show that

$$\bar{\beta} = \liminf_{m,k\to+\infty} \beta_{m,k} \le \beta .$$

This is a consequence of the upper semicontinuity of the limit in the monotonicity formula.

Indeed, if $\bar{\beta} = 0$, there is nothing to prove. Thus let $\beta_{m,k} > 0$. Fix $r > 0$ and let

$$J_r(x_{m,k}, w_k) = \frac{1}{r^4}\int_{B_r(x_{m,k})} \frac{|\nabla w_k^+|^2}{|x - x_{m,k}|^{n-2}}\,dx \int_{B_r(x_{m,k})} \frac{|\nabla w_k^-|^2}{|x - x_{m,k}|^{n-2}}dx.$$

Then from Corollary 12.6, for every r

(6.5) $$J_r(x_{m,k}, w_k) \ge c(n)\alpha_{m,k}^2\beta_{m,k}^2 .$$

On the other hand, we have

(6.6) $$\lim_{k\to+\infty} J_r(x_{m,k}, w_k) = J_r(x_m, u)$$

since $x_{m,k} \to x_m$ and $w_k \to u$ uniformly. For the same reason

(6.7) $$\lim_{m\to+\infty} J_r(x_m, u) = J_r(x_0, u) .$$

From Corollary 12.6 we also have

$$\lim_{r\to0} J_r(x_0, u) = c(n)\alpha^2\beta^2 .$$

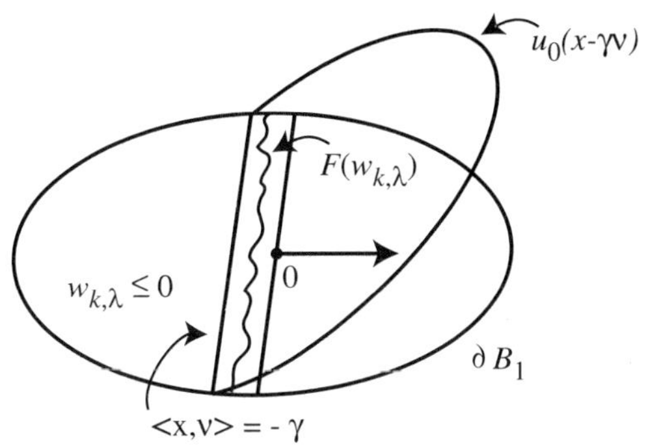

Figure 6.4

Hence, given $\varepsilon > 0$, there exists r such that

$$J_r(x_0, u) \leq c(n)\alpha^2\beta^2 + \varepsilon$$

and, from (6.5), (6.6), (6.7), there are m, k such that

$$c(n)\alpha_{m,k}^2\beta_{m,k}^2 \leq c(n)\alpha^2\beta^2 + 2\varepsilon \ .$$

Since $\liminf_{m,k\to+\infty} \alpha_{m,k} \geq \alpha$, it follows that $\bar{\beta} \leq \beta$. □

6.6. u is a viscosity subsolution

We now want to show that u is a subsolution in the viscosity sense, that is, u satisfies condition (ii)(b) of Definition 2.4: if $x_0 \in F(u)$ is a regular point from the left with touching ball $B \subset \Omega^-(u)$, then near x_0

$$u^-(x) = \beta\langle x - x_0, \nu\rangle^- + o(|x - x_0|), \qquad \beta \geq 0,$$

in B, and

$$u^+(x) = \alpha\langle x - x_0, \nu\rangle^+ + o(|x - x_0|), \qquad \alpha > 0,$$

in $\mathcal{C}B$, with

$$\alpha \geq G(\beta) \ .$$

Actually, due to the nondegeneracy of u^+, even if $\beta = 0$, both $\Omega^+(u)$ and $\Omega^-(u)$ are tangent to $\{\langle x - x_0, \nu\rangle = 0\}$ at x_0 and therefore u has a full asymptotic development, as in the next lemma.

Lemma 6.11. *Assume that near* x_0

$$u(x) = \alpha\langle x - x_0, \nu\rangle^+ - \beta\langle x - x_0, \nu\rangle^- + o(|x - x_0|)$$

with $\alpha > 0$, $\beta \geq 0$. *Then*

$$\alpha \geq G(\beta) \ .$$

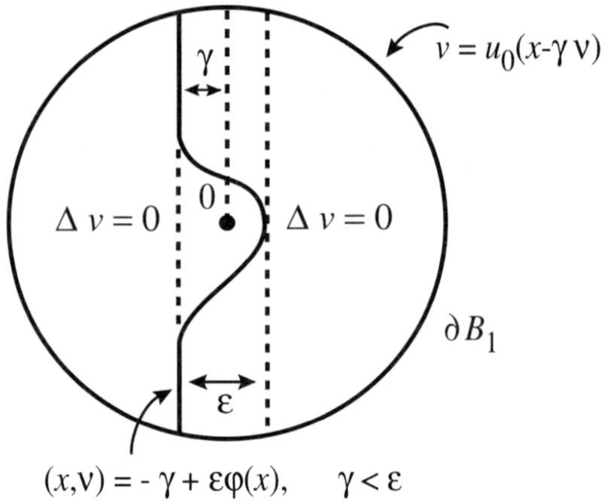

Figure 6.5

Proof. We use a perturbation argument to show that if $\alpha < G(\beta)$, we may construct a function $w \in \mathcal{F}$, smaller than u at some point, contradicting the minimality of u.

Let

$$u_0(x) = \lim_{\lambda \to 0} \frac{1}{\lambda} u(x_0 + \lambda x) = \alpha\langle x, \nu\rangle^+ - \beta\langle x, \nu\rangle^- .$$

Assume $\alpha \leq G(\beta) - \delta_0$, with $\delta_0 > 0$, and choose $w_k \searrow u$ uniformly so that (Corollary 6.8) $w_k \leq 0$ on any $K \subset\subset \Omega^-(u)$. Then for any $\gamma = \gamma(\delta) > 0$ to be chosen, we can find λ, k so that

$$w_{k,\lambda}(x) = \frac{1}{\lambda} w_k(x_0 + \lambda x)$$

satisfies the following.

(a) If $\beta > 0$, $w_{k,\lambda}(x) \leq u_0 + \gamma \min\{\alpha, \beta\}$ on ∂B_1.

(b) If $\beta = 0$, $w_{k,\lambda}(x) \leq u_0 + \alpha\gamma$ on ∂B_1 and

$$w_{k,\lambda}(x) \leq 0 \quad \text{in} \quad \{\langle x, \nu\rangle < -\gamma\} \cap \bar{B}_1 .$$

Equivalently, for any $\gamma > 0$,

$$w_{k,\lambda}(x) \leq u_0(x + \gamma\nu) \quad \text{in} \quad \bar{B}_1 .$$

We now make a standard perturbation of u_0 by changing its free boundary, still keeping it a supersolution:

Define v as follows:

$$\begin{cases} v(x) = u_0(x + \gamma\nu) & \text{on } \partial B_1, \\ v(x) = 0 & \text{on } \langle x, \nu\rangle = -\gamma + \varepsilon\varphi(x), \\ \Delta v = 0 & \text{if } \langle x, \nu\rangle \lessgtr -\gamma + \varepsilon\varphi(x) \end{cases}$$

where $\varphi \geq 0$ is a cut-off function, $\varphi \equiv 0$ outside $B_{1/2}$, $\varphi \equiv 1$ inside $B_{1/4}$. Along the new free boundary

$$F(v) = \{\langle x, \nu \rangle = -\gamma + \varepsilon\varphi(x)\}$$

we have by standard estimates

$$|v_\nu^+ - \alpha| \leq c(\varepsilon + \gamma) \ , \quad |v_\nu^- - \beta| \leq c(\varepsilon + \gamma) \ .$$

Hence, if we choose ε, γ small, depending only on δ_0, and rescale back, the function

$$\bar{w}_k = \begin{cases} \min\{w_k, \lambda v(\frac{1}{\lambda}(x - x_0))\} & \text{in } B_\lambda(x_0), \\ \quad w_k & \text{in } \Omega \setminus B_\lambda(x_0) \end{cases}$$

becomes an element of $\mathcal{F}$, since

$$v_\nu^+ < G(v_\nu^-)$$

along $F(v)$.

Moreover, if $\gamma \ll \varepsilon$, the set

$$\{\langle x, \nu \rangle \leq -\gamma + \varepsilon\varphi(x)\}$$

contains a neighborhood of the origin, which implies, after rescaling back, $x_0 \in \Omega^-(\bar{w}_k)$, a contradiction since $\Omega^+(\bar{w}_k) \supseteq \Omega^+(u)$. □

Remark. Let us stress once more that the minimal viscosity solution does not necessarily belong to $\mathcal{F}$.

6.7. Measure-theoretic properties of $F(u)$

The minimal viscosity solution constructed in Theorem 6.1 is locally Lipschitz and u^+ is nondegenerate. Thus, we can apply Theorem 3.4 to u^+ and conclude that $F(u)$ *has locally finite* $(n-1)$*-dimensional Hausdorff measure; moreover*, for H^{n-1}-a.e. point $x \in F(u)$ and any ball $B_r(x) \subset\subset \Omega$,

$$H^{n-1}(F(u) \cap B_r(x)) \sim r^{n-1} \ .$$

In particular, $\Omega^+(u) \cap B_r(x)$ is a set of finite perimeter. As we pointed out in the introduction, there is always the danger of ending up, locally, with a solution u that behaves like $u \sim |x_1|$ for which our "flatness implies $C^{1,\alpha}$ theorem" will not apply in a nonnegligible part of the free boundary. We show now that this is not the case for our u. This is because the minimality feature of u implies that the perimeter of this set is also equivalent to r^{n-1}. Precisely,

Theorem 6.12. *The reduced boundary of* $\Omega^+(u)$ *has positive density in* H^{n-1}*-measure at any point of* $F(u)$*, i.e.,*

$$H^{n-1}(F^*(u) \cap B_r(x)) \geq c r^{n-1}$$

for every $x \in F(u)$*.*

Proof. Let $x_0 \in F(u)$. By rescaling, letting

$$u_r(x) = \frac{1}{r} u(r(x - x_0)),$$

it is enough to consider $r = 1$ and $x_0 = 0$. Let v be the auxiliary function (see the proof of Theorem 3.4) satisfying ($\sigma < 1$)

$$\begin{cases} \Delta v = -\dfrac{1}{|B_\sigma|} \chi_{B_\sigma} & \text{in } B_1, \\ v = 0 & \text{on } \partial B_1. \end{cases}$$

Now take $w \in \mathcal{F}$. Since $w > u$ in $\bar{\Omega}^+(u)$, $\Omega^+(u) \cap B_1 \subset\subset \Omega^+(w) \cap B_1$, so that ∇w is a continuous vector field in $\overline{\Omega^+ \cap B_1}$ and we can use it to test for perimeter.

Therefore we can write

$$\text{(6.8)} \qquad \int_{\Omega^+(u)\cap B_1} (v\Delta w - w\Delta v) = \int_{F^*(u)\cap B_1} (v w_\nu - w v_\nu)\, dH^{n-1} - \int_{\partial B_1 \cap \Omega^+(u)} w v_\nu\, dH^{n-1}.$$

Now choose $w_k \searrow u$ uniformly in $\bar{B}_1$, $\Delta w_k = 0$ in $\Omega^+(u)$. Since the w_k are uniformly lipschitz, $|v_\nu| \sim c_0$ on ∂B_1 and $0 \le v < c\sigma^{2-n}$ outside B_σ, we have

$$\Big| \int_{F^*(u)\cap B_1} (w_k)_\nu v\, dH^{n-1} \Big| \le c\sigma^{2-n} H^{n-1}(F^*(u) \cap B_1) .$$

Moreover, as $k \to +\infty$,

$$\int_{F^*(u)\cap B_1} w_k v_\nu\, dH^{n-1} \to 0 , \qquad \int_{\partial B_1 \cap \Omega^+(u)} w_k v_\nu\, dH^{n-1} \to \int_{\partial B_1} u^+ v_\nu\, dH^{n-1}$$

and

$$-\int_{\Omega^+(u)\cap B_1} w_k \Delta v = \frac{1}{|B_\sigma|} \int_{\Omega^+(u)\cap B_\sigma} w_k \to ⨍_{B_\sigma} u^+ .$$

Thus, from (6.8) we get

$$\text{(6.9)} \qquad ⨍_{B_\sigma} u^+ + \int_{\partial B_1} u^+ v_\nu\, dH^{n-1} \le c\sigma^{2-n} H^{n-1}(F^*(u) \cap B_1) .$$

By the nondegeneracy of u^+ we can write

$$\int_{\partial B_1} u^+ v_\nu\, dH^{n-1} \ge c_1 > 0$$

while by Lipschitz continuity,

$$⨍_{B_\sigma} u^+ \le c_2 \sigma .$$

Choosing $\sigma = c_1/2c_2$, the proof is complete. □

Remark. In general, if $\partial\Omega$ has finite $(n-1)$-dimensional measure, $H^{n-1}_{\lfloor\partial^*\Omega}$ is absolutely continuous w.r.t. H^{n-1} with density f, $0 \le f \le 1$, H^{n-1}-a.e.

The theorem shows that in our case f is strictly positive on $F^*(u)$:

$$0 < \mu \le f(x) \le 1\ , \qquad H^{n-1}\text{-a.e. on } F^*(u)\ .$$

In particular

$$H^{n-1}(F(u) \setminus F^*(u)) = 0\ .$$

6.8. Asymptotic developments

We want to show that H^{n-1}-a.e. along $F(u)$ and in particular at every point x_0 of the reduced free boundary $F^*(u)$, u has an appropriate asymptotic development.

First of all, recall that if Ω is a set of finite perimeter and if $\partial^*\Omega$ is its reduced boundary, the following structural theorem holds [G]:

Theorem 6.13. *Let Ω be a set of finite perimeter. Then for every point $x \in \partial^*\Omega$, there exists a normal unit vector $\nu(x)$ in the* measure-theoretic sense, *i.e.*,

$$\lim_{r\to 0} \frac{|\Omega \cap B_r(x) \cap \{\langle y-x, \nu\rangle \ge 0\}|}{|B_r(x)|} = \frac{1}{2} \tag{6.10}$$

and

$$\lim_{r\to 0} \frac{|C\Omega \cap B_r(x) \cap \{\langle y-x, \nu\rangle \le 0\}|}{|B_r(x)|} = \frac{1}{2}. \tag{6.11}$$

In other words, (setting $x = 0$) if

$$\Omega_r = \{x : rx \in \Omega\}\ ,$$

(6.10) and (6.11) are equivalent to

$$\frac{|\Omega_r \cap B_1 \cap \{\langle y, \nu\rangle \ge 0\}|}{|B_1|} \to \frac{1}{2}$$

and

$$\frac{C\Omega_r \cap B_1 \cap \{\langle y, \nu\rangle \le 0\}}{|B_1|} \to \frac{1}{2}\ ,$$

respectively. That is, $\Omega_r \cap B_1$ and $C\Omega_r \cap B_1$ converge in measure to the half ball $B_1 \cap \{\langle y, \nu\rangle \ge 0\}$ and $B_1 \cap \{\langle y, \nu\rangle \le 0\}$, respectively.

We have

Lemma 6.14. *If $x_0 \in F^*(u)$, u has at x_0 the asymptotic development ($\nu = \nu(x_0)$)*

$$u(x) = \alpha\langle x - x_0, \nu\rangle^+ - \beta\langle x - x_0, \nu\rangle^- + o(|x - x_0|)$$

with

$$\alpha = G(\beta)\ .$$

Before proving Lemma 6.14, let us point out some general facts, regarding perturbations of two-plane functions.

Let $\alpha > 0$, $\beta \geq 0$ and let

$$u_0(x) = \alpha x_n^+ - \beta x_n^- \ . \tag{6.12}$$

Let φ be a cut-off function in B_1, $\varphi \equiv 1$ in $B_{1/3}$, $\varphi \equiv 0$ outside $B_{2/3}$. For $\varepsilon > 0$, set

$$\Omega^+ = B_1 \cap \{x_n > \gamma\varphi(x)\} \ , \qquad \Omega^- = B_1 \setminus \bar{\Omega}^+ \ ,$$

and consider the function v_γ so defined:

$$\begin{aligned} &\Delta v_\gamma = 0 \ \text{ in } \ \Omega^+ \ \text{ and } \ \Omega^-, \\ &\quad v_\gamma = \alpha x_n^+ \text{ on } \partial B_1 \cap \{x_n \geq 0\}, \quad v_\gamma = -\beta x_n^- \text{ on } \partial B_1 \cap \{x_n < 0\}, \\ &\quad v_\gamma = 0 \ \text{ on } \ B_1 \cap \{x_n = \gamma\varphi(x)\} \ . \end{aligned}$$

Then we have the following lemma, whose proof comes from standard estimates.

Lemma 6.15. *Let u_0 and v_γ be as above. Then if $\gamma \to 0$,*

$$(v_\gamma^+)_\nu \to \alpha \qquad \textit{and} \qquad (v_\gamma^-)_\nu \to \beta$$

uniformly in $B_{7/8} \cap \{x_n = \gamma\varphi(x)\}$.

As a consequence,

Corollary 6.16. *Suppose that u is a minimal viscosity solution of our free boundary problem such that $|u - u_0| \leq \varepsilon$ in B_1. Then if ε is small enough, we must have in* (6.12)

$$|\alpha - G(\beta)| \leq \delta(\varepsilon)$$

where $\delta(\varepsilon) \to 0$ if $\varepsilon \to 0$.

Proof. Suppose that this is not true. Then $\alpha - G(\beta) > \bar{\delta} > 0$ for a sequence $\varepsilon_j \to 0$. Consider the auxiliary function v_γ, with $\gamma < 0$. A translation $v_{\gamma,s}(x) = v_\gamma(x - se_n)$, for some s_j, $|s_j| < \varepsilon_j$, will touch the free boundary of u at a point $x_j \in B_{2/3}$, regular from the right and inside the strip $|x_n| < \varepsilon$. If γ and ε_j are small, by Lemma 6.15, $v_{\gamma,s}$ is a classical subsolution. Contradiction.

The case $\alpha - G(\beta) < -\bar{\delta}$ is treated analogously. □

Suppose now that $0 \in F^*(u)$, $\nu(0) = e_n$ and rescale defining

$$u_R(x) = \frac{1}{R} u(Rx) \ .$$

Then since $\Omega^+(u_R) = \Omega_R^+(u)$, the structural Theorem 6.13 says that $B_1 \cap \Omega^+(u_R)$ and $B_1 \cap \Omega^-(u_R)$ converge in measure to the half balls

$$B_1^+ = B_1 \cap \{x_n > 0\} \quad \text{ and } \quad B_1^- = B_1 \cap \{x_n < 0\}.$$

We have

Lemma 6.17. *Given* $\varepsilon > 0$, *there exists* R_0 *such that if* $R < R_0$,

$$\Omega^+(u_R) \cap B_1 \subset \{x_n > -\varepsilon\} \cap B_1 \ . \tag{6.13}$$

Proof. From the uniform positive density of $\Omega^+(u)$ along $F(u)$, if $R_j \to 0$ and $x_j \in \partial\Omega^+(u_{R_j}) \cap \{x_n \le -\varepsilon\} \cap B_1$, then $|B_{\varepsilon/2}(x_j) \cap \Omega^+(u_{R_j})| \ge c\varepsilon$, contradicting the convergence in measure of $\Omega^+(u_{R_j})$ to B_1^+. □

Notice that Lemma 6.17 implies that $B_1 \cap \Omega^-(u_R) \supset \{x_n \le -\varepsilon\} \cap B_1$.

The proof of Lemma 6.14 is an immediate consequence of the next lemma, which deals with the asymptotic behavior of the sequence of rescalings $\{u_R\}$, as $R \to 0$.

Lemma 6.18. *There exist unique* $\alpha > 0$, $\beta \ge 0$ *such that every sequence* $\{R_k\}$, $R_k \to 0$, *has a subsequence* $\{R_{k_j}\}$ *such that*

$$u_{R_{k_j}} \to \alpha x_n^+ - \beta x_n^-$$

uniformly in any compact subset of $\mathbb{R}^n$, *with* $\alpha = G(\beta)$.

Proof. Let $R_k \to 0$. Since $\{u_{R_k}\}$ is an equilipschitz family, we can extract a subsequence $\{u_{R_{k_j}}\}$ converging uniformly in larger and larger balls to a limiting function u_0.

We now use the monotonicity formula and the results in Chapter 12. Let

$$J_R(u) = \frac{1}{R^4} \int_{B_R} \frac{|\nabla u^+|^2}{|x|^{n-2}}\, dx \int_{B_R} \frac{|\nabla u^-|^2}{|x|^{n-2}}\, dx \ .$$

Since $J_R(u)$ is increasing (Theorem 12.3),

$$J_R(u) \downarrow J_0(u) \equiv \gamma_0 \ .$$

Assume $\gamma_0 > 0$. Since $J_R(u) = J_1(u_R)$ and $u_{R_{k_j}} \to u_0$ uniformly in B_2 (say), we can write

$$J_1(u_{R_{k_j}}) \to J_1(u_0) = \gamma_0 \ge J_0(u_0) \ .$$

It follows that $J_0(u_0) = \gamma_0$. In fact, if $J_0(u_0) < \gamma_0$, we have $J_r(u_0) < \gamma_0$ for r small. On the other hand, if $(u_0)_r(x) = u_0(rx)/r$,

$$J_r(u_0) = J_1((u_0)_r)$$

and, since for r fixed

$$\frac{1}{r} u_{R_{k_j}}(rx) = u_{rR_{k_j}}(x) \to (u_0)_r(x)$$

uniformly, we have

$$\gamma_0 \le J_1(u_{rR_{k_j}}) \to J_1((u_0)_r) .$$

Contradiction.

Therefore $J_1(u_0) = J_0(u_0)$ which implies that $J_r(u_0)$ is constant. Formula (12.14) and Corollary 12.4 force u_0^+ and u_0^- to be linear functions.

From the nondegeneracy of u_0^+ and Lemma 6.17, we deduce $\Omega^+(u_0) \subset \{x_n \ge 0\}$ so that we can write

$$u_0(x) = \alpha x_n^+ - \beta x_n^- . \tag{6.14}$$

Then from Corollary 6.16 it must be that

$$\alpha = G(\beta) .$$

The case $\gamma_0 = 0$ reduces to $\gamma_0 > 0$ considering $\tilde{u} = u - \varepsilon x_n^-$ instead of u. Then we obtain a limiting function $\tilde{u}_0$ such that

$$\tilde{u}_0^+(x) = \alpha x_n^+ .$$

Since $\Omega^+(\tilde{u}_0) \subset \{x_n \ge 0\}$ if u_0 is as before, $\tilde{u}_0^+ = u_0^+$ and the nondegeneracy of u_0^+ implies that $\beta = 0$ in (6.14).

It only remains to prove that α, β are independent of the particular selected subsequence. This follows as in the end of Section 3.3, from Lemma 12.5 and the strict monotonicity of G. □

6.9. Regularity and compactness

It is now easy to see that if u is a minimal viscosity solution, then $F(u)$ is a $C^{1,\gamma}$-surface around H^{n-1}-a.e. point.

Precisely, we have

Theorem 6.19. *For every $x_0 \in F^*(u) \cap B_{1/2}$, $F(u)$ is a $C^{1,\gamma}$-surface in a neighborhood of x_0.*

Proof. Let $0 \in F^*(u)$ and let $u_{R_j} = \frac{1}{R_j} u(R_j x)$ be a sequence of rescaling of u, converging to

$$u_0(x) = \alpha x_n^+ - \beta x_n^-$$

as in the proof of Lemma 6.14. From the same proof, it follows that given ε, for $R_j < R_0(\varepsilon)$, $F(u_{R_j}) \cap B_1$ is contained in the strip $|x_n| < \varepsilon$. Thus, since u_R^+ is nondegenerate, u_R^+ falls under the hypotheses of Theorem 5.10.

This ends the proof. □

As a final result we give the following compactness results.

Theorem 6.20. *Let u_k be a sequence of minimal viscosity solutions to the free boundary problems*

$$\begin{aligned} &\Delta u_k = 0 \quad \text{in} \quad \Omega^+(u_k) \text{ and } \Omega^-(u_k), \\ &(u_k^+)_\nu = G_k((u_k^-)_\nu), \\ &\underline{u}_k \le u_k \le \bar{u}_k \ . \end{aligned}$$

Assume $G_k \to G$ and $\underline{u}_k \to \underline{u}$ uniformly and that the assumptions on G_k and $\underline{u}_k$ are satisfied uniformly (in particular, the uniform one-side regularity of the free boundary of $\underline{u}_k$).

Then if $u_k \to u$ uniformly in a domain D, u is a viscosity solution of the limiting free boundary problem in D.

The proof is an adaptation of those of Lemmas 6.10 and 6.11 and we leave it as an exercise. (Hint: let $x_0 \in F(u)$ be regular from the right (left) and near x_0

$$u(x) = \alpha\langle x - x_0, \nu\rangle^+ - \beta\langle x - x_0, \nu\rangle^- + o(|x - x_0|) \ .$$

Let $x_k \in F(u_k)$ be regular from the right (left) and near x_k

$$w_k(x) = \alpha_k\langle x - x_k, \nu_k\rangle^+ - \beta_k\langle x - x_0, \nu_k\rangle^- + o(|x - x_0|) \ .$$

Prove that if $x_k \to x_0$, then

$$\alpha_0 \le \liminf \alpha_k \qquad (\beta_0 \ge \limsup \beta_k) \ .)$$

Part 2

Evolution Problems

Chapter 7

Parabolic Free Boundary Problems

7.1. Introduction

In this part we will discuss several free boundary problems of evolution type.

The typical problem consists mainly of finding a function u in a space time domain that satisfies some parabolic equation (usually the heat equation) when u is different from zero (one or two phases), and across the surface where u vanishes, there is a balance condition between the speed of the interphase, that is, of the zero level surface, and may be the flux discontinuity, i.e., the jump on the normal derivative of u, and may also be some geometrical quantity along the free boundary, like its curvature.

As in the elliptic case, and different from, for instance, conservation laws, where the conservation law or constitutive relation is a smooth function across a shock, and where it is the solution itself that jumps from an admissible value to another, in the case of a parabolic free boundary problem, it is the law itself that changes discontinuously.

Some examples are Stefan-like problems, that is, problems of melting or solidification, where caloric energy changes discontinuously across the melting temperature (see, e.g., [D], [F1], [F2], [K]), problems of flame propagation, where the existence of a sharp flame front is assumed (sometimes constructed as a limit of singular perturbation problems as in our example of Chapter 1 ([CV], [CLW1], [CLW2])), and more recently problems arising from financial mathematics where the edging strategy changes when the present value of an option goes through a certain threshold.

Mathematically, the general lines of attack of the problems follow naturally those of the elliptic case, and the necessary tools are similar: Harnack type inequalities in the interior and at the boundary of one of the phases, optimal regularity, monotonicity formulas, and the perturbation techniques, through continuous families of supersolutions.

But one soon realizes that the study of these problems entails new, serious difficulties.

The first obvious difficulty is the role of time, and it is already present in the standard Harnack inequality, which says that for a nonnegative solution, in a parabolic cylinder, the past controls the future only from below, since one can obviously "inject" as much heat as one likes from the sides of the cylinder, something that the bottom will never see.

This implies that on one hand stronger hypotheses have to be made in the geometry of the problem, or the starting configuration, and that the very strong local conclusions we obtained in the elliptic case do not necessarily hold.

An example of this phenomenon is the waiting time counterexample to instantaneous regularization of the free boundary for a two-phase Stefan problem, which we discuss below.

A second difficulty comes from the different homogeneities corresponding on one hand to the evolution equation that u satisfies away from the transition surface (being a parabolic equation, it remains invariant under dilations that are linear in space and quadratic in time), and on the other hand to that of the free boundary that, relating speed with flux, is of a Hamilton-Jacobi nature and as such scales homogeneously of degree one, both in time and space.

Hence we are faced with the dilemma that parabolic scaling will keep the heat equation but will, generically, make the free boundary vertical and information will be lost while hyperbolic scaling will preserve the asymptotic geometry of the free boundary but we will lose the time derivative in the parabolic part, disconnecting u in time away from the free boundary.

From these difficulties comes the very delicate balance in the intermediate rescaling that we will use in the Stefan problem, which allows us (an almost miraculous fact) to reconstruct a Dini domain out of our iteration process.

Finally, the third ingredient we are missing is that of the very strong, local geometric measure-theoretical properties of the free boundary, and this can usually be seen in the case in which there is focusing (total melting of a solid region or extinction of a flame) and all hopes of universal regularity or nondegeneracy are broken.

This is one more reason why we can only treat geometries for which one can assert a priori that, at least after finite time, focusing is ruled out (for instance, flatness hypotheses).

7.2. A class of free boundary problems and their viscosity solutions

Perhaps the best known example of parabolic two-phase free boundary problem is the Stefan problem, a simplified model describing the melting (or solidification) of a material with a solid-liquid interphase.

The concept of solution can be stated in several ways: classical solutions, weak solutions on divergence form or viscosity solutions.

Locally, a classical solution of the Stefan problem may be described as follows. In the unit cylinder $\mathcal{C}_1 = B_1' \times (-1,1) \subset R^m \times R$ we have two complementary domains, Ω^+ and $\Omega^- = \mathcal{C}_1 \setminus \bar{\Omega}^+$, separated by a smooth surface $F = \partial\Omega^+ \cap \mathcal{C}_1$. In Ω^+ and Ω^-, respectively, we have two solutions u_1 and u_2 of the heat equations

$$\Delta u_1 - a_1 \partial_t u_1 = 0 = \Delta u_2 - a_2 \partial_t u_2$$

with $u_2 \le 0 \le u_1$.

The functions u_1 and u_2 are C^1 up to F, and along F both $u_1 = u_2 = 0$ and the interphase energy balance condition

$$\frac{\partial_t u_i}{|\nabla u_i|} = |\nabla u_1| - |\nabla u_2| \tag{7.1}$$

are satisfied. Note that the ratios (for $i = 1, 2$) on the left-hand side of (7.1) represent the speed of F in the direction $-\nu$, $\nu = \nabla u^+/|\nabla u^+|$.

What can be constructed for all times are weak solutions to the equation

$$\Delta u \in \beta(u)_t$$

with $\beta(u) = a_1 u^+ - a_2 u^- + \operatorname{sign} u$, subject to proper initial and boundary conditions ([K], [F1], [F2]).

From [CE], u is continuous in $\mathcal{C}_1$ and heuristically, $u_1 = u^+$ in $\Omega^+ = \{u > 0\}$, $u_2 = -u^-$ in $\Omega^- = \{u \le 0\}^0$ and $F = \partial\{u > 0\} \cap \mathcal{C}_1$ becomes the *free boundary*.

Modeled on the example of the Stefan problem, we formally introduce the following class of free boundary problems (f.b.p. in the sequel): *to find a function u, continuous in $\mathcal{C}_1 = B_1' \times (-1,1)$, such that the following hold.*

(a) $\Delta u - a_1 u_t = 0$ in $\Omega^+(u) = \{u > 0\} \cap \mathcal{C}_1$, and
$\Delta u - a_2 u_t = 0$ in $\Omega^-(u) = \{u \le 0\}^0 \cap \mathcal{C}_1$.

(b) On $F(u) = \partial\Omega^+(u) \cap \mathcal{C}_1$, the (*free boundary*) condition,

$$V_\nu = -G(u_\nu^+, u_\nu^-) \tag{7.2}$$

must be satisfied, where $a_1 > 0$, $a_2 > 0$ and $V_\nu(\cdot, \tau)$ is the speed of the surface $F(u) \cap \{t = \tau\}$ in the direction $\nu = \nabla u^+/|\nabla u^+|$.

The basic requirements on the function $G : [0, \infty)^2 \mapsto \mathbb{R}$ follow.

(i) G is continuous in $[0, \infty)^2$.

(ii) G is strictly increasing in u_ν^+ and strictly decreasing in u_ν^-.

(iii) $G \to +\infty$ when $u_\nu^+ - u_\nu^- \to \infty$.

Note that there is no nondegeneracy property of G, since this is the case of the Stefan problem. In fact, harmonic functions are stationary solutions of a Stefan problem and we cannot hope for nondegeneracy properties in this case.

Let us now define what we mean by a C^k-classical ($k \geq 1$) subsolution, supersolution and solution.

Definition 7.1. A function $v \in C(\mathcal{C}_1)$ is a C^k-classical subsolution if the following hold.

(a) $v \in C^k(\bar{\Omega}^+(v)) \cap C^k(\bar{\Omega}^-(v))$.

(b) $\Delta v - a_1 v_t \geq 0$ in $\Omega^+(v)$ and $\Delta v - a_2 v_t \geq 0$ in $\Omega^-(v)$.

(c) The free boundary $F(v) = \partial\Omega^+(v) \cap C_1$ is a C^k-surface, $|\nabla v^+| > 0$ on $F(v)$ and

$$-V_\nu = \frac{v_t^+}{v_\nu^+} \leq G(v_\nu^+, v_\nu^-) \tag{7.3}$$

where $\nu = \frac{\nabla v^+}{|\nabla v^+|}$.

The free boundary condition (7.3) indicates that the speed of $F(v)$ towards the "solid phase" $\{u \leq 0\}^0$ is "smaller" for a subsolution than for a solution. If the inequality in (7.3) is *strict*, we call v a *strict subsolution.*

A C^k-classical supersolution is defined by reversing the inequalities in (b) and (7.3). A C^k-classical solution is both a C^k-classical subsolution and supersolution.

As in the elliptic case, we use classical strict subsolutions and supersolutions as test functions to define viscosity solutions.

Again, from now on, it is understood that $k = 2$. Let

$$Q_r(x_0, t_0) = B_r'(x_0) \times (t_0 - r^2, t_0] \ .$$

Definition 7.2. A function $u \in C(\mathcal{C}_1)$ is a viscosity subsolution (supersolution) to f.b.p. if, whenever $(x_0, t_0) \in \mathcal{C}_1$ and v is a classical strict supersolution (subsolution) in $Q_r(x_0, t_0) \subset \mathcal{C}_1$, then v cannot touch u from above (below) at (x_0, t_0); u is a viscosity solution if it is both a subsolution and a supersolution.

In other words, if u is a subsolution and v is a classical strict supersolution in $Q_r(x_0, t_0) = Q_r$, and $(x_0, t_0) \in F(u)$, *it is not possible that* $v > u$ in $Q_r \setminus \{(x_0, t_0)\}$ and $v(x_0, t_0) = u(x_0, t_0)$. Analogously if u is a supersolution and v is a subsolution in $Q_r(x_0, t_0)$, it is not possible that $v < u$ in $Q_r \setminus \{(x_0, t_0)\}$ and $v(x_0, t_0) = u(x_0, t_0)$.

It is easy to check that a C^2-classical (sub, super) solution is a solution in the viscosity sense and that a viscosity solution u which is of class C^2, with its free boundary, in $\bar{\Omega}^+(u)$ and $\bar{\Omega}^-(u)$ is also a C^2-classical solution.

Sometimes it is more convenient to use the definition of viscosity solution in the following equivalent form.

Definition 7.2′. A function $u \in C(\mathcal{C}_1)$ is a *viscosity subsolution (supersolution)* to f.b.p. if, for any cylinder Q, $Q \subset\subset \mathcal{C}_1$, and for every classical strict supersolution (subsolution) v in Q, $u < v$ on $\partial_p Q$ implies $u < v$ in $\bar{Q}$; u is a *viscosity solution* if it is both a subsolution and supersolution.

Indeed Definitions 7.2 and 7.2′ are equivalent.

(a) Definition 7.2′ $\Rightarrow$ Definition 7.2. Let u be a viscosity subsolution, $(x_0, t_0) \in \mathcal{C}_1$, $Q_r = Q_r(x_0, t_0) \subset \mathcal{C}_1$ and let v be a classical strict supersolution in Q_r such that $v > u$ in $Q_r \setminus \{(x_0, t_0)\}$. If $u(x_0, t_0) = v(x_0, t_0)$, we have a contradiction to Definition 7.2′ (in a slightly smaller cylinder $Q_{r'}(x_0, t_0)$, $r' < r$).

(b) Definition 7.2 $\Rightarrow$ Definition 7.2′. Let u be a viscosity subsolution, $Q \subset\subset \mathcal{C}_1$, and let v be a classical strict supersolution in Q such that $v > u$ on $\partial_p Q$. Suppose $v > u$ in $\bar{Q}$ is not true. Then there is a first time τ and a point $(x_0, \tau) \in F(u) \cap F(v)$ such that $u(x_0, \tau) = v(x_0, \tau)$. Assume e_n is the normal direction to $F(v)$ at (x_0, τ). Then near (x_0, τ) we can write

$$\begin{aligned} v(x,t) = &\left(\alpha^+(x - x_0)_n + \beta^+(t - \tau)\right)^+ \\ &- \left(\alpha^-(x - x_0)_n + \beta^-(t - \tau)\right)^- + o(|x - x_0| + |t - \tau|) \end{aligned}$$

where $\alpha^\pm = v^\pm_{x_n}$, $\beta^\pm = v^\pm_t$, with $\alpha^+ > 0$, $\alpha^- > 0$ and

$$\frac{\beta^-}{\alpha^-} = \frac{\beta^+}{\alpha^+} > G(\alpha^+, \alpha^-) \, .$$

On the other hand, for $t \le \tau$, $u(x,t) < v(x,t)$, and therefore, using Theorem 7.1, we deduce that

$$\frac{\beta^-}{\alpha^-} \le C(\alpha^+, \alpha^-) .$$

Contradiction. □

Remark. Weak solutions of the two-phase Stefan problem are viscosity solutions. This follows from the following comparison theorem in [F1]: if u and v are a subsolution and supersolution, respectively, in $\mathcal{C}_1$ and $u > v$ on $\partial_p \mathcal{C}_1$, then $u > v$ in $\mathcal{C}_1$.

7.3. Asymptotic behavior and free boundary relation

From the results in Section 13.3, and in particular from Lemma 13.19, we can deduce asymptotic inequalities at regular points of the free boundary and give a weak formulation of the free boundary condition. First, let us clarify what we mean by regular point in this case.

Definition 7.3. A point $(0,0)$ on $F(u)$ is a regular point from the right (from the left) if it has a touching ball $B \subset \Omega^+(u)$ $(B \subset \Omega^-(u))$ with tangent plane, say $\alpha e_n + \beta t = 0$, of finite slope β/α.

The following theorem holds ([ACS1]):

Theorem 7.1. *Let u be a viscosity solution in $\mathcal{C}_1$ of the f.b.p., according to either Definition 7.2 or 7.2′. Suppose $(0,0) \in F(u)$ and that near $(0,0)$, for $t \le 0$, the following asymptotic inequality holds.*

(a) $u(x,t) \ge (\beta^+ t + \alpha^+ \langle x, \nu \rangle)^+ - (\beta^- t + \alpha^- \langle x, \nu \rangle)^- + o(d(x,t))$
with $\alpha^+, \alpha^-, \beta^+, \beta^- \in \mathbb{R}$, $\alpha^+ > 0$, $\alpha^- \ge 0$

or

(b) $u(x,t) \le (\beta^+ t + \alpha^+ \langle x, \nu \rangle)^+ - (\beta^- t + \alpha^- \langle x, \nu \rangle)^- + o(d(x,t))$
with $\alpha^+, \alpha^-, \beta^+, \beta^- \in \mathbb{R}$, $\alpha^+ \ge 0$, $\alpha^- > 0$

where ν denotes the inward spatial direction to $\Omega^+(u)$ at $(0,0)$ and $d(x,t) = |x| + |t|$.

Then

$$\frac{\beta^+}{\alpha^+} \ge G(\alpha_+, \alpha_-) \qquad \text{(supersolution condition)} \tag{7.4}$$

in case (a), while

$$\frac{\beta^-}{\alpha^-} \le G(\alpha_+, \alpha_-) \qquad \text{(subsolution condition)} \tag{7.5}$$

in case (b).

Proof. We give it only for case (a), the other case being completely analogous.

Let $\mathcal{N}$ denote the neighborhood in which (a) is valid and assume $\nu = e_n$. Define, for $\lambda > 0$,

$$u_\lambda(x,t) = \frac{1}{\lambda} u(\lambda x, \lambda t) \ .$$

Then, in $\mathcal{N}$,

$$u_\lambda(x,t) \geq (\alpha^+ x_n + \beta^+ t)^+ - (\alpha^- x_n + \beta^- t) + o(1)$$

where $o(1) \to$ uniformly in $\mathcal{N}$, when $\lambda \to 0$.

Now if (7.4) were false, there would exist $\eta > 0$ such that

$$\frac{\beta^+}{\alpha^+} \leq G(\alpha^+, \alpha^-) - \eta \ .$$

We show that this leads to a contradiction. Let R be a small parabolic neighborhood of $(0,0)$ contained in $\mathcal{N}$, i.e.,

$$R = \{(x', x_n, t) : |x'| < a, \ |x_n| < b, \ -t_0 < t \leq 0\}$$

with a, b, t_0 small. Define

$$\psi(x', x_n, t) = \bar{\alpha}^+ x_n + \bar{\beta}^+ t - ct^2 + x_n^2 - \frac{|x'|^2}{2(n-1)}$$

where $\bar{\alpha}^+ = \alpha^+ + \varepsilon$, $\bar{\beta}^+ = \beta^+ + \varepsilon$ for some ε, positive and small, to be determined later. Choose $C > 0$, large, so that the level surface $\{\psi = 0\}$ is strictly convex and $\{\psi > 0\} \cap R \subset \Omega^+(u)$.

Observe that

$$\Delta\psi - \lambda a_1 \psi_t = 1 - \lambda a_1(\bar{\beta}^+ - 2Ct) > 0$$

in R if λ is small enough.

Claim. *If ε, a, b, t_0 are small enough, depending on η, G, then the function*

$$\varphi = \psi^+ - \frac{\bar{\alpha}^-}{\bar{\alpha}^+}\psi^- \quad \textit{with} \quad \bar{\alpha}^- = \alpha^- + \varepsilon$$

is a classical strict subsolution to the f.b.p. in R.

To prove the claim, it is enough to show that

$$\frac{\varphi_t^+}{|\nabla\varphi^+|} \leq G(|\nabla\varphi^+|, |\nabla\varphi^-|) \tag{7.6}$$

on $\{\varphi = 0\} \cap R$. By continuity, $\varphi_t^+/|\nabla\varphi^+|$ is close to

$$\frac{\varphi_t^+(0,0)}{|\nabla\varphi^+(0,0)|} = \frac{\bar{\beta}^+}{\bar{\alpha}^+}$$

uniformly in R. On the other hand

$$\frac{\bar{\beta}^+}{\bar{\alpha}^+} = \frac{\beta^+}{\alpha^+} + O(\varepsilon) \leq G(\alpha^+, \alpha^-) - \eta/2$$

if $\varepsilon = \varepsilon(\eta)$ is small enough.

Since G is continuous in all of its arguments, we obtain (7.6) if the size of R is small enough.

If we now choose λ very small, we see that

$$u_\lambda(x,t) \geq (\alpha^+ x_n + \beta^+ t)^+ - (\alpha^- x_n + \beta^- t)^- + o(1) > \varphi(x,t)$$

on the parabolic boundary of R, hence in $R \setminus \{(0,0)\}$. Since u_λ is a viscosity solution of the f.b.p., we must have $u_\lambda > \varphi$ in $\bar{R}$. But then $0 = u_\lambda(0,0) > \varphi(0,0) = 0$ gives a contradiction. □

Remark. Let u be a viscosity solution in $\mathcal{C}_1$ and let $(0,0) \in F(u)$. If $(0,0)$ is a regular point from the right (resp. left), then (a) (resp. (b)) holds near $(0,0)$. Precisely, let B an $(n+1)$-dimensional ball touching $F(u)$ at $(0,0)$ from the positive side. Let $\nu = e_n$ be the spatial inward unit normal to B at $(0,0)$ and let $\bar{\alpha} x_n + \bar{\beta} t = 0$ be the equation of the tangent plane. Assume that the slope $-\bar{\beta}/\bar{\alpha}$ is finite ($\bar{\alpha} > 0$). Then apply Lemma 13.19 to u^+ and u^-. We obtain

$$u^+(x,t) \geq (\alpha^+ x_n + \beta^+ t)^+ + o(d(x,t)) \tag{7.7}$$

in B, with $\alpha^+ > 0$ (which we may assume finite) and $\beta^+ \in \mathbb{R}$, and

$$u^-(x,t) \leq (\alpha^- x_n + \beta^- t)^- + o(d(x,t)) \tag{7.8}$$

in $\mathcal{C}B$, with $\alpha^- > 0$ (finite) and $\beta^+ \in \mathbb{R}$.

Then the conclusion of Theorem 7.1 follows. In particular, note that

$$\infty > \frac{\bar{\beta}}{\bar{\alpha}} = \frac{\beta^+}{\alpha^+} \geq G(\alpha_+, \alpha_-)$$

and therefore $\sup\{\alpha : (7.7) \text{ holds}\} < \infty$.

Regular points from the left can be handled in the same way. Also, we know, from Lemma 13.19, that equality holds in (7.7) and (7.8), along paraboloids of the form $t = -\gamma x_n^2$, $\gamma > 0$.

7.4. R-subsolutions and a comparison principle

As in Section 2.3, for comparison purposes, it is useful to introduce another class of subsolutions. We say that $v \in C(\mathcal{C}_1)$ is an *R-subsolution* if the following hold.

(i) $\Delta v - a_1 v_t \geq 0$ in $\Omega^+(v)$, $\Delta v - a_2 v_t \geq 0$ in $\Omega^-(v)$.

(ii) Whenever $(x_0, t_0) \in F(v)$ has a touching $(n+1)$-dimensional ball $B \subset \Omega^+(v)$, then near (x_0, t_0), in B,

$$v^+(x,t) \geq (\alpha^+\langle x - x_0, \nu\rangle + \beta^+(t-t_0))^+ + o(|x-x_0| + |t-t_0|) \tag{7.9}$$

and in $\mathcal{C}B$

$$v^-(x,t) \leq (\alpha^-\langle x - x_0, \nu\rangle + \beta^-(t-t_0))^- + o(|x-x_0| + |t-t_0|) \tag{7.10}$$

where $\alpha^\pm \geq 0$, $\beta^\pm \in \mathbb{R}$, $\nu = \nu(x_0, t_0)$ is the spatial inward unit normal to B at (x_0, t_0),

$$\beta^+ \leq \alpha^+ G(\alpha^+, \alpha^-) \qquad (\text{or } \beta^- \leq \alpha^- G(\alpha^+, \alpha^-)) \,. \tag{7.11}$$

Here

$$\alpha^+\langle x - x_0, \nu\rangle + \beta^+(t-t_0) = 0 = \alpha^-\langle x - x_0, \nu\rangle + \beta^-(t-t_0)$$

are equations for the tangent plane to B at (x_0, t_0).

The following analog of Lemma 2.1 holds.

Theorem 7.2. *Let u, v be a viscosity solution and an R-subsolution in $\mathcal{C}_1$, respectively, with $u \geq v$. Moreover let $u > v$ in $\Omega^+(v)$ and $(0,0) \in F(v) \cap F(u)$. Then $(0,0)$ cannot be a regular point from the right.*

Proof. Suppose $B \subset \Omega^+(v)$ touches $(0,0)$, with tangent plane $\alpha x_n + \beta t = 0$, $-\beta/\alpha$ finite. Then near $(0,0)$ in B, we have

$$v^+(x,t) \geq (\alpha^+ x_n + \beta^+ t)^+ + o(d(x,t)) \qquad (\alpha^+ > 0) \tag{7.12}$$

and in $\mathcal{C}B$,

$$v^-(x,t) \leq (\alpha^- x_n + \beta^- t)^- + o(d(x,t)) \,. \tag{7.13}$$

From Theorem 7.1 and the Remark after it, we know that near $(0,0)$

$$u^+(x,t) \geq (\bar\alpha^+ x_n + \bar\beta^+ t)^+ + o(d(x,t)) \tag{7.14}$$

in B and

$$u^-(x,t) \geq (\bar\alpha^- x_n + \bar\beta^- t)^- + o(d(x,t)) \tag{7.15}$$

in $\mathcal{C}B$, with equality along any paraboloid of the form $t = -\gamma x_n^2$, $\gamma > 0$.

We have the following.

(a) Computing along the paraboloid $t = -\gamma x_n^2$, $\gamma > 0$, from (7.12)–(7.15) and $u \geq v$, we deduce that

$$\bar\alpha^+ \geq \alpha^+ \,, \quad \bar\alpha^- \leq \alpha^-$$

and therefore

$$G(\bar\alpha^+, \bar\alpha^-) \geq G(\alpha^+, \alpha^-) \,. \tag{7.16}$$

(b) From (7.11) and (7.4) of Theorem 7.1, since $\alpha^+ > 0$,

$$G(\bar{\alpha}^+, \bar{\alpha}^-) \leq \frac{\bar{\beta}^+}{\bar{\alpha}^+} = \frac{\beta^+}{\alpha^+} \leq G(\alpha^+, \alpha^-) \tag{7.17}$$

which implies, by the strict monotonicity of G,

$$\bar{\alpha}^+ = \alpha^+ \,, \quad \bar{\alpha}^- = \alpha^- \,.$$

On the other hand $u - v$ is supercaloric in $\Omega^+(v)$, so that the Hopf principle along $t = -\gamma x_n^2$, gives, for some $\varepsilon > 0$,

$$u - v \geq \varepsilon x_n^+ \,.$$

Contradiction. □

Chapter 8

Lipschitz Free Boundaries: Weak Results

8.1. Lipschitz continuity of viscosity solutions

As we already mentioned in the introduction of Chapter 7, regularity theory follows the lines of elliptic theory, at least for the class of parabolic problems that we deal with.

Thus, as a first step we assume that the free boundary is given by the graph of a Lipschitz function in space and time. Although Lipschitz regularity in time versus Lipschitz regularity in space is not, of course, the natural homogeneity balance for the study of general parabolic equations, it is so for the study of phase transition relations of the form

$$F(u_\nu^+, u_\nu^-, V_\nu) = 0$$

along the free boundary, which are invariant precisely under hyperbolic scaling.

In this chapter we first prove optimal regularity (Lipschitz continuity) for a solution and then we refine the results in Section 8.2, giving a more precise free boundary behavior. In the last section, two counterexamples, one in spatial dimension $n = 2$ for a one-phase Stefan problem, the other in dimension $n = 3$ for a two-phase Stefan problem, indicate that, in general, there is no instantaneous regularization of the free boundary.

We now prove the optimal regularity of a viscosity solution, having Lipschitz free boundary ([ACS1]).

Theorem 8.1. *Let u be a viscosity solution to f.b.p. in $\mathcal{C}_1 = B_1' \cap (-1,1)$. If $F(u) = \partial\Omega^+(u)$ is Lipschitz in some space direction ν, then u is Lipschitz continuous in $\mathcal{C}_{1/2}$.*

Proof. Let d_0 be so small that all the results in Section 13.2 can be applied in a d_0-neighborhood of $F(u)$, from both sides. In particular, from Corollary 13.14 we have $|u_t| \le C|\nabla u|$. Thus it is enough to show that the spatial gradient is bounded across $F(u)$.

We show that $|\nabla u^+|$ is bounded; a similar proof can be done for $|\nabla u^-|$. Take $(x_0, t_0) \in \Omega^+(u) \cap \mathcal{C}_{1/2}$ at distance d from $F(u)$, with $d \le d_0$. Suppose that the $(n+1)$-dimensional ball $B_d(x_0, t_0)$ touches $F(u)$, say, at $(0,0)$. Let $h = \operatorname{dist}((x_0, 0), (0,0))$ and note that $cd \le h \le d$, for $c = c(L,n)$. Set $u(x_0, 0) = Ah$. We want to show that A is bounded. By interior estimates this proves the theorem.

As in Lemma 13.15, in $\Omega^+(u)$ set

$$w_+ = u + u^{1+\varepsilon}, \quad w_- = u - u^{1+\varepsilon}.$$

We know that $\Delta w_+ \ge 0$ and $\Delta w_- \le 0$ in $\Omega^+(u) \cap \{t = 0\}$. Moreover

$$w_+ \le c w_- .$$

At time $t = 0$, we have

$$Ah = u(x_0, 0) \le w_+(x_0, 0) \le \fint_{B'_{h/4}(x_0)} w_+ \le c \fint_{B'_{h/4}(x_0)} w_- \le c_0 \inf_{B'_h(x_0)} w_-$$

where the constant c_0 depends only on n, L, a_1 and the value of u at, say, $(\frac{3}{4}\nu, -\frac{3}{4})$.

Considering $n > 2$, set

$$H(x) = \frac{Ah}{c_0(r^{n-2} - 1)} \left(\frac{h^{n-2}}{r^{n-2}} - 1 \right), \qquad r = |x - x_0| .$$

Then $\Delta H = 0$ in the ring $B'_h(x_0) \setminus \bar{B}'_{h/4}(x_0)$, $H = 0$ on $\partial B_h(x_0)$ and $H = Ah/c_0$ on $\partial B'_{h/4}(x_0)$. By the maximum principle, $u > w_- > H$ in the ring. Choose the coordinate system so that $x_0 = he_n$. Then near 0, inside the ring,

$$u(x) > w_-(x) \ge cAx_n + o(|x|) . \tag{8.1}$$

We now construct a classical subsolution in a small cylinder around $(0,0)$, smaller than u on the parabolic boundary.

Let $C_\rho = B'_\rho \times (-\rho, \rho)$, $\rho \le \frac{h}{2}$ and define on the hyperplane $\{t = 0\}$, in $B'_h(x_0) \cap B'_\rho$,

$$v = \frac{1}{2} H .$$

Fix ρ small so that $v < 1$ and define in C_ρ

$$z^+(x,t) = \frac{1}{2}\left[v(x+t\beta e_n) + (v(x+t\beta e_n))^2\right]^+ .$$

Then in $\{z^+ > 0\}$

$$\begin{aligned}\Delta z^+ - z_t^+ &= |\nabla v|^2 - \frac{1}{2}\beta v_{x_n}(1 - 2v_{x_n}) \\ &\geq v_{x_n}(v_{x_n} - \bar{c}\beta) > (\bar{c}A - c\beta) > 0\end{aligned}$$

if $0 < \beta < \frac{\bar{c}}{2c}A$. Also let $\beta > 2L$, L the Lipschitz constant of $F(u)$ and, if necessary, lower ρ to make sure that $u > z^+$ in $\{z^+ > 0\} \cap \{t < 0\}$.

Let $Q^+ = C_\rho \cap \{z^+ > 0\}$, $Q^- = C_\rho \setminus \bar{Q}^+$, and $Q_{-\rho} = \bar{Q}^- \cap \{t = -\rho\}$. In $Q_{-\rho}$ define w as

$$\begin{cases} \Delta w = 0 & \text{in } Q_{-\rho}, \\ w = 0 & \text{on } \partial Q_{-\rho} \cap F(z^+), \\ w = 2u^- + \eta g & \text{on } \partial Q_{-\rho} \setminus F(z^+) \end{cases}$$

where η is a small positive number and $g = g(x,t)$ is a smooth function, positive on $\partial_p Q^- \setminus F(z^+)$, vanishing on $\partial Q_{-\rho} \cap F(z^+)$.

Now let z^- be defined in Q^- such that

$$\begin{cases} \Delta z^- - z_t^- = 0 & \text{in } Q^-, \\ z^- = 2u^- + \eta g & \text{on } \partial_p Q^- \setminus F(z^+), \\ z^- = 0 & \text{on } \partial_p Q^- \cap F(z^+), \\ z^- = w & \text{in } Q_{-\rho}. \end{cases}$$

At $t = 0$, near the origin, we have

$$z^-(x,0) = \alpha^- x_n^- + o(|x|)$$

with

$$\alpha^- \leq \frac{c_1}{A\rho^2}$$

by the monotonicity formula, and $c_1 = c_1(n, \|z^\pm\|_{L^\infty(C_\rho)})$. On the other hand, if we choose β such that

$$\frac{1}{3}\min\left\{\frac{\bar{C}M}{2c}, G\left(CA, \frac{c_1}{A\rho^2}\right)\right\} < \beta < \min\left\{\frac{\bar{C}A}{2c}, g\left(CA, \frac{c_1}{A\rho^2}\right)\right\},$$

at $(0,0)$ we have

$$\frac{z_t^+(0,0)}{z_{e_n}^+(0,0)} = \beta < G\left(CA, \frac{c_1}{A\rho^2}\right) \leq G(CA, \alpha^-) .$$

By continuity, this inequality propagates in a small subcylinder $C_{\rho'}$ of C_ρ.

Thus, inside $C_{\rho'}$, the function

$$z = \begin{cases} z^+ & \text{in } Q^+, \\ -z^- & \text{in } Q^- \end{cases}$$

is a classical strict subsolution. Now, if A is very large, β is also very large, and $\partial\{z^+ > 0\} \cap C_\rho$ stays to the right of $F(u)$ for $t < 0$, since the free boundary of u is Lipschitz, that is, $F(u)$ has bounded speed.

Then by construction, $u > z$ on $\partial_p C_\rho$ so that it must be $u > z$ in $C_{\rho'}$. Since $u(0,0) = z(0,0) = 0$, we have a contradiction. Therefore A is controlled by a constant depending on the Lipschitz constant of $F(u)$ and the maximum of $|u|$ in $\bar{C}_{2/3}$. □

8.2. Asymptotic behavior and free boundary relation

At a regular point of $F(u)$, the asymptotic behavior from both sides of the free boundary follows from Theorem 8.1. When $F(u)$ is Lipschitz, it is possible to show a more precise result near a point which is both of differentiability of the free boundary and of Lebesgue differentiability with respect to surface measure, for instance, for $\nabla u^\pm$ and $D_t u^\pm$.

Notice that almost every point on $F(u)$ is of this kind. In fact, from Section 13.2, we know that caloric and surface measures are mutually absolutely continuous and any partial derivative of $u^\pm$ is a bounded solution of the heat equation on each side of $F(u)$. Therefore, these derivatives have a well-defined L^∞-trace on the free boundary, in the sense of nontangential convergence.

Theorem 8.2. *Let u be as in Theorem 8.1 and let $(0,0)$ be a common point of differentiability for the free boundary and of Lebesgue differentiability for $\nabla u^\pm$ and $D_t u^\pm$ w.r.t. surface measure.*

Then in a neighborhood of $(0,0)$,

$$u(x,t) = (\alpha^+\langle x,\nu\rangle + \beta^+ t)^+ - (\alpha^-\langle x,\nu\rangle + \beta^- t)^- + o(d(x,t)) \tag{8.2}$$

where $\alpha^+\langle x,\nu\rangle + \beta^+ t = \alpha^-\langle x,\nu\rangle + \beta^- t = 0$ is the equation of the tangent plane to $F(u)$ at $(0,0)$. Moreover, $\alpha^+ \geq 0$, $\alpha^- \geq 0$ and

$$\beta^+ = \alpha^+ G(\alpha^+,\alpha^-),$$
$$\beta^- = \alpha^- G(\alpha^+,\alpha^-)\,.$$

Proof. Consider u^+. The proof for u^- is similar. Since u_t^+, u_ν^+ are bounded solutions of $H_{a_1} v = 0$ and since $(0,0)$ is a Lebesgue differentiability point for both functions, the nontangential limits of u_t^+ and u_ν^+ exist when $(x,t) \to$

(0, 0). Therefore, we can write

$$\alpha^+ = \text{N.T.} \lim_{(x,t)\to(0,0)} u_\nu^+(x,t) , \qquad \beta^+ = \text{N.T.} \lim_{(x,t)\to(0,0)} u_t^+(x,t) .$$

(N.T. stands for nontangential).

Now, the asymptotic development of u^+ follows easily and, since we have an equality sign in it, by Theorem 7.1, the equality sign must hold also in the free boundary relations (7.4) and (7.5). □

8.3. Counterexamples

The natural question now is: suppose u is a viscosity solution in $\mathcal{C}_1$. If the free boundary $F(u)$ is a Lipschitz graph (in space and time), can we deduce further regularity, that is, that $F(u)$ is actually a $C^{1,\gamma}$ or a C^1-graph?

The answer is no, in general, as the following counterexamples show.

Counterexample 1. *A bidimensional one-phase problem.*

Consider the function

$$w(\rho, \theta, t) = \rho^{g(t)}\{\cos[g(t)\theta]\}^+$$

where ρ, θ are polar coordinates in the plane and g is a smooth decreasing function greater than 2.

Then

$$\Delta w - w_t < 0$$

in $\Omega^+(w) = \{(\rho, \theta, t) : |\theta| < \pi/2g(t), \rho > 0\}$ and on

$$F(w) = \{(\rho, \theta, t) : |\theta| - \pi/2g(t), \rho \geq 0\} ,$$

$$w_t^+ = -\frac{\pi}{2}\frac{g^1}{g}\rho^g , \qquad |\nabla w^+|^2 = \rho^{2g-2}g^2 .$$

Therefore, if $R \leq R_0$ is sufficiently small, in the cylinder $C_R = B'_R \times (0, R^2)$, we have

$$-\frac{\pi g^1}{2g^3} \leq \rho^{g(t)-2}$$

which makes w a supersolution of a one-phase Stefan problem. Now if u is a solution in C_R of the same problem (in a weak or viscosity sense) with $u = w$ on $\partial_p C_R$, then $u \leq w$ in C_R.

But at the origin, $F(w)$ has a persistent corner singularity. Since $(0, 0) \in F(u)$ initially with zero speed, $(0, 0) \in F(u)$ for $0 \leq t < R_0^2$ and since $u \leq w$, the origin is a persistent corner for $F(u)$ also.

Counterexample 2. *A three-dimensional two-phase problem* ([ACS2]).

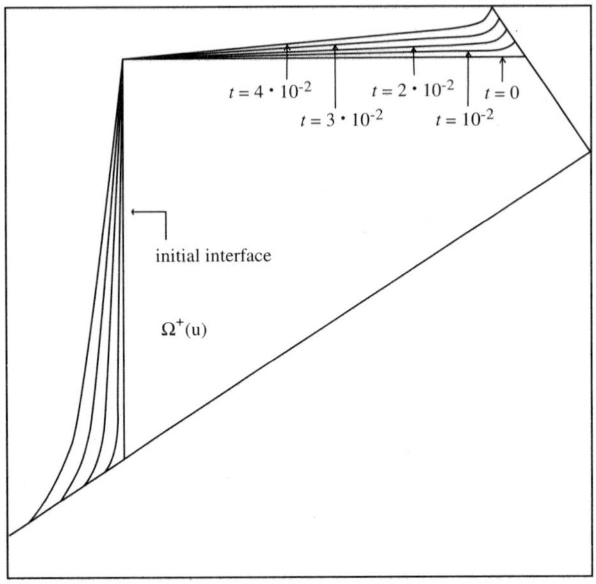

Figure 8.1. $\Omega = (-10^{-1}, 10^{-1}) \times (-10^{-1}, 10^{-1}$, $g(t) = 2.1 - 10t$

The same kind of hyperbolic behavior as in Counterexample 1 can occur in two-phase problems also, in spite of the fact that both phases have nonzero temperature.

To construct the example, we need the following lemma on spherical harmonics.

Lemma 8.3. *In* $\mathbb{R}^3$, *let* $D_{\lambda,\mu}$ *denote the domain*

$$D_{\lambda,\mu} = \{x_3 > r(\lambda \cos \lambda\theta - \mu)\}$$

where (r, θ) *are polar coordinates in the* (x_1, x_2)*-plane and* λ *is an integer. Let* $\rho = |x|$, $\sigma = x/|x|$ *and*

$$h_{\lambda,\mu} = \rho^\alpha f_{\lambda,\mu}(\sigma)$$

be the positive harmonic function in $D_{\lambda,\mu}$, *homogeneous of degree* $\alpha(\lambda,\mu)$, *vanishing on* $\partial D_{\lambda,\mu}$, *with* $\|f_{\lambda,\mu}\|_{L^2(S_{\lambda,\mu})} = 1$ *where*

$$S_{\lambda,\mu} = D_{\lambda,\mu} \cap \partial B_1 \ .$$

Then for $\lambda > \lambda_0$, $|\mu| < 1$,

$$\alpha \geq 3 \ .$$

Proof. The function h is constructed by separation of variables by computing

$$\Delta h = \rho^{\alpha-2}[\alpha(\alpha+1)f + \Delta_\sigma f]$$

where Δ_σ is the Laplace-Beltrami operator on the unit sphere ∂B_1 of $\mathbb{R}^3$. Thus f is chosen to be the first normalized eigenfunction of $-\Delta_\sigma$ for the spherical domain $S_{\lambda,\mu}$ and α must be chosen so that $\alpha(\alpha+1)$ equals the corresponding first eigenvalue.

If we fix $|\mu| \leq 1$ and let λ go to infinity, the domain $D_{\lambda,\mu}$ becomes very narrow; for instance, any disc of radius c/λ cuts the complement of $S_{\lambda,\mu}$ in a fixed proportion so that

$$\alpha \sim C\lambda$$

if $\lambda \to +\infty$.

This proves the lemma. □

Lemma 8.4. *Let*

$$\begin{cases} w(x,t) = (1+Mt)h_{\lambda_0+1,t}(x) & \text{in } D_{\lambda_0+1,t}, \\ w(x,t) = 0 & \text{otherwise.} \end{cases}$$

Then for ε small and M properly chosen, w is a supersolution of the Stefan problem in $B'_\varepsilon(0) \times (0,\varepsilon)$.

Proof. We must show that

$$\Delta w - w_t \leq 0 \quad \text{in} \quad D_{\lambda_0+1,t}$$

and that the speed of the free boundary V_ν (ν pointing towards the positive region) is smaller than $-w_\nu$, on the free boundary $\partial D_{\lambda_0+1,t}$.

Observe that $h_{\lambda,\mu}$ is smooth with respect to λ, μ since $D_{\lambda,\mu}$ changes smoothly (in fact, analytically) in λ, μ. Moreover, $D_{\lambda_0+1,t}$ is increasing in t, so that by choosing M large, we can make $(1+Mt)h_{\lambda_0+1,t}$ monotone increasing in t. Therefore

$$\Delta w = -D_t[(1+Mt)h_{\lambda_0+1,t}] < 0 \ .$$

On the free boundary,

$$w_\nu = (1+Mt)r^\alpha f_{\nu^*}(\sigma) \qquad (\alpha = \alpha(\lambda_0+1,t))$$

where f_{ν^*} denotes the (smooth) normal inward derivative of f along $\partial S_{\lambda_0+1,t}$. Thus, since for t small, $\alpha \geq 3$, for r small we have

$$w_\nu \leq cr^3 \ .$$

The speed of the free boundary $\partial D_{\lambda_0+1,t}$ is minus the ratio between the time and the space components of the gradient of

$$g(x,t) = x_3 - r[(\lambda_0+1)\cos(\lambda_0+1)\theta - t] \ ,$$

that is,

$$V_\nu = -\frac{r}{|\nabla g|} \leq -cr \ .$$

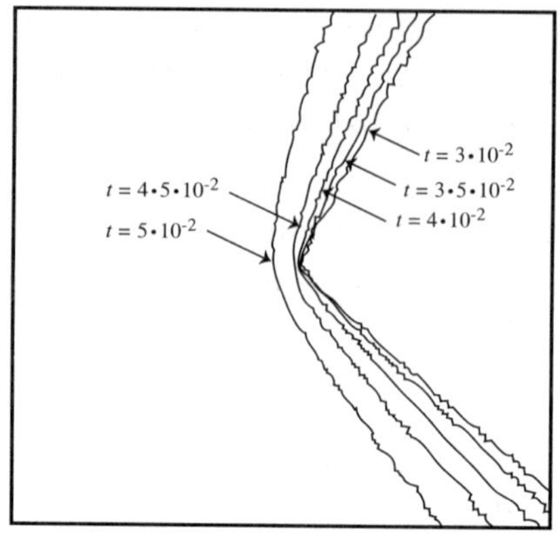

Figure 8.2. Regularization effect after a waiting time

Therefore, for $r, t < \varepsilon$, small,

$$V_\nu \leq -w_\nu$$

which makes w a supersolution in $B'_\varepsilon(0) \times (0, \varepsilon)$. □

Observe now that $\mathcal{C}D_{\lambda,0}$, the complement of $D_{\lambda,0}$, is obtained from $D_{\lambda,0}$ by a rotation R. Therefore the same construction can be used to construct a negative subsolution in $B'_\varepsilon(0) \times (0, \varepsilon)$.

We define for $t = 0$, in $\mathcal{C}D_{\lambda_0+1,t}$,

$$v(x, 0) = -w(Rx, 0)$$

and we let v evolve according to the formula

$$\begin{cases} v(x,t) = -(1 + Mt)h_{\lambda_0+1,-t}(Rx) & \text{in } \mathcal{C}D_{\lambda_0+1,-t}, \\ v(x,t) = 0 & \text{otherwise.} \end{cases}$$

It is easy to check that v is indeed a subsolution of the Stefan problem in $B'_\varepsilon \times (0, \varepsilon)$. As a consequence,

Theorem 8.5. *Let u be the solution of the two-phase Stefan problem in $B'_\varepsilon \times (0, \varepsilon)$ with initial and lateral data satisfying*

$$v \leq u \leq w \, .$$

Then $F(u)$ is contained in

$$\left\{x_n \geq r\left[(\lambda_0 + 1)\cos(\lambda_0 + 1)\theta - t\right]\right\} \cap \left\{x_n \leq r\left[(\lambda_0 + 1)\cos(\lambda_0 + 1)\theta + t\right]\right\}$$

and, in particular, has a persistent corner at the origin.

Thus, Lipschitz free boundaries in evolution problems *do not enjoy*, in general, *instantaneous regularization.*

A closer look at both counterexamples reveals that two main factors seem to prevent immediate smoothing:

(a) the simultaneous vanishing of the heat fluxes from both sides of the free boundary,

(b) the size of the Lipschitz constant (too large).

Points where (a) and (b) occur do not move at least for short time and are able to carry a singularity. Concerning condition (b), in Counterexample 1 the function w ceases to be a supersolution when g becomes smaller than 2 (which seems to indicate that $\pi/2$ is a critical angle) and one expects that the singularity will eventually disappear, showing *finite time regularization.*

Numerical evidence that this is indeed the case is shown in Figures 8.1 and 8.2 (courtesy of Nochetto, Schmidt and Verdi).

We will examine this question in Chapter 10.

According to previous observations, a regularity theory can be developed only under additional conditions, able to prevent (a) and/or (b) above.

The next chapter deals with regularization under an appropriate nondegeneracy condition that prevents (a).

Chapter 9

Lipschitz Free Boundaries: Strong Results

9.1. Nondegenerate problems: main result and strategy

We start our study of the regularity of the free boundary. According to the final comments in Section 8.3, we shall concentrate on a class of problems that we call *nondegenerate*, for which the regularity of the free boundary can be pushed to be C^1. The nondegeneracy condition states, roughly speaking, that the heat fluxes from the two sides of the free boundary are not vanishing simultaneously. In some cases, the validity of this condition can be inferred by global consideration (see for instance [N]). On the other hand, we expect this situation to be generic in a sense that will be explained in the next chapter, when we deal with *flat free boundaries*.

The main result can be stated in the following way ([ACS2]):

Theorem 9.1. *Let u be a viscosity solution to a f.b.p in $\mathcal{C}_1 = B_1' \times (-1,1)$ whose free boundary, $F(u)$, is given by the graph of a Lipschitz function $x_n = f(x',t)$ with Lipschitz constant L. Assume that $M = \sup_{\mathcal{C}_1} |u|$, $u(e_n, -\frac{2}{3}) = 1$, $(0,0) \in F(u)$ and that the following hold.*

(i) *$G = G(a,b) : \mathbb{R}^2_+ \to \mathbb{R}$ is a locally Lipschitz function and, for some positive number c^*,*

$$\partial_a G \geq c^* , \qquad \partial_b G \leq -c^* .$$

(ii) *(Nondegeneracy condition) There exists $m > 0$ such that if $(x_0, t_0) \in F(u)$ is a regular point (from the right or from the left), then for any small r,*

$$\fint_{B'_r(x_0)} |u| \geq mr \ .$$

Then the following conclusions hold.

(1) *In $\mathcal{C}_{1/2}$, the free boundary is a C^1-graph in space and time. Moreover, for any small $\eta > 0$, there exists a positive constant $c_0 = c_0(n, M, L, c^*, m, a_1, a_,\eta)$ such that, for every (x', x_n, t), $(y', y_n, t) \in F(u) \cap \mathcal{C}_{1/2}$,*

$$\begin{aligned} |\nabla_{x'} f(x', t) - \nabla_{x'} f(y', t)| &\leq c_0(-\log|x' - y'|)^{-3/2+\eta}, \\ |\partial_t f(x', b) - \partial_t f(x', s)| &\leq c_0(-\log|t - s|)^{-1/2+\eta} \ . \end{aligned} \tag{9.1}$$

(2) *$u \in C^1(\bar{\Omega}^+(u)) \cap C^1(\bar{\Omega}^-(u))$ and, on $F(u) \cap \mathcal{C}_{1/2}$,*

$$u_\nu^+ \geq c_1 > 0$$

with $c_1 = c_1(n, M, L, c^*, m, a_1, a_2, \eta)$.

Strategy of the proof. Although the proof follows the general lines of the elliptic case, new features occur here that we like to emphasize. Thus, let us examine and comment on the various steps in the proof.

Step 0. From Chapter 8 and Section 1.3, we know that u is Lipschitz continuous in $\mathcal{C}_{2/3}$ (say) and that, near $F(u)$, u is monotone increasing along every direction τ belonging to a cone $\Gamma(e_n, \theta)$ (cone of monotonicity) with axis e_n and opening θ, $\theta = \frac{1}{2}\cot^{-1}(L)$. Equivalently, for any $\varepsilon > 0$,

$$v_\varepsilon(p) = \sup_{B_{\varepsilon \sin\theta}(p)} u(q - \varepsilon e_n) \leq u(p) \qquad (p = (x, t)) \ . \tag{9.2}$$

After a hyperbolic scaling we may assume that u is monotone in all of $\mathcal{C}_1$.

Step 1. We show that, away from the free boundary, on both sides and in parabolic homogeneity, the enlargement of the space section of the cone of monotonicity (with a suitable choice of new axes) is a simple consequence of Harnack's inequality.

However this enlargement could occur in opposite directions in $\Omega^+(u)$ and $\Omega^-(u)$, as shown in Figure 9.1, preventing a common enlarged cone.

In the parabolic case, the role of the two phases is more symmetrical than in the elliptic case, so that we do not know which one of the two is the commanding one or even if they may alternate in being the commanding one. Therefore we must rule out the situation described in Figure 9.1 and show that there is a common axis of improvement in both phases. This amounts to an improvement of the Lipschitz constant in space of the level sets of u, away from $F(u)$.

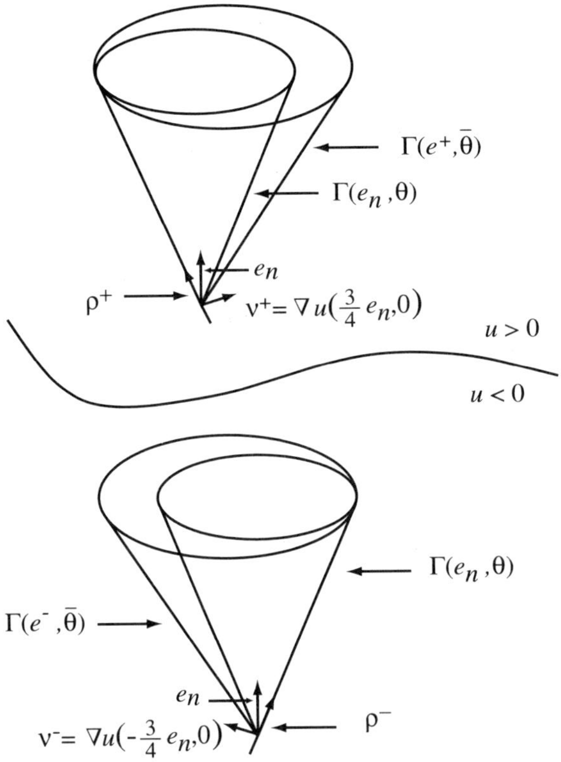

Figure 9.1

Step 2. Due to the underlying double homogeneity (parabolic for the heat equation, hyperbolic for the free boundary condition) the parabolically scaling gain in Step 1 is not well suited for iteration purposes (it is too short in time). Thus, we need to prove a gain in monotonicity, in a hyperbolically scaling region.

Step 3. The interior improvement of the Lipschitz constant in time of the level sets of u requires new ideas. At this stage one realizes that the derivatives of u along purely spatial directions behave differently from those involving a time component. This different behavior will of course result in a different opening speed between the spatial section of the cone $\Gamma_x(e, \theta^x)$ and the space-time section $\Gamma_t(\nu_1, \theta^t)$, where ν_1 belongs to the plane spanned by e, e_t. Clearly, initially $e = e_n = \nu$, $\theta^x = \theta^t = \theta$. In other words cones cease to be circular in space and time and the decay rate of the spatial defect will be faster than that of the time defect.

We will denote by $\delta = \frac{\pi}{2} - \theta^x$ the *defect angle in space* and by $\mu = \frac{\pi}{2} - \theta^t$ the *defect angle in time*.

Steps 4, 5. Iteration of Step 3 requires $\delta \ll \mu^3$. This implies that in carrying the opening gain to the free boundary, we have to prove regularity first in space and then in time. As in the elliptic case, this can be done by constructing a continuous family of R-subsolutions, by varying the radius where the supremum in (9.2) is taken. It is at this stage that the nondegeneracy condition plays its role.

We first prove (Step 4) that for each time level τ, $F_\tau = F(u) \cap \{t = \tau\}$ is a C^1-surface with a rough modulus of continuity of $\nabla_{x'} f$. Using this partial regularity in space, it is possible to obtain (Step 5) a gain in space and time described by the relations

$$\delta_1 \leq \delta - c_1 \frac{\delta^2}{\mu}, \tag{9.3}$$

$$\mu_1 \leq \mu - c_2 \delta \tag{9.4}$$

as long as $\delta \ll \mu^3$. The improvements (9.3) and (9.4) hold in a cylinder of the form $B'_{1/2} \times (-\frac{1}{2}\frac{\delta}{\mu}, \frac{1}{2}\frac{\delta}{\mu})$, which shows the right scaling, intermediate between parabolic and hyperbolic.

Step 6. We now iterate Steps 1 to 5. We obtain that in a sequence of contracting cylinders $B'_{2^{-k}} \times (-\frac{\delta_k}{2^k \mu_k}, \frac{\delta_k}{2^k \mu_k})$, u is increasing along all the directions in a sequence of corresponding spatial cones $\Gamma^x(\nu_k, \theta_k^x)$ and space-time cones $\Gamma^t(\bar{\nu}_k, \theta_k^t)$, whose respective defect angles satisfy the recurrence relations

$$\delta_{k+1} \leq \delta_k - c_1 \frac{\delta_k^2}{\mu_k}, \tag{9.5}$$

$$\mu_{k+1} \leq \mu_k - c_2 \delta_k \tag{9.6}$$

with $\delta_k \ll \mu_k^3$, $|\nu_{k+1} - \nu_k| \leq c_3(\delta_k - \delta_{k+1})$, $|\bar{\nu}_{k+1} - \bar{\nu}_k| \leq c_4(\bar{\nu}_k - \bar{\nu}_{k+1})$.

From (9.5) and (9.6) we get the asymptotic behavior

$$\delta_k \sim \frac{c_1(\eta)}{k^{3/2-\eta}}, \qquad \mu_k \sim \frac{c_2(\eta)}{k^{1/2-\eta}}$$

for any small $\eta > 0$. According to the discussion at the end of Section 4.1 this asymptotic behavior corresponds exactly to the modulus of continuity of $\nabla_{x'} f$ and $\partial_t f$ in Theorem 9.1. In particular, if we set

$$\omega(r) = \sup_{|x'-y'|<r} |\nabla_{x'} f(x') - \nabla_{x'} f(y')|,$$

then $\omega(r) \leq c_0 |\log r|^{-4/3}$ (say, choosing $\eta = 1/6$) and in particular

$$\int_0 \frac{\omega(r)}{r} \, dr \leq c_D < \infty. \tag{9.7}$$

The condition (9.7), known as the *Dini-condition*, appears in several instances in the regularity theory for partial differential equations. An important question we may ask is under which minimal conditions does one have C^1-regularity up to the boundary for, say, a harmonic function in a domain $\Omega \subset \mathbb{R}^n$?

From [W] it turns out that a necessary and sufficient condition for this to happen is that Ω be a Liapunov-Dini domain, i.e., a bounded domain in $\mathbb{R}^n$, whose boundary is given locally by the graph of a function $x_n = f(x')$ satisfying (9.7) uniformly.

More precisely, we will use the following consequence of the results in [W]: if v is harmonic in the C^1 domain Ω and if it vanishes on the surface disc $\Delta_R = \partial\Omega \cap B_R(x_0)$, $x_0 \in \partial\Omega$, then u is C^1 up to $\Delta_{R/2}$ and

$$v_\nu \geq c > 0 \quad \text{on} \quad \Delta_{R/2}$$

if and only if (9.7) holds on Δ_R.

But for the solution u of our f.b.p., (9.7) tells us that $\Omega^\pm(u) \cap \{t = \tau\}$ is a Liapunov-Dini domain for each $\tau \in (-\frac{1}{2}, \frac{1}{2})$. Applying the results of [W] and the free boundary condition, we obtain the remaining assertions of Theorem 9.1.

As usual, we do not explicitly express the dependence of a constant when its arguments are some of the relevant parameters $n, L, M, c^*, m, a_1, a_2$.

9.2. Interior gain in space (parabolic homogeneity)

According to the comments in Step 0, we assume from now on that u is a viscosity solution to f.b.p. in $\mathcal{C}_1 = B_1' \times (-1, 1)$, with $(0,0) \in F(u)$, monotone increasing along every $\tau \in \Gamma(e_n, \theta)$. We also assume that the coefficients of u_t in the heat equations $\Delta u - a_i u_t = 0$ in both phases are much smaller than $\delta = \frac{\pi}{2} - \theta$ and that $B'_{1/4}(\frac{3}{4}e_n) \times (-1,1) \subset \Omega^+(u)$, $B'_{1/4}(-\frac{3}{4}e_n) \times (-1,1) \subset \Omega^-(u)$. There is no loss of generality, since we can start from a very small neighborhood of a free boundary point, say $(0,0)$, and perform an initial fixed scaling of the type

$$u_\lambda(x,t) = \frac{u(\lambda x, \lambda t)}{\lambda}\,.$$

The following lemma parallels Lemma 4.5.

Lemma 9.2 (Harnack's principle). *Let $u_1 \leq u_2$ be two viscosity solutions to f.b.p. in $\mathcal{C}_1$, monotone increasing along every $\tau \in \Gamma(\nu, \theta)$, $\nu \in \mathbb{R}^{n+1}$. Let $a = \min\{a_1, a_2\}$ and suppose that for $\varepsilon > 0$, $\sigma > 0$, small, the following hold.*

(a) $v_\varepsilon(x,t) = \sup_{B_\varepsilon(x,t)} u_1 \leq u_2(x,t)$ *in* $\mathcal{C}_{1-\varepsilon}$.

(b) *If* $p_0 = (\frac{3}{4}e_n, -\frac{1}{4}a^2)$,

$$u_2(p_0) - v_\varepsilon(p_0) \geq \sigma\varepsilon u_2(p_0).$$

(c) $B'_{1/4}(\frac{3}{4}e_n) \times (-1,1) \subset \Omega^+(u_1)$.

Then there exist positive constants C, h *such that, in* $B'_{1/8}(\frac{3}{4}e_n) \times (0, \frac{1}{4}a^2)$, *we have*

$$u_2(x,t) - v_{(1+h\sigma)\varepsilon}(x,t) \geq C\sigma\varepsilon u_2(p_0) \ . \tag{9.8}$$

Proof. For any unit vector $\bar{\nu}$, write ($p = (x,t)$)

$$u_2(p) - u_1(p + \varepsilon\bar{\nu}(1+h\sigma)) = w(p) + z(p)$$

where

$$w(p) = u_2(p) - u_1(p+\varepsilon\bar{\nu}) \ , \quad z(p) = u_1(p+\varepsilon\bar{\nu}) - u_1(p+\varepsilon\bar{\nu}(1+h\sigma)) \ .$$

Now, w is a nonnegative solution of $\Delta w - a_1 w_t = 0$ in

$$B'_{1/4-\varepsilon}\left(\frac{3}{4}e_n\right) \times (-1+\varepsilon, 1-\varepsilon).$$

By Harnack's inequality, for $p \in B_{1/8}(\frac{3}{4}e_n) \times (0, \frac{1}{4}a^2)$, we have, from (9.8)

$$w(p) \geq c w(p_0) \geq c\sigma\varepsilon u_2(p_0) \ . \tag{9.9}$$

From Corollary 13.14 we have, in $B'_{1/8}(\frac{3}{4}e_n) \times (0, a^2)$

$$|\nabla u_1(p)| \leq C u_1(p) \leq C u_1(p_0) \leq C u_2(p_0) \ . \tag{9.10}$$

Hence, from (9.8) and (9.9),

$$w(p) + z(p) \geq C\sigma\varepsilon u_2(p_0) + Ch\sigma\varepsilon u_2(p_0) \geq \bar{C}\sigma\varepsilon u_2(p_0)$$

for h small enough. □

The above Harnack's principle remains valid if the supremum in (a) and (b) is taken only over n-dimensional spatial balls, for every time level t. Precisely, with the same proof we have

Corollary 9.3. *Let* u_1, u_2 *as in Lemma 9.2. Suppose that for* $\varepsilon > 0$, $\sigma > 0$ *small,*

(a) $\sup_{y \in B'_\varepsilon(x)} u_1(y,t) \leq u_2(x,t)$ *in* $\mathcal{C}_{1-\varepsilon}$,

(b) $u_2(p_0) - \sup_{y \in B'_\varepsilon(\frac{3}{4}e_n)} u_1(y_1, -\frac{1}{4}a^2) \geq \sigma\varepsilon u_2(p_0)$,

(c) $B'_{1/4}(\frac{3}{4}e_n) \times (-1,1) \subset \Omega^+(u_1)$.

Then there exist constants $c > 0$, $h > 0$ *such that, in* $B'_{1/8}(\frac{3}{4}e_n) \times (0, \frac{1}{4}a^2)$,

$$u_2(x,t) - \sup_{y \in B'_{(1+h\sigma)\varepsilon}(x)} u_1(y,t) \geq c\sigma\varepsilon u_2(p_0) \ . \tag{9.11}$$

We now apply Corollary 9.3 to the positive part of our viscosity solution in u. Denote by $\Gamma_x(e_n, \theta)$ the section in space of $\Gamma(e_n, \theta)$. Let $\tau \in \Gamma_x(e_n, \frac{\theta}{2})$ be a small vector and let $\varepsilon = |\tau| \sin \frac{\theta}{2}$.

Setting $u_1(x,t) = u(x-\tau, t)$, for $y \in B'_{1/4}(\frac{3}{4}e_n)$ we have

$$\begin{aligned} u_1\left(y, -\frac{1}{4}a^2\right) &= u\left(y-\tau, -\frac{1}{4}a^2\right) = u\left(\frac{3}{4}e_n - \bar{\tau}, -\frac{1}{4}a^2\right) \\ &= u\left(\frac{3}{4}e_n, -\frac{1}{4}a^2\right) - D_{\bar{\tau}}u\left(y^*, -\frac{1}{4}a^2\right) \end{aligned} \tag{9.12}$$

where $\bar{\tau} = \frac{3}{4}e_n - y + \tau$.

Since $\alpha(\tau, \bar{\tau}) \le \frac{1}{2}\theta$, $D_{\bar{\tau}}u$ is a nonnegative solution of the heat equation in $\Omega^+(u)$. Hence, from Harnack's inequality,

$$\begin{aligned} \inf_{y \in B'_{|\tau|+\frac{1}{4}}(3e_n/4)} D_{\bar{\tau}}u\left(y, -\frac{1}{4}a^2\right) &\ge cD_{\bar{\tau}}u\left(\frac{3}{4}e_n, -\frac{1}{2}a^2\right) \\ &= c\left|\nabla u\left(\frac{3}{4}e_n, -\frac{1}{2}a^2\right)\right| \cdot |\bar{\tau}| \cos\alpha(\bar{\nu}, \bar{\tau}) \end{aligned} \tag{9.13}$$

where $\bar{\nu} = \nabla u(\frac{3}{4}e_n, -\frac{1}{2}a^2)$. Since $|\bar{\tau}| \ge c|\tau|$ and $\alpha(\bar{\nu}, \bar{\tau}) \le \alpha(\bar{\nu}, \tau) + \frac{\theta}{2}$, by Corollary 13.14, we have,

$$\left|\nabla u\left(\frac{3}{4}e_n, -\frac{1}{2}a^2\right)\right| \ge cu(p_0)$$

and therefore, from (9.12), for $y \in B'_{1/4}(\frac{3}{4}e_n)$

$$u_1\left(y, -\frac{1}{4}a^2\right) \le (1 - \sigma\varepsilon)u(p_0)$$

with $\sigma = \sigma(\tau) = c(\frac{\pi}{2} - (\alpha(\bar{\nu}, \tau) + \frac{\theta}{2}))$. Applying Corollary 9.3, we have proved

Lemma 9.4. *There exist $c > 0$, $h > 0$ such that*

$$\sup_{y \in B'_{(1+h\sigma)\varepsilon}(x)} u(y-\tau, t) \le u(x,t) - c\sigma\varepsilon u(p_0)$$

for every $(x,t) \in B'_{1/8}(\frac{3}{4}e_n) \times (0, \frac{1}{4}a^2)$.

An analogous lemma holds for the negative part of u, in $B'_{1/8}(-\frac{3}{4}e_n) \times (0, \frac{1}{4}a^2)$. Now applying the intermediate cone Theorem 4.2, we conclude that

Corollary 9.5. *In $B'_{1/8}(\pm\frac{3}{4}e_n) \times (0, \frac{1}{4}a^2)$, $u^\pm$ is monotone increasing along every $\tau \in \Gamma_x(e^\pm, \theta^\pm) \supset \Gamma_x(e_n, \theta)$, respectively, with*

$$\delta^\pm \le \bar{b}\delta$$

where $\bar{b} < 1$.

9.3. Common gain

At this point we have enlarged the spatial section $\Gamma_x(e_n,\theta)$ of the monotonicity cone in $B'_{1/8}(\frac{3}{4}e_n)\times(0,\frac{1}{4}a^2)$ and in $B'_{1/8}(-\frac{3}{4}e_n)\times(0,\frac{1}{4}a^2)$ to new cones $\Gamma_x^+=\Gamma_x(e^+,\theta^+)$ and $\Gamma_x^-=\Gamma_x(e^-,\theta^-)$, respectively.

Now, if $\nu^\pm=\nabla u(\pm\frac{3}{4}e_n,-\frac{1}{2}a^2)$ and

$$\alpha(\nu^+,e_n)<\delta \quad \text{or} \quad \alpha(\nu^-,e_n)<\delta, \tag{9.14}$$

we can use ν^+ or ν^-, respectively, as our special direction ν in Theorem 4.2 and obtain a gain on both sides with the same corrected cone $\Gamma_x(e^*,\theta^*)$, $e^*=\nu^+/|\nu^+|$ or $e^x=\nu^-/|\nu^-|$.

The problem arises when both angles in (9.14) are close to δ, say,

$$\delta-\eta=\alpha(\nu^+,e_n)\approx\alpha(\nu^-,e_n)\le\delta$$

with $\eta\ll\delta$, because the new cone axis e^+, necessary for the gain on the positive side of u, may be far from that on the other side. For instance, it could be that

$$\alpha(\nu^+,e_n)=\alpha(\nu^-,e_n)=\delta$$

and, if $\nu^+\ne\nu^-$, the corresponding cones Γ_x^+ and Γ_x^- are tangent to the original cone $\Gamma_x(e_n,\theta)$ along different directions, making it impossible to have a common enlarged cone.

This is precisely the situation described in Figure 9.1. We rule out this situation by estimating the distance between e^+ and e^- in terms of δ, showing that we can always choose a common enlarged cone on both sides.

In the following lemma we find a good candidate for the axis of the common cone.

Let $\rho^\pm$ denote the unit vectors on the boundary of $\Gamma_x(e_n,\theta)$, opposite to $\nu^\pm$, respectively (that is, $\langle\rho^\pm,\nu^\pm\rangle$ is minimum among all $\langle\tau,\nu^\pm\rangle$, $\tau\in\Gamma_x(e_n,\theta)$), and thus $\rho^\pm$ give us the directions for which expected gain is minimum.

Lemma 9.6. *There exists a (universal) constant K and a normal unit vector ν at some point of differentiability of the free boundary such that simultaneously*

$$\langle\nu,\rho^+\rangle\le K[\delta-\alpha(\nu^+,e_n)]\equiv K\gamma^+,$$
$$\langle\nu,\rho^-\rangle\le K[\delta-\alpha(\nu^-,e_n)]\equiv K\gamma^-\ .$$

Proof. Normalize u^+ so that

$$\langle\nu^+,e_n\rangle=D_{e_n}u\left(\frac{3}{4}e_n,-\frac{1}{2}a^2\right)=1\ .$$

We have

(1) $D_{\rho^+}u \geq 0$,

(2) $D_{\rho^+}u\left(\frac{3}{4}e_n, -\frac{1}{2}a^2\right) = \langle \nu^+, \rho^+\rangle \leq |\nu^+|\cos\alpha(\nu^+, \rho^+) \leq c\gamma^+$.

Denote by ω the caloric measure in $\Omega^+(u) \cap \{t \geq -20a_1^2\}$, evaluated at $(\frac{3}{4}e_n, -\frac{1}{2}a^2)$ and set

$$E = \{(x,t) \in F(u) : -10a_1^2 < t < -a_1^2\} \,.$$

Then from (1) and (2),

$$0 \leq \int_E D_{\rho^+}u^+ \, d\omega \leq c\gamma^+ \,.$$

Since caloric and surface measure on $F(u)$ are mutually absolutely continuous, the set

$$E \cap \{D_{\rho^+}u^+ \geq K\gamma^+\}$$

has surface measure as small as we want depending only on K.

Also, since $D_{e_n}u^+$ is equivalent to caloric measure on $F(u)$, the set

$$E \cap \left\{D_{e_n}u^+ < \frac{1}{K}\right\}$$

has surface measure as small as we want. Similarly for u^-, ν^-, ρ^-.

Thus, for K large, there is a point of differentiability of $F(u)$ where

$$D_{e_n}u^+ \geq \frac{1}{K}\,, \quad D_{e_n}u^- \geq \frac{1}{K}$$

and

$$D_{\rho^+}u^+ \leq K\gamma^+\,, \quad D_{\rho^-}u^- \leq K\gamma^+ \,.$$

If ν is the normal unit vector to $F(u)$ at that point, inward to $\Omega^+(u)$, we have

$$\frac{\langle \nu, \rho^+\rangle}{\langle \nu, e_n\rangle} = \frac{D_{\rho^+}u^+}{D_{e_n}u^+} \leq K^2\gamma^+ \,.$$

The same inequality holds for $\langle \nu, \rho^-\rangle$. This completes the proof. □

We want now to show that we can associate to ν, by the gain process indicated in Theorem 4.2, a cone $\bar{\Gamma}_x(\nu, \bar{\theta})$ that, for a small enough gain, is contained in both the cones Γ_x^+ and Γ_x^- associated to ν^+ and ν^-.

The cones Γ_x^+ and Γ_x^- are constructed on the basis that, after normalization, any directional derivative $D_\tau u$, with $\tau \in \partial\Gamma - x(e_n, \theta)$, has in a neighborhood of $(\pm\frac{3}{4}, 0)$ a gain of

$$D_\tau u \sim \langle \tau, \nu^\pm\rangle \,.$$

This is a consequence of formula (9.13). Therefore, to obtain a common enlarged cone on both sides, we need to control $\langle \tau, \nu^\pm\rangle$ from below by $\langle \tau, \nu\rangle$.

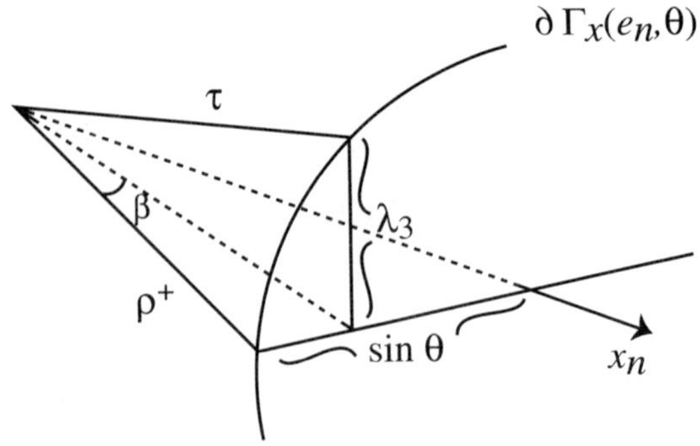

Figure 9.2. $\sin\beta \approx \delta\lambda_3^2$, $\delta = \frac{\pi}{2} - \theta$, $|\rho^+| = |\tau| = 1$

This is obtained in the following corollary to Lemma 9.6.

Corollary 9.7. *There exists a constant $A = A(n, \theta)$ such that*

$$\langle \tau, \nu\rangle \le A \min\{\langle \tau, \bar\nu^+\rangle, \langle \tau, \bar\nu^-\rangle\}$$

for every unit vector $\tau \in \partial\Gamma_x(e_n, \theta)$, where $\bar\nu^+ = \nu^+/|\nu^+|$, $\bar\nu^- = \nu^-/|\nu^-|$.

Proof. We show that $\langle \tau, \bar\nu^+\rangle \ge A\langle \tau, \nu\rangle$. It is enough to consider $\tau \in \partial\Gamma_x(e_n, \theta)$ in a small neighborhood of ρ^+. That is, we may suppose

$$\tau = \lambda_1 e_n + \lambda_2 \rho^+ + \lambda_3 \tau^*$$

with $\tau^* \perp \operatorname{span}\{e_n, \rho^+\}$ and $\lambda_3 \ll 1$. Then (see Figure 9.2)

$$\alpha(\lambda_1 e_n + \lambda_2 \rho^+, \rho^+) \le c\delta\lambda_3^2 \tag{9.15}$$

On the other hand, we have simultaneously

$$\alpha(\nu, e_n) \le \delta$$

just because $\nu = \nabla u^+/|\nabla u^+|$ at a free boundary point, and

$$\alpha(\nu, \rho^+) \ge \frac{\pi}{2} - \bar K\gamma^+$$

since $\langle \nu, \rho^+\rangle \le K\gamma^+$. Therefore, the component ν_1 of ν in the (ρ^+, e_n)-plane satisfies

$$\alpha(\nu_1, \bar\nu^+) \le c\bar K\gamma^+ \tag{9.16}$$

and the normal component ν_2 satisfies

$$|\nu_2| \le (c\delta\gamma^+)^{1/2} \tag{9.17}$$

(see Figure 9.3).

Now, we have

$$\langle \nu, \tau\rangle = \langle \nu - \bar\nu^+, \tau\rangle + \langle \bar\nu^+, \tau\rangle\ . \tag{9.18}$$

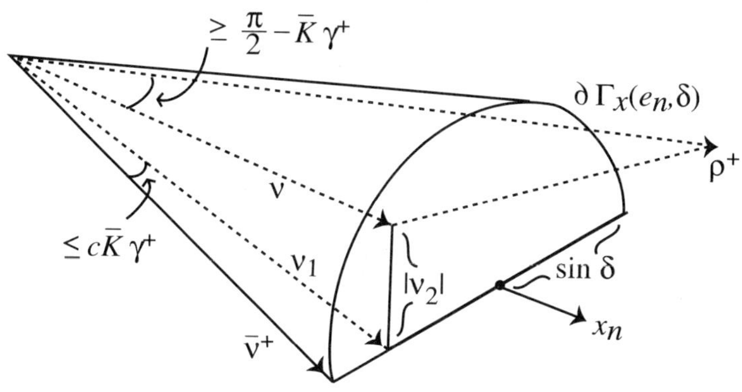

Figure 9.3. $\alpha(\nu_1, \bar{\nu}^+) \le c\bar{K}\gamma^+$, $|\nu_2| \le c(\delta\gamma^+)^{1/2}$

Since $\bar{\nu}^+ \perp \tau^*$,

$$\langle \bar{\nu}^+, \tau\rangle = \langle \bar{\nu}^+, \lambda_1 e_n + \lambda_2 \rho^+\rangle \ge c \cos \alpha(\bar{\nu}^+, \lambda_1 e_n + \lambda_2 \rho^+) \ .$$

From (9.15) we get

$$\begin{aligned} \alpha(\bar{\nu}^+, \lambda_1 e_n + \lambda_2 \rho^+) &\le \alpha(\bar{\nu}^+, \rho^+) + \alpha(\lambda_1 e_n + \lambda_2 \rho^+, \rho^+) \\ &\le \frac{\pi}{2} - \gamma^+ + c\delta\lambda_3^2. \end{aligned}$$

Hence

$$\langle \bar{\nu}^+ \tau\rangle \ge c(\gamma^+ + \delta\lambda_3^2) \ . \tag{9.19}$$

Moreover

$$\begin{aligned} \langle \nu - \bar{\nu}^+, \tau\rangle &= \langle \nu_1 - \bar{\nu}^+, \tau\rangle + \langle \nu_2, \tau\rangle \le |\nu_1 - \bar{\nu}^+| + \lambda_3 |\nu_2| \\ &\le c\bar{K}\gamma^+ + c\lambda_3(\delta\gamma^+)^{1/2} \le \tilde{c}(\gamma^+ + \delta\lambda_3^2) \ . \end{aligned} \tag{9.20}$$

Finally, (9.18), (9.19) and (9.20) give

$$\langle \nu, \tau\rangle \le A\langle \bar{\nu}^+, \nu\rangle \ . \qquad \square$$

Remark. This indicates that the "good direction ν" to improve defect can be detected from the distribution of the normal vector along the free boundary.

9.4. Interior gain in space (hyperbolic homogeneity)

By the results in Section 9.3, a common enlarged cone of monotonicity $\Gamma_x(\nu, \bar{\theta})$ exists on both sides of $F(u)$. As was already mentioned, this gain is valid only in a parabolic region away from the free boundary. Now, if we are willing to give up a portion of $\Gamma_x(\nu, \bar{\theta})$, we can always have a gain in a hyperbolic region. This can be obtained in the following way.

For any unit vector $\bar{\tau} \in \Gamma_x(\nu, \bar{\theta})$, $D_{\bar{\tau}} u \geq 0$ is equivalent to

$$D_\tau u \geq \frac{\beta}{\alpha} D_{e_n} u$$

where $\bar{\tau} = \alpha\tau - \beta e_n$, $\tau \in \Gamma_x(e_n, 0)$, $|\tau| = 1$, with $\beta \geq 0$ and

$$1 \leq \alpha \leq \sin(2\bar{\theta} - \theta)/\sin\theta \ .$$

Note that if the cones touch along τ, then $\alpha = 1$, $\beta = 0$.

If we delete a small percentage of $\Gamma_x(e_n, \theta)$ around the contact line, we can have

$$\frac{\beta}{\alpha} \geq c\delta$$

where c depends on the size of the deleted neighborhood. Thus, the inequality $D_\tau u \geq c\delta D_{e_n} u$ holds in $B'_{1/8}(\pm\frac{3}{4}e_n) \times (0, \frac{1}{2}a^2)$.

We will show that this inequality propagates in time to an interval of order δ/μ, where we recall that μ is the defect angle in time. Although we start with an $(n+1)$-dimensional circular cone, so that $\delta = \mu$, in the iterative process we will use later, the defect angle goes to zero much faster in space than in time. Therefore we assume from the beginning $\delta \leq \mu$.

Lemma 9.8. *Let u be a viscosity solution to f.b.p. in $\mathcal{C}_1$. Let $0 < \delta \leq \mu < \frac{\pi}{2}$. If for $\tau \in \Gamma_x(e_n, \theta)$*

$$D_\tau u(x,t) \geq c\delta D_{e_n} u(x,t)$$

in $B'_{1/8}(\pm\frac{3}{4}e_n) \times (0, \frac{1}{4}a^2)$, then there exist $\bar{c}$ and c_0 such that

$$D_\tau u(x,t) \geq \bar{c}\delta D_{e_n} u(x,t)$$

in $B'_{1/8}(\pm\frac{3}{4}e_n) \times (-c_0\delta/\mu, c_0\delta/\mu)$.

Proof. Let $\gamma = ae_n + be_t$, $a^2 + b^2 = 1$, be normal to the axis of the full cone of monotonicity. Note that

$$|D_\gamma u(x,t)| \leq c\mu D_{e_n} u(x,t)$$

for any (x,t) where the derivative exists.

By interior a priori estimates and Corollary 13.14,

$$|D_{\gamma\tau} u(x,t)| \leq c\mu D_{e_n} u(x,t) \tag{9.21}$$

if (x,t) stays uniformly away from $F(u)$ and $\partial_p \mathcal{C}_1$.

Now, for any $p = (x, 0)$, $x \in B'_{1/8}(\pm\frac{3}{4}e_n)$, we have

$$D_\tau u(p + \lambda\gamma) \geq c\delta D_{e_n} u(p) + \lambda D_{\gamma\tau} u(p + \tilde{\lambda}\gamma)$$

for some $0 < |\tilde{\lambda}| < |\lambda|$. Using (9.21) and Corollary 13.14 again, we obtain

$$D_\tau u(p + \lambda\gamma) \geq (c\delta - C|\lambda|\mu) D_{e_n} u(p + \lambda\gamma) \ .$$

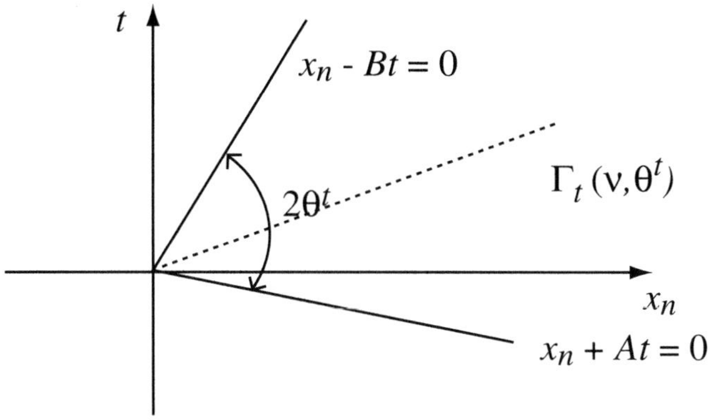

Figure 9.4. The space-time section of the monotonicity cone

Choosing $|\lambda| \leq \frac{1}{2}\frac{c\delta}{c\mu}$ and using Harnack's inequality for $D_\tau u$, we see that

$$D_\tau u(q) \geq c\delta D_{e_n} u(q)$$

if q belongs, at least, to $B_{1/8}(\pm\frac{3}{4}e_n) \times (-c_0\delta/\mu, c_0\delta/\mu)$. □

The convex envelope of the union of the directions in Lemma 9.8 and those in the original cone $\Gamma_x(e_n, \theta)$ is easily seen to contain a cone $\Gamma_x(\nu, \bar{\theta})$ such that u is monotone increasing along every $\tau \in \Gamma_x(\nu, \bar{\theta})$ and

$$\bar{\delta} \leq \bar{b}_0 \delta$$

with $\bar{b}_0 < 1$.

9.5. Interior gain in time

The method used in the previous sections to enlarge the cone in space fails for the enlargement in time. We will proceed in a different way. The idea is the following.

Let e_n be the projection in space of the axis of the (full) monotonicity cone with defect angle in time $\mu = \frac{\pi}{2} - \theta^t$. This means that there are numbers $A, B \in \mathbb{R}$, $A \leq B$, such that $B - A \approx \mu$ and ($D_n = \partial_{x_n}$)

$$A \leq -\frac{D_t u^+(x,t)}{D_n u^+(x,t)} \leq B \quad \text{or} \quad A \leq -\frac{D_t u^-(x,t)}{D_n u^-(x,t)} \leq B \tag{9.22}$$

for (x, t) a.e. on $F(u)$ and everywhere outside $F(u)$.

To increase the cone in time means to *lower* B or to *increase* A.

From the first one of (9.22), the functions

$$w = D_t u + BD_n u \quad \text{or} \quad v = -D_t u - AD_n u$$

satisfy the heat equation and are nonnegative in $\Omega^+(u)$.

We manage to enlarge Γ_t in some region if in that region we can prove that

$$w \geq c\mu D_n u \quad \text{or} \quad v \geq c\mu D_n u \ .$$

The same kind of consideration can be done in $\Omega^-(u)$.

We start by estimating in terms of δ/μ the oscillation of $D_n u$, pointwise in the interior and in measure on $F(u)$. The first step is

Lemma 9.9. *Let u be a viscosity solution of f.b.p. in $\mathcal{C}_1$. Then*

$$u(x,t) = u\left(\pm\frac{3}{4}e_n, 0\right) + \alpha_\pm\left(x_n \pm \frac{3}{4}\right) + \alpha_\pm O(\delta/\mu) \tag{9.23}$$

for every $(x,t) \in B'_{1/8}(\pm\frac{3}{4}e_n) \times (-c_0\delta/\mu, c_0\delta/\mu)$, where

$$\alpha_\pm = D_n u\left(\pm\frac{3}{4}e_n, 0\right) \ .$$

Proof. For any $x \in B'_{1/8}(\frac{3}{4}e_n)$, we have ($D_i = \partial_{x_i}$, $D_{ij} = \partial_{x_i x_j}$)

$$u(x,t) = u\left(\frac{3}{4}e_n, t\right) + D_n u\left(\frac{3}{4}e_n, t\right)\left(x_n - \frac{3}{4}\right) + \sum_{i=1}^{n-1} D_i u\left(\frac{3}{4}e_n, t\right) x_i$$
$$+ \int_0^1\int_0^s \sum_{i,j=1}^{n} D_{ij}u\left((1-r)\frac{3}{4}e_n + rx, t\right)\left(x - \frac{3}{4}e_n\right)_i \left(x - \frac{3}{4}e_n\right)_j \, dr\, ds \ .$$

We estimate each term on the right, separately. For the first one, we write

$$u\left(\frac{3}{4}e_n, t\right) = u\left(\frac{3}{4}e_n, 0\right) + \int_0^t D_t u\left(\frac{3}{4}e_n, s\right) ds \ .$$

Since $|D_t u| \leq cD_n u$, using Corollary 13.14, we have, for $|t| \leq c\delta/\mu$,

$$u\left(\frac{3}{4}e_n, t\right) = u\left(\frac{3}{4}e_n, 0\right) + \alpha_+ O\left(\frac{\delta}{\mu}\right) \ . \tag{9.24}$$

For the second term, write

$$D_n u\left(\frac{3}{4}e_n, t\right) = D_n u\left(\frac{3}{4}, 0\right) + \int_0^t D_{nt} u\left(\frac{3}{4}e_n, s\right) ds \ .$$

Using interior estimates and Corollary 13.14, we see that $|D_t u| \leq cD_n u$ implies

$$\left|D_{nt} u\left(\frac{3}{4}e_n, t\right)\right| \leq c\alpha_+ \ .$$

Therefore, for $|t| \leq c\delta/\mu$,

$$D_n u\left(\frac{3}{4}e_n, t\right) = \alpha_+\left(1 + O\left(\frac{\delta}{\mu}\right)\right) \ . \tag{9.25}$$

Now the third term. Since for $i = 1, \ldots, n-1$, $|D_i u| \le c\delta D_n u$, by the previous estimates, we have, for $|t| \le c\delta/\mu$,

$$\left| D_i u \left(\frac{3}{4} e_n, t \right) \right| = \alpha_+ O(\delta) \ .$$

Finally, for the fourth term, observe that since $|D_i u(p)| \le c\delta D_n u(p)$, $i = 1, \ldots, n-1$, then interior estimates and Corollary 13.14 give

$$|D_{ij}(p)| \le c\delta D_n u(p)$$

whenever p is away from $F(u)$ and $\partial_p C_1$.

Thus, using the equation $\Delta u = a_1 D_t u$ and $a_1 < \delta$, (by hypothesis)

$$|D_{nn} u(p)| \le c\delta D_n u(p) + c a_1 D_n u(p) \le c\delta D_n u(p) \ .$$

Once more, Corollary 13.14 gives

$$|D_{ij}(p)| \le c\delta\alpha_+, \qquad i, j = 1, \ldots, n \ .$$

Collecting all the estimates above, we get (9.23) for the positive side of u. In a similar fashion we prove the result for the negative side too. □

Using Lemma 9.9, it is possible to estimate the measure of the set of points on $F(u)$ where $D_n u^+$ ($D_n u^-$) is close to α_+ (α_-). We do this by estimating their difference in the L^2-sense at all time levels.

Lemma 9.10. *Let F_t denote the t-section of $F(u)$. Then for all $|t| \le c_0\delta/\mu$*

$$\fint_{B'_{1/8} \cap F_t} |D_n u^\pm - \alpha_\pm|^2 \, dS \le c\alpha_\pm^2 \frac{\delta}{\mu} \ .$$

Proof. We prove only the estimate for u^+, since the negative part can be treated in a similar way. We do this in two steps.

Step 1. The first step is to prove

$$\left| \fint_{B'_{1/8} \cap F_t} (D_n u^+ - \alpha_+) \, dS \right| \le c\alpha_+ \frac{\delta}{\mu} \ .$$

For $0 \le r \le \frac{1}{8}$ set

$$D_r = \left\{ (x', x_n) : |x'| < r, \ |x_n| < \frac{3}{4} \right\} \cap \Omega^+(u) \ .$$

We know that, after an initial rescaling if necessary, for a suitable $\varepsilon > 0$, the function $w_+ = u + \delta u^{1+\varepsilon}$ is subharmonic in D. Then for a.e. t, $|t| \le c\delta/\mu$,

$$0 \le \int_{D_{1/8}} \Delta w_+(x, t) \, dx = \int_{\partial D_{1/8}} D_\nu w_+(x, t) \, dS_x$$

or

$$\int_{F_t \cap \partial D_{1/8}} |\nabla u^+(x,t)|\, dS_x \le \int_{S_{1/8}} D_\nu w_+(x,t)\, dS_x + \int_{T_{1/8}} D_\nu w_+(x,t)\, dS_x$$

where

$$S_r = \left\{ |x'| = r,\ |x_n| < \frac{3}{4} \right\} \cap \Omega^+(u) , \quad T_r = \left\{ |x'| < r,\ x_n = \frac{3}{4} \right\} \cap \Omega^+(u) .$$

Since

$$|D_j w_+| \le (1 + c\delta)|D_j u|, \qquad j = 1, \dots, n, \tag{9.26}$$

$$|D_i u| \le c\delta D_n u, \qquad j = 1, \dots, n-1, \tag{9.27}$$

$$|\nabla u^+| \ge D_n u^+,$$

we have, from Lemma 9.9,

$$\int_{F_t \cap \partial D_{1/8}} D_n u^+\, dS_x \le c\delta \int_{S_{1/8}} D_n u\, dS_x + (1 + O(\delta/\mu)) \int_{T_{1/8}} \alpha_+\, dS_x$$

or

$$\int_{F_t \cap \partial D_{1/8}} (D_n u^+ - \alpha_+)\, dS_x \le (c\delta + O(\delta/\mu)) \int_{T_{1/8}} \alpha_+\, dS_x \le \alpha_+ O(\delta/\mu) .$$

Similarly, using $w_- = u - \delta u^{1+\varepsilon}$, superharmonic, we obtain

$$\int_{F_t \cap \partial_{1/8}} (D_n u^+ - \alpha_+)\, dS_x \ge -\alpha_+ O(\delta/\mu) .$$

Step 2. The second step is to prove

$$\left| \fint_{B'_{1/10} \cap F_t} \left((D_n u^+)^2 - \alpha_+^2 \right) dS \right| \le c\alpha_+^2 \frac{\delta}{\mu} .$$

As in Step 1, for a.e. t, $|t| \le c\delta/\mu$,

$$\int_{D_r} \nabla(D_n w_+) \nabla w_+\, dx \le \int_{D_r} \left[\nabla(D_n w_+) \nabla w_+ + D_n w_+ \Delta w_+ \right] dx$$
$$= \int_{\partial D_r} D_n w_+ D_\nu w_+\, dS_x$$

or

$$\int_{\partial D_r \cap F_t} D_n u |\nabla u^+|\, dS_x + \frac{1}{2} \int_{D_r} D_n(|\nabla w_+|^2)\, dx$$
$$\le \int_{S_r} D_n w_+ D_\nu w_+\, dS_x + \int_{T_r} (D_n w_+)^2\, dS_x .$$

Integrating the second term on the left with respect to x_n and using $|\nabla u_+| \geq D_n u$, we obtain

$$\frac{1}{2}\int_{\partial D_n \cap F_t} (D_n u)^2 \, dS_x \leq \int_{S_r} D_n w_+ D_\nu w_+ \, dS_x + \frac{1}{2}\int_{T_r} (D_n w_+)^2 \, dS_x \, .$$

Since $|D_i u| \leq c\delta D_n u$ for $i = 1, \ldots, n-1$, also $|D_i w_+| \leq c\delta D_n w_+$, so that

$$\int_{\partial D_r \cap F_t} (D_n u)^2 \, dS_x \leq c\delta \int_{S_R} (D_n w_+)^2 \, dS_x + \int_{T_r} (D_n w_+)^2 \, dS_x \, .$$

Integrate the last inequality with respect to r from $1/10$ to $R \leq 1/8$. We get, since $D_{1/10} \subset D_r \subset D_R$,

$$\begin{aligned}\left(R - \frac{1}{10}\right) &\int_{\partial D_{1/10} \cap F_t} (D_n u)^2 \, dS_x \\ &\leq c\delta \int_{1/10}^{R} \int_{S_R} |\nabla w_+|^2 \, dS_x \, dr + \left(R - \frac{1}{10}\right) \int_{T_R} (D_n w_+)^2 \, dS_x \, .\end{aligned}$$

Integrating by parts the first term on the right and using the subharmonicity of w_+, we obtain

$$\begin{aligned}\left(R - \frac{1}{10}\right) \int_{\partial D_{1/10} \cap F_t} (D_n u)^2 \, dS_x &\leq \left(R - \frac{1}{10}\right) \int_{T_R} (D_n w_+)^2 \, dS_x \\ + c\delta \left\{ \int_{S_R} w_+ D_\nu w_+ \, dS_x + \int_{S_{1/10}} w_+ D_\nu w_+ \, dS_x \right. &+ \left. \int_{T_R \setminus T_{1/10}} w_+ D_n w_+ \, dS_x \right\} .\end{aligned}$$

Using (9.26) and (9.27), we can write, using Lemma 9.9,

$$\begin{aligned}\left(R - \frac{1}{10}\right) \int_{\partial D_{1/10} \cap F_t} (D_n u)^2 \, dS_x &\leq \left(R - \frac{1}{10}\right) \int_{T_R} (D_n u)^2 \, dS_x \\ + c\delta \left\{ \int_{S_R} \delta D_n(u^2) \, dS_x + \int_{S_{1/10}} \delta D_n(u^2) \, dS_x \right. &+ \left. \int_{T_R \setminus T_{1/10}} u D_n u \, dS_x \right\} .\end{aligned}$$

Integrating the two middle terms on the right and using Corollary 13.14, we get

$$\begin{aligned}\left(R - \frac{1}{10}\right) \int_{\partial D_{1/10} \cap F_t} (D_n u)^2 \, dS_x &\leq \left(R - \frac{1}{10}\right) \int_{T_R} (D_n u)^2 \, dS_x \\ + c\delta \left\{ \int_{\partial S_R \cap \{x_n = 3/4\}} \delta (D_n u)^2 \, dS_x \right. & \\ + \int_{\partial S_{1/10} \cap \{x_n = 3/4\}} \delta (D_n u)^2 \, dS_x &+ \left. \int_{T_R \setminus T_{1/10}} (D_n u)^2 \, dS_x \right\}\end{aligned}$$

and, using Lemma 9.9,

$$\leq \left(R - \frac{1}{10}\right) \int_{T_R} (D_n u)^2 \, dS_x + c\delta^2 \alpha_+^2 + c\delta \left(1 + O\Big(\frac{\delta}{\mu}\Big)\right) \alpha_+^2 |T_R - T_{1/10}| \, .$$

Now choose $R = \frac{1}{10} + \delta$. We have, then

$$\int_{\partial D_{1/10} \cap F_t} \left((D_n u)^2 - \alpha_+^2\right) dS_x \leq c\delta\alpha_+^2 \leq c\frac{\delta}{\mu}\alpha_+^2 \, . \tag{9.28}$$

If we use w_- instead of w_+, we get $\geq -c\frac{\delta}{\mu}\alpha_+^2$ in (9.28). Combining Steps 1 and 2, we arrive at the desired result. □

With Lemma 9.10 at hand we can now show how to increase in time the cone of monotonicity. Actually, the enlargement can be done simultaneously on both sides of $F(u)$, if we stay away from $F(u)$ itself.

Precisely, we have

Lemma 9.11. *Let $b = \frac{A+B}{2}$ and assume*

$$G(\alpha_+, \alpha_-) \geq -b \qquad (\textit{resp. } G(\alpha_+, \alpha_-) \leq -b) \, .$$

There exist positive constants c, c_0 such that if δ is small, $\delta \ll \mu^3$, then

$$-\frac{D_t u}{D_n u} \leq B - c\mu \qquad \left(\textit{resp.} \quad -\frac{D_t u}{D_n u} \geq A + c\mu\right)$$

in $B'_{1/8}(\pm\frac{3}{4}e_n) \times (-c_0\delta/\mu, c_0\delta/\mu)$.

Proof. Assume $G(\alpha_+, \alpha_-) \geq -b$. The other case is analogous. Observe that $w = D_t u + B D_n u$ satisfies $\Delta w - a_1 D_t w = 0$ in $\Omega^+(u)$ and $\Delta w - a_2 D_t w = 0$ in $\Omega^-(u)$.

Let $|t| \leq c_0\delta/\mu$, c_0 as in Lemmas 9.9 and 9.10. Moreover, let

$$R_t = \Omega^+(u) \cap (-a_1 + t, t)$$

and let $\omega^{(x,t)}$ be the caloric measure in R_t evaluated at (x, t).

Now, on the free boundary, a.e. with respect to surface measure, we have

$$\frac{D_t u^+}{D_n u^+} = \frac{D_t u^+}{|\nabla u^+|}(1 + O(\delta)) = (1 + O(\delta)) G(|\nabla u^+|, |\nabla u^-|) \, . \tag{9.29}$$

By Lemma 9.10, if $\bar{c}$ is suitably chosen, since $\delta^{1/3} \leq c\mu$,

$$\Sigma_t = \left\{p \in F \cap \bar{R}_t : |D_n u^\pm(p) - \alpha_\pm| \leq \bar{c}\delta^{1/3}\alpha_\pm\right\} ,$$

then

$$|\Sigma_t| \geq \frac{1}{3}|F \cap \bar{R}_t|$$

for any $t \in (-c_0\delta/\mu, c_0\delta/\mu)$.

From the results in Section 13.5, the restriction of $\omega^{(x,t)}$ to $F \cap \bar{R}_t$ is an A_∞ weight with respect to surface measure, so that we have

$$\omega^{(x,t)}(\Sigma_t) \geq c > 0 \tag{9.30}$$

for every $(x,t) \in B_{1/8} \times (-c_0\delta/\mu, c_0\delta/\mu)$.

On the other hand, on Σ_t, by the local Lipschitz continuity of G,

$$\begin{aligned} G(|\nabla u^+|, |\nabla u^-|) &= G(\alpha_+, \alpha_-) + O(\delta^{1/3}) \geq -b + O(\delta^{1/3}) \\ &\geq -B + c\mu + O(\delta^{1/3}) \geq -B + \bar{c}\mu \ . \end{aligned}$$

Therefore, if $(x,t) \in B'_{1/8}(\frac{3}{4}e_n) \times (c_0\delta/\mu, c_0\delta/\mu)$, we can write

$$w(x,t) = \int_{\partial_p R_t} w \, d\omega^{(x,t)} \geq \int_{\Sigma_t} w \, d\omega^{(x,t)} \geq \bar{c}\mu\alpha_+ \omega^{(x,t)}(\Sigma_t) \geq \tilde{c}\mu\alpha_+ \ .$$

Since $D_n u(x,t) = \alpha_+(1 + O(\delta/\mu)) \leq c\alpha_+$ in $B'_{1/8}(\frac{3}{4}e_n) \times (-c_0\delta/\mu, c_0\delta/\mu)$, we obtain finally

$$w(x,t) \geq \tilde{c}\mu D_n u(x,t) \ .$$

In $\Omega^-(u)$ the same inequality holds in $B'_{1/8}(-\frac{3}{4}e_n) \times (-c_0\delta/\mu, c_0\delta/\mu)$. □

9.6. A continuous family of subcaloric functions

As in the elliptic case, starting from a viscosity solution u, we construct a particular family of subcaloric functions that plays a major role in carrying to the free boundary the interior gains obtained in Sections 9.4 and 9.5.

These functions are again constructed by taking the supremum of u over $(n+1)$-dimensional balls (thus, space-time balls) with variable radius $\varphi = \varphi(x,t)$. The main question is to find the right condition on φ that makes the supremum a subcaloric function in its positivity and negativity set. This is the content of the following lemma.

Lemma 9.12. *Let u be a viscosity solution to f.b.p. in $\mathcal{C}_3$, monotone increasing along every $\tau \in \Gamma(e,\theta)$. Suppose φ is a C^2-function such that $1 \leq \varphi \leq 2$ and such that it satisfies*

$$D_t\varphi \geq 0, \quad \Delta\varphi - c_1 D_t\varphi - C\frac{|\nabla\varphi|^2}{\varphi} - c_2|\nabla\varphi| \geq 0$$

in $\mathcal{C}_3$, for some positive constant $c_1, c_2, C, C > 1$, depending on n, θ. Then the function

$$v_\varphi(x,t) = \sup_{B_{\varphi(x,t)}(x,t)} u \tag{9.31}$$

satisfies $\Delta v_\varphi - a_1 D_t v_\varphi \geq 0$ in $\{v_\varphi > 0\} \cap \mathcal{C}_1$ and $\Delta v_\varphi - a_2 D_t v_\varphi \geq 0$ in $\{v_\varphi < 0\} \cap \mathcal{C}_1$.

Proof. It is enough to show that the expression

$$\liminf_{r\to 0}\left\{\frac{2n(n+2)}{r^2}\fint_{B'_r(\xi)}[v(x,\tau)-v(\xi,\tau)]\,dx\right\} - a_j \limsup_{h\to 0}\frac{1}{h}[v_\varphi(\xi,\tau+h)-v_\varphi(\xi,\tau)] \tag{9.32}$$

($j=1,2$) is nonnegative for every $(\xi,t)\in\{v_\varphi>0\}\cap\mathcal{C}_1$. For simplicity, assume $(\xi,\tau)=(0,0)$, $v_\varphi(0,0)>0$ and $\varphi(0,0)=1$. To estimate the first term from below, we proceed as in Lemma 4.7. Choose the system of coordinates so that

(1) $v_\varphi(0,0)=u(p)$ where $p=\varepsilon e_n+\eta e_t$, $\quad|\bar\nu|=1$, $\varepsilon>0$,

(2) $\nabla\varphi(0,0)=\alpha e_1+\beta e_n$.

By the definition of v_φ

$$v_\varphi(x,0)\ge u(y(x),\eta)$$

for

$$y(x)=x+\sqrt{\varphi^2(x,0)-\eta^2}\cdot\frac{\nu_x}{|\nu_x|}$$

where

$$\nu_x=e_n+\frac{\beta x_1-\alpha x_n}{\varepsilon^2}e_1+\frac{\gamma}{\varepsilon}\sum_{i=2}^{n-1}x_ie_i \tag{9.33}$$

with γ such that

$$(1+\gamma)^2=\left(1+\frac{\beta}{\varepsilon}\right)^2+\frac{\alpha^2}{\varepsilon^2}\,. \tag{9.34}$$

Expand and collect terms to get

$$y(x)=y^*(x)+q(x)e_n+O(|x|^2)\tau^*+O(|x|^3)e^*$$

where

$$y^*(x)=\varepsilon e_n+x+\frac{1}{\varepsilon}(\beta x_1-\alpha x_n)e_1+\frac{1}{\varepsilon}(\alpha x_1+\beta x_n)e_n+\gamma\sum_{i=2}^{n-1}x_ie_i$$

is εe_n plus a first order term,

$$q(x)=\frac{1}{2\varepsilon}\left\{\sum_{i,j=1}^{n}D_{ij}\varphi(0,0)x_ix_j-\frac{(\beta x_1-\alpha x_n)^2}{\varepsilon^2}-\frac{\delta^2(\alpha x_1+\beta x_n)^2}{\varepsilon^2}-\gamma^2\sum_{i=2}^{n-1}x_i^2\right\}$$

is the quadratic part, and $\tau^*,e^*\in\mathbb{R}^n$, $|\tau^*|=|e^*|=1$, $\gamma^*\cdot e_n=0$.

By the choice of γ in (9.34), the transformation

$$x\mapsto y^*(x)-\varepsilon e_n$$

is a rotation in the (e_1, e_n)-plane (see Lemma 4.7). Therefore

$$\lim_{r\to 0} \frac{2n(n+2)}{r^2} \fint_{B_r(0)} \Big[u(y^*(x),\eta) - u(p)\Big]\, dx = (1+\gamma)^2 \Delta u(p)\ .$$

Now evaluate $u(y(x),\eta) - u(y^*(x),\eta)$, observing that $\nabla u(p) = \nabla u(y(0),\eta)$ must point in the direction of e_n. We have

$$\begin{aligned} u(y,\eta) - u(y^*,\eta) &= \nabla u(y^*,\eta)\cdot(y-y^*) + O(|y-y^*|^2) \\ &= \nabla u(p)\cdot(y-y^*) + O(|y-y^*|^2) \\ &= |\nabla u(p)|\{q(x) + O(|x|^3)\}. \end{aligned}$$

Therefore

$$\begin{aligned} &\lim_{r\to 0} \frac{2n(n+2)}{r^2} \fint_{B_r(0)} [u(y(x),\eta) - u(y^*(x),\eta)]\, dx \\ &\quad = \frac{|\nabla u(p)|}{\varepsilon}\left\{\Delta\varphi(0,0) - \frac{1+\delta^2}{\varepsilon^2}(\alpha^2+\beta^2) - (n-2)\gamma^2\right\} \\ &\quad \ge \frac{|\nabla u(p)|}{\varepsilon}\left\{\Delta\varphi(0,0) - \frac{n}{\varepsilon^2}|\nabla\varphi(0,0)|^2\right\}\ . \end{aligned}$$

From the above calculations, we have ($v(0,0) = u(p)$)

$$\begin{aligned} &\liminf_{r\to 0}\left\{\frac{2n(n+2)}{r^2} \fint_{B_r(0)} (v(x,0) - v(0,0))\, dx\right\} \\ &\quad \ge (1+\gamma)^2\Delta u(p) + \frac{|\nabla u(p)|}{\varepsilon}\left\{\Delta\varphi(0,0) - \frac{n}{\varepsilon^2}|\nabla\varphi(0,0)|^2\right\}. \end{aligned}$$

The second term in (9.32) can be easily bounded from below by

$$-a_1\Big\{\varepsilon|\nabla u(p)|\cdot D_t\varphi(0,0) + D_t u(p)(1+\delta D_t\varphi(0,0))\Big\}\ .$$

Hence, using the equation $\Delta u(p) = a_1 D_t u(p)$, we see that the full expression in (9.32) is greater than or equal to

$$\begin{aligned} &\frac{1}{\varepsilon}|\nabla u(p)|\cdot\left\{\Delta\varphi(0,0) - a_1\varepsilon^2 D_t\varphi(0,0) - \frac{n}{\varepsilon^2}|\nabla\varphi(0,0)|^2\right\} \\ &\qquad + a_1 D_t u(p)\left\{\frac{2\beta}{\varepsilon} + \frac{|\nabla\varphi(0,0)|^2}{\varepsilon^2} - \delta D_t\varphi(0,0)\right\}\ . \end{aligned} \tag{9.35}$$

Since u is monotone along every $\tau \in \Gamma(e,\theta)$, $|u_t| \le \cot\theta|\nabla u|$ and $\varepsilon \ge \sin 2\theta$. Therefore

$$(9.35) \ge \frac{1}{\varepsilon}|\nabla u(p)|\Big\{\Delta\varphi(0,0) - c_1 D_t\varphi(0,0) - C|\nabla\varphi(0,0)|^2 - c_2|\nabla\varphi(0,0)|\Big\}$$

where c_1, c_2 depend only on θ, while c depends on n, θ. □

We now have to examine which kind of condition v_φ satisfies on $F(v_\varphi) = \partial\Omega^+(v_\varphi)$ (compare with Lemma 4.9). First, the asymptotic behavior.

Lemma 9.13. *Let u be a viscosity solution to f.b.p. in $\mathcal{C}_3$ and define in $\mathcal{C}_1$*

$$v_\varphi(x,t) = \sup_{B_{\varphi(x,t)}(x,t)} u$$

with φ as in Lemma 9.12 and $|D_t\varphi|, |\nabla\varphi| \ll 1$. If $(x_0,t_0) \in F(v_\varphi) \cap C_1$, $(y_0,s_0) \in F(u) \cap \partial B_{\varphi(x_0,t_0)}(x_0,t_0)$, then the following hold.

(i) *$F(v_\varphi)$ has a tangent $(n+1)$-dimensional ball at (x_0,t_0) from the right (i.e., there is $B \subset \Omega^+(v_\varphi)$ such that $\partial B \cap F(v_\varphi) = \{(x_0,t_0)\}$).*

(ii) *If $F(u)$ is a Lipschitz graph and $|\nabla\varphi|$, $D_t\varphi$ are small enough (depending on the Lipschitz constant L of $F(u)$), the set $F(v_\varphi)$ is a Lipschitz graph with Lipschitz constant*

$$L' \leq L + C\sup(|\nabla\varphi| + D_t\varphi).$$

(iii) *If near (y_0,s_0), at the s_0-level, u has the asymptotic expansion*

$$u(y,s_0) = \alpha_+\langle y-y_0,\nu\rangle^+ - \alpha_-\langle y-y_0,\nu\rangle^- + o(|y-y_0|)$$

where $\nu = (y_0-x_0)/|y_0-x_0|$, then, near (x_0,t_0) at the t_0-level,

$$\begin{aligned} v_\varphi(x,t_0) \geq \alpha_+ &\left\langle x-x_0, \nu+\frac{\varphi(x_0,t_0)}{|y_0-x_0|}\nabla\varphi(x_0,t_0)\right\rangle^+ \\ &-\alpha_-\left\langle x-x_0, \nu+\frac{\varphi(x_0,t_0)}{|y_0-x_0|}\nabla\varphi(x_0,t_0)\right\rangle^- + o(|x-x_0|)\,. \end{aligned}$$

Proof. The proofs of (i) and (ii) are identical to thc proofs of (a) and (c) in Lemma 4.9.

To prove (iii), for x near x_0, set $y = x + \nu\bar{\varphi}(x)$ where

$$\bar{\varphi}(x) = \sqrt{\varphi^2(x,t_0) - (s_0-t_0)^2} \leq \varphi(x,t_0)\,.$$

Then $v(x,t_0) \geq u(y,s_0)$. Hence, since

$$\bar{\varphi}(x) = \bar{\varphi}(x_0) + \left\langle x-x_0, \frac{\varphi(x_0,t_0)}{|y_0-x_0|}\nabla\varphi(x_0,t_0)\right\rangle + o(|x-x_0|)\,,$$

we have ($\bar{\varphi}(x_0) = |x_0-y_0|$)

$$\langle y-y_0,\nu\rangle = \left\langle x-x_0, \nu + \frac{\varphi(x_0,t_0)}{|x_0-y_0|}\nabla\varphi(x_0,t_0)\right\rangle + o(|x-x_0|)\,,$$

from which we easily conclude the proof. □

If u, φ, v_φ are as in the preceding lemmas, it is not difficult to show that v_φ is "almost an R-subsolution", in the sense explained in Section 7.4.

In fact, if (x_0,t_0), (y_0,s_0) are as in (ii) of Lemma 9.13, at (y_0,s_0) $F(u)$ has a tangent ball from the left with slope $(s_0-t_0)/|y_0-x_0|$.

Therefore, u has the asymptotic expansion in (iii) with

$$\frac{s_0 - t_0}{|y_0 - x_0|} \leq G(\alpha_+, \alpha_-) .$$

On the other hand if $\nu_0 = \alpha_0 \nu + \beta_0 e_t$ is the normal unit vector to $F(v_\varphi)$ at (x_0, t_0), we have (see the proof of (c) in Lemma 4.9)

$$\frac{\beta_0}{\alpha_0} = \left(\frac{s_0 - t_0}{|y_0 - x_0|} + \frac{\varphi(x_0, t_0)}{|y_0 - x_0|} D_t \varphi(x_0, t_0) \right) \cdot |\tau^*|^{-1}$$

where

$$\tau^* = \frac{y_0 - x_0}{|y_0 - x_0|} + \frac{\varphi(x_0, t_0)}{|y_0 - x_0|} \nabla \varphi(x_0, t_0) .$$

Then

$$\frac{\beta_0}{\alpha_0} \leq \left(\frac{s_0 - t_0}{|y_0 - x_0|} + \frac{\varphi(x_0, t_0)}{|y_0 - x_0|} D_t \varphi(x_0, t_0) \right) \cdot \left(1 + \frac{c\varphi(x_0, t_0)}{|y_0 - x_0|} |\nabla \varphi(x_0, t_0)| \right)$$

or

$$\frac{\beta_0}{\alpha_0} \leq G(\alpha_+, \alpha_-) + \frac{c\varphi(x_0, t_0)}{|y_0 - x_0|} \left(D_t \varphi(x_0, t_0) + |\nabla \varphi(x_0, t_0)| \right) . \tag{9.36}$$

From Lemma 9.13(iii), if x is near x_0, we can write

$$v_\varphi(x, t_0) \geq \alpha_+^* \langle x - x_0, \nu^* \rangle^+ - \alpha_-^* \langle x - x_0, \nu^* \rangle^- + o(|x - x_0|) \tag{9.36'}$$

where

$$\alpha_+^* = \alpha_+ |\tau^*| , \quad \alpha_-^* = \alpha_- |\tau^*| , \quad \nu^* = \tau^* / |\tau^*| .$$

Since $D_t \varphi$, $|\nabla \varphi|$ are small and $\varphi(x_0, t_0)/|y_0 - x_0| \leq c = c(n, \theta)$, we have

$$\begin{aligned} |\alpha_\pm - \alpha_\pm^*| \leq \alpha_\pm (1 - |\tau^*|) &\leq c \frac{\varphi(x_0, t_0)}{|y_0 - x_0|} (|\nabla \varphi(x_0, t_0)| + D_t \varphi(x_0, t_0)) \\ &\leq c(|\nabla \varphi(x_0, t_0)| + D_t \varphi(x_0, t_0)) \end{aligned}$$

and, using the Lipschitz continuity of G, from (9.36) we get

Corollary 9.14. *Let v_φ as in Lemma 9.13. Then at $(x_0, t_0) \in F(v_\varphi)$,*

$$\frac{\beta_0}{\alpha_0} \leq G(\alpha_+^*, \alpha_-^*) + c(D_t \varphi(x_0, t_0) + |\nabla \varphi(x_0, t_0)|) . \tag{9.37}$$

9.7. Free boundary improvement. Propagation lemma

In this section we show how an interior gain can be carried to the free boundary. Following the same lines of the elliptic case, we first construct a family of functions φ_η, depending on the parameter η, $0 \leq \eta \leq 1$, and satisfying the requirements in Lemma 9.12. In the end we want the $v_{\varepsilon \varphi_\eta}$, defined as in (9.31), to carry the monotonicity gain from $B'_{1/8}(\pm \frac{3}{4} e_n) \times (-c_0 \delta / \mu, c_0 \delta / \mu)$ to the free boundary as η goes from 0 to 1. This means that we must have $\varphi_\eta \le 1$ along $\partial_p C_{1-\varepsilon}$, $\varphi_\eta \approx 1 + \eta b$ on the sides and top of the internal cylinders and $\varphi_\eta \approx 1 + c\eta b$ ($c > 0$, $b > 0$ both small) in $C_{1/2}$.

Lemma 9.15. *Let $0 < T_0 \le T$ and $C > 1$. There exist positive constants $\bar{C}, k$ and h_0, depending only on C and T_0, such that, for any h, $0 < h < h_0$, there is a family of C^2-continuous functions φ_η, $0 \le \eta \le 1$, defined in the closure of*

$$D = \left[B' \setminus \left\{\bar{B}'_{1/8}\left(\frac{3}{4}e_n\right) \cup \bar{B}'_{1/8}\left(-\frac{3}{4}e_n\right)\right\}\right] \times (-T, T)$$

such that

(1) $1 \le \varphi_\eta \le 1 + \eta h$ *in* $\bar{D}$,

(2) $\Delta\varphi_\eta - c_1 D_t\varphi_\eta - C\frac{|\nabla\varphi_\eta|^2}{\varphi_\eta} - c_2|\nabla\varphi| \ge 0$ *in* D,

(3) $\varphi_\eta \equiv 1$ *outside* $B'_{8/3} \times (-\frac{7}{8}T, T)$,

(4) $\varphi_\eta \ge 1 + \kappa\eta h$ *in* $B'_{1/2} \times (-\frac{T}{2}, \frac{T}{2})$,

(5) $D_t\varphi_\eta, |\nabla\varphi| \le \bar{C}\eta h$ *in* $\bar{D}$,

(6) $D_t\varphi_\eta \ge 0$ *in* D,

provided c_1 and c_2 are small enough.

Proof. Set

$$\psi_\eta = -1 + \varphi_\eta^{1-c} .$$

If $c > 1$ and ψ_η satisfies

$$\Delta\psi_\eta - c_1 D_t\psi_\eta - c_1|\nabla\psi_\eta| \le 0, \tag{9.38}$$

then φ_η satisfies (2). Now, if c_1 and c_2 are small enough, it is not difficult to construct a C^2-function ψ satisfying (9.38) in D, such that

$$\begin{cases} -a \le \psi \le 0 & \text{in } \bar{D}, \\ \psi \equiv 0 & \text{outside } B'_{8/9} \times (-\frac{7}{8}T, T), \\ \psi \le -\kappa b & \text{in } \bar{B}'_{1/2} \times [-\frac{T}{2}, \frac{T}{2}], \\ |D_t\psi|, |\nabla\psi| \le \tilde{c} & \text{in } \bar{D}, \\ D_t\psi \le 0 & \text{in } \bar{D} . \end{cases}$$

Choose h small and $a, b, \tilde{c}$ ($b < a$) so that

$$1 - b\kappa\eta h \le (1 + \kappa\eta h)^{1-c}, \quad 1 - a\eta h \le (1 + \eta h)^{1-c}, \quad \text{and} \quad \tilde{c}(c-1) < 2^c\bar{c} .$$

Then if $\psi_\eta = \eta h \psi$, the function

$$\varphi_\eta = (1 + \psi_\eta)^{1/1-c}$$

satisfies (1)–(6). □

The following lemma is fundamental.

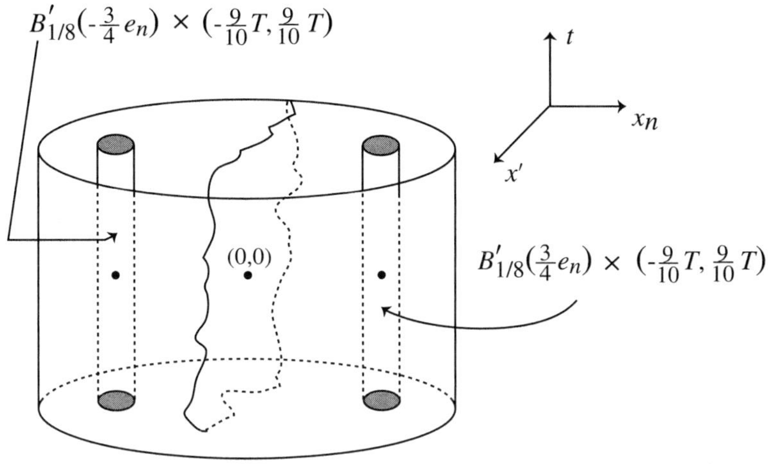

Figure 9.5. The domain D in (9.39)

Lemma 9.16. *Let $u_1 \le u_2$ be two viscosity solutions to f.b.p. in C_2, with Lipschitz free boundaries. Assume u_2 satisfies the hypotheses of Theorem 9.1, in particular the nondegeneracy condition (ii), and G satisfies condition (i) of the same theorem. Suppose further that*

$$v_\varepsilon(x,t) = \sup_{B_\varepsilon(x,t)} u_1 \le u_2$$

in $B'_1 \times (-T,T)$ and, for some positive $\varepsilon, h, \bar{C}$ and $\sigma \ge 0$,

$$v_{(1+\sigma h)\varepsilon}(x,t) \le u_2(x,t) - \bar{C}\sigma\varepsilon u_2\left(\frac{3}{4}e_n, 0\right)$$
$$\forall\ (x,t) \in B'_{1/8}\left(\frac{3}{4}e_n\right) \times (-T,T) \subset \Omega^+(u_2),$$

$$v_{(1+\sigma h)\varepsilon}(x,t) \le u_2(x,t) + \bar{C}\sigma\varepsilon u_2\left(-\frac{3}{4}e_n, 0\right)$$
$$\forall\ (x,t) \in B'_{1/8}\left(-\frac{3}{4}e_n\right) \times (-T,T) \subset \Omega^-(u_2)\ .$$

Then if ε and h are small enough, there exists γ, $0 < \gamma < 1$, such that

$$v_{(1+\gamma\sigma h)\varepsilon} \le u_2 \quad \text{in} \quad B'_{1/2} \times \left(-\frac{T}{2}, \frac{T}{2}\right)\ .$$

Proof. We construct a continuous family of functions $\bar{v}_\eta \le u_2$ for $0 \le \eta \le 1$, such that $\bar{v}_1 \ge v_{(1+\gamma h\sigma)\varepsilon}$ in $B'_{1/2} \times (-T,T)$. Let φ_η be the family of radii constructed in Lemma 9.15 and set

$$\bar{v}_\eta = v_{\varepsilon\varphi_{\sigma\eta}} + c\sigma\varepsilon w$$

with c to be chosen later and w the continuous function in

$$D = \left[B'_{9/10} \setminus \left\{B'_{1/8}\left(\frac{3}{4}e_n\right) \cup B'_{1/8}\left(\frac{3}{4}e_n\right)\right\}\right] \times \left(-\frac{9}{10}T, \frac{9}{10}T\right) \tag{9.39}$$

defined as follows:

$$\begin{cases} \Delta w - a_1 D_t w = 0 & \text{in } D \cap \Omega^+(u_2), \\ \Delta w - a_2 D_t w = 0 & \text{in } D \cap \Omega^-(u_2), \\ w = \pm u_2(\pm\frac{3}{4}e_n, 0) & \text{on } \partial B'_{1/8}(\pm\frac{3}{4}e_n) \times (-\frac{9}{10}T, \frac{9}{10}T), \\ w = 0 & \text{on } F(u_2) \text{ and on the rest of } \partial_p D\ . \end{cases}$$

We prove now that the set E of η's for which $\bar{v}_\eta \le u_2$ is open and closed in $[0,1]$. Notice first that by hypothesis and the maximum principle, the set E is nonempty. Also, it is clearly closed. To show that it is open, assuming that $\bar{v}_{\eta_0} \le u_2$ for some $\eta_0 \in [0,1]$, it is enough to show that $D \cap \Omega^+(\bar{v}_{\eta_0})$ is compactly contained in $D \cap \Omega^+(u_2)$.

Replacing ε, if necessary, by any smaller ε', we have that $\bar{v}_{\eta_0} < u_2$ on $\partial_p D$. Thus, if $D \subset\subset \Omega^+(\bar{v}_{\eta_0})$ is not true, there exists $(x_0, t_0) \in F(\bar{v}_{\eta_0}) \cap F(u_2) \cap D$.

Claim. *If h is small enough, at (x_0, t_0), $\bar{v}_{\eta_0}$ satisfies a subsolution condition.*

Indeed, let $(y_0, s_0) \in F(u_2) \cap \partial B_{\varepsilon\varphi_{\eta_0}}(x_0, t_0)$. Then if $\nu = (y_0 - x_0)/|y_0 - x_0|$, at the s_0-level, near y_0,

$$u_1(y, s_0) = \alpha_+^{(1)} \langle y - y_0, \nu\rangle^+ - \alpha_-^{(1)} \langle y - y_0, \nu\rangle^- + o(|y - y_0|)$$

with $\alpha_-^{(1)} > 0$.

On the other hand, (x_0, t_0) is a regular point from the right for both $F(\bar{v}_{\eta_0})$ and $F(u_2)$. If $\nu_0 = \alpha_0 \nu + \beta_0 e_t$ is the normal unit vector to both free boundaries, inward to the positive side, we have, from Lemma 9.13 and Corollary 9.14,

$$v_{\varepsilon\varphi_{\eta_0}(x,t_0)}(x, t_0) \ge \alpha_+^* \langle x - x_0, \nu^*\rangle^+ - \alpha_-^* \langle x - x_0, \nu^*\rangle^- + o(|x - x_0|)$$

with

$$\alpha_+^* = \alpha_+^{(1)} |\tau^*|\ , \quad \alpha_-^* = \alpha_-^{(1)} |\tau^*|\ , \quad \tau^* = \nu + \frac{\varepsilon^2 \varphi_{\sigma\eta_0}(x_0, y_0)}{|y_0 - x_0|} \nabla \varphi_{\sigma\eta_0}(x_0, t_0)\ ,$$

and

$$\frac{\beta_0}{\alpha_0} \le G(\alpha_+^*, \alpha_-^*) + c\varepsilon\sigma h\ . \tag{9.40}$$

Notice that if (x_0, t_0) belongs to the region of constant $\varphi_{\sigma\eta_0}$, then we can take $h = 0$ in (9.40) and $\bar{v}_{\eta_0}$ satisfies a subsolution condition. If this is not the case, (x_0, t_0) stays away from $\partial_p D$ and we can apply the comparison Theorem 12.3 and use the monotonicity of u_2 along a space-time cone to get

$$w \geq c_0 u_2 \text{ in } \Omega^+(u_2) \text{ and } w \geq -c_0 u_2 \text{ in } \Omega^-(u_2) , \tag{9.41}$$

near (x_0, t_0).

Now, at the t_0-level, near x_0 we have

$$u_2(x, t_0) = \alpha_+^{(2)} \langle x - x_0, \nu^* \rangle^+ - \alpha_-^{(2)} \langle x - x_0, \nu^* \rangle^- + o(|x - x_0|) \tag{9.42}$$

with

$$\frac{\beta^0}{\alpha^0} \geq G(\alpha_+^{(2)}, \alpha_-^{(2)}) . \tag{9.43}$$

Thus, we have, near x_0, at the t_0-level, from (9.41) and (9.42)

$$\bar{v}_\eta(x, t_0) \geq \bar{\alpha}_+ \langle x - x_0, \nu^* \rangle^+ - \bar{\alpha}_- \langle x - x_0, \nu^* \rangle^- + o(|x - x_0|)$$

where

$$\bar{\alpha}_+ = \alpha_+^* + c_0 \varepsilon \sigma \alpha_+^{(2)} , \quad \bar{\alpha}_- = \alpha_-^* - c_0 \varepsilon \sigma \alpha_-^{(2)} .$$

Since $u_2 - \bar{v}_{\eta_0} \geq 0$ and it is supercaloric in $\Omega^+(\bar{v}_{\eta_0})$, we have $\alpha_-^{(2)} \leq \bar{\alpha}_-$ and, by the Hopf maximum principle, $\alpha_+^{(2)} > \bar{\alpha}_+$. Therefore

$$G(\bar{\alpha}_+, \bar{\alpha}_-) < G(\alpha_+^{(2)}, \alpha_-^{(2)}) \tag{9.44}$$

by the strict monotonicity of G.

On the other hand, from (9.40) and the hypotheses on G,

$$\begin{aligned} \frac{\beta^0}{\alpha^0} &\leq G(\bar{\alpha}_+, \bar{\alpha}_-) - c^* c_0 \varepsilon \sigma (\alpha_+^{(2)} + \alpha_-^{(2)}) + c \varepsilon \sigma h \\ &\leq G(\bar{\alpha}_+, \bar{\alpha}_-) - \varepsilon \sigma (c^* c_0 m - ch) \\ &\leq G(\bar{\alpha}_+, \bar{\alpha}_-) \end{aligned}$$

if $h < c^* c_0 m / c$.

But then from (9.43) and (9.44) we get

$$G(\bar{\alpha}_+, \bar{\alpha}_-) < \frac{\beta^0}{\alpha^0} \leq G(\bar{\alpha}_+, \bar{\alpha}_-) .$$

We have reached a contradiction that concludes the proof. □

9.8. Regularization of the free boundary in space

In this section we apply the *propagation lemma* to show that the free boundary is a C^1 domain in space. This preliminary step is necessary to say that the defect angle in space is as small as we want in order to be able to apply the results of Sections 9.4 and 9.5, in particular Lemmas 9.8 and 9.11 which require $\delta \ll \mu$.

We use the symbol $\Gamma(\nu, \theta, \theta^t)$ to denote an "elliptic" cone with axis ν and opening θ in space and θ^t in space-time.

Lemma 9.17. *Let u be a viscosity solution to f.b.p. in $\mathcal{C}_1$ monotone increasing along every $\tau \in \Gamma(e_n, \theta, \theta^t)$, for some $0 < \theta_0 \le \theta^t \le \theta \le \frac{\pi}{2}$. Then there exist positive constants $c, \bar{c}$ and a unit vector ν_1 such that, in $B'_{1/2} \times (-1, 1)$, the function*

$$u^*(x,t) = u(x, \bar{c}\delta^2 t)$$

is monotone increasing along every direction $\tau \in \Gamma(\nu_1, \theta_1, \theta_1^t)$ with

$$\theta_1 - \theta_0 \ge c\delta^3 , \qquad \theta_1^t \ge \theta_0 .$$

Proof. From Corollary 9.5 and Lemma 9.6, u is monotone increasing along every $\tau \in \Gamma_x(\nu, \bar{\theta})$, with $\bar{\theta} - \theta \le \bar{b}\delta$, $b < 1$, in $B_{1/8}(\pm\frac{3}{4}e_n) \times (-\frac{1}{16}a^2, \frac{1}{16}a^2)$.

The cone $\Gamma_x(\nu, \bar{\theta})$ contains $\Gamma_x(e_n, \theta)$. Consider all spatial unit vectors $\tau \in \Gamma_x(e_n, \theta) \setminus \mathcal{N}$, where $\mathcal{N}$ denotes a neighborhood of the line $\partial\Gamma_x(e_n, \theta) \cap \partial\Gamma_x(\nu, \bar{\theta})$ in the case where the two cones touch. From Lemma 9.8,

$$D_\tau u \ge \tilde{c}\delta D_{e_n} u$$

inside the internal cylinders $B'_{1/8}(\pm\frac{3}{4}e_n) \times (-\frac{c_0\delta}{\mu}, \frac{c_0\delta}{\mu})$.

Now choose a number $c_1 \ll 1$, set $\bar{c} = \frac{Cc_1}{\mu}$ and perform a dilation in time of order $\bar{c}\delta^2$, that is, set

$$u^*(x,t) = u(x, \bar{c}\delta^2 t).$$

Observe that the coefficient of the t-derivative of u_1 in the heat equations for u_1 as well as in the free boundary condition is multiplied by the factor $1/\bar{c}\delta^2$. Also, the region $B'_1 \times (-\frac{c_0\delta}{\mu}, \frac{c_0\delta}{\mu})$ is mapped into $B'_1 \times (-\frac{1}{c_1\delta}, \frac{1}{c_2\delta})$ and the cone $\Gamma(e_n, \theta, \theta^t)$ is transformed into a cone which certainly contains a circular cone $\Gamma(e_n, \theta)$, if c_1 is small enough.

Now take a direction in space and time $\rho \in \Gamma(e_n, \theta)$ of the form $\rho = \lambda_1\sigma + \lambda_2 e_t$, with σ unit vector in $\Gamma_x(e_n, \theta) \setminus \mathcal{N}$, $\lambda_1 + \lambda_2^2 = 1$ and $|\lambda_2| \le \frac{1}{2}\frac{\lambda_1}{\bar{c}\delta}$. We have

$$D_\rho u^* \ge c\delta D_{e_n} u^* \tag{9.45}$$

in $B'_{1/8}(\pm\frac{3}{4}e_n) \times (-\frac{1}{c_1\delta}, \frac{1}{c_1\delta})$, since $|D_t u^*| \le \bar{c}\delta^2 D_{e_n} u^*$. Now let $\tau \in \Gamma(e_n, \frac{\theta}{2})$ be a small vector of the form

$$\tau = \eta\rho \qquad (0 < \eta \le 1)$$

and $(x_0, t_0) \in B'_{1/8}(\pm\frac{3}{4}e_n) \times (-\frac{1}{c_1\delta}, \frac{1}{c_1\delta})$. Then if $\varepsilon = |\tau| \sin\frac{\theta}{2}$, $(y, s) \in B_\varepsilon(x_0, t_0)$ and $\bar{\tau} = \tau + (x_0 - y, t_0 - s)$, we have

$$u_1(y,s) \equiv u^*((y,s) - \bar{\tau}) = u^*((x_0, t_0) - \tau) = u^*(x_0, t_0) - D_\tau u^*(\bar{x}, \bar{t}) .$$

From (9.45) and the fact that $D_{e_n} u \sim u/d_{x,t}$, we obtain

$$\bar{v}_\varepsilon(x_0, t_0) \equiv \sup_{B_\varepsilon(x_0,t_0)} u_1 \le u^*(x_0, t_0) - c\delta\varepsilon u^*(x_0, t_0) \ . \tag{9.46}$$

Then Lemma 9.2 (with $u_2 = u^*$) gives us that there exist $h > 0$, $c > 0$, such that

$$\bar{v}_{(1+h\delta)\varepsilon}(x, t) \le u^*(x, t) - c\delta\varepsilon u^*(x_0, t_0)$$

for every (x, t) in a parabolic neighborhood of (x_0, t_0).

But since for

$$(x_0, t_0) \in B'_{1/8}\left(\pm\frac{3}{4}e_n\right) \times \left(-\frac{1}{c_1\delta}, \frac{1}{c_1\delta}\right)$$

we have $u^*(x_0, t_0) \sim u^*(\pm\frac{3}{4}e_n, 0)$, respectively, we can write

$$\bar{v}_{(1+h\delta)\varepsilon}(x, t) \le u^*(x, t) - c\delta\varepsilon u^*\left(\pm\frac{3}{4}e_n, 0\right)$$

for every $(x, t) \in B'_{1/8}(\pm\frac{3}{4}e_n) \times (-\frac{1}{c_1\delta}, \frac{1}{c_1\delta})$, respectively.

Since u_1 and u^* are two viscosity solutions to f.b.p. in the region $B'_{1-2\varepsilon} \times (-\frac{1}{c_1\delta}, \frac{1}{c_1\delta})$ and since they satisfy the hypotheses of the propagation lemma, taking into account the modifications in the free boundary relation due to the time scaling, we deduce that for a small positive constant γ,

$$\bar{v}_{(1+\gamma h\delta^3)\varepsilon} \le u^*$$

in $B'_{1/2} \times (-\frac{1}{2c_1\delta}, \frac{1}{2c_1\delta})$. Notice that in order to apply the propagation lemma, the coefficient of $D_t u^*$ in the heat equation must be kept small. This can be obtained by a previous hyperbolic scaling and will produce at the end of our iteration procedure a very weak modulus of continuity of $\nabla_{x'} f$.

The last inequality implies that along any direction of the form $\tau + (1 + \gamma h\delta^3)\tilde{\nu}$, $\tilde{\nu} \in \mathbb{R}^{n+1}$, $|\tilde{\nu}| = 1$, u^* is monotone increasing. Using again the geometric Theorem 4.2, it is readily seen that the convex envelope of this family of directions and the cone $\Gamma(e_n, \theta)$ contains a new cone

$$\Gamma(\nu_1, \theta_1, \theta^t) \text{ with } \theta^t \ge \theta_2 \text{ and, for some constants } b_1, b_2,$$
$$\theta_1 - \theta \ge b_1\delta^3 \ , \quad |\nu_1 - e_n| \le b_2\delta \ .$$

Iteration of Lemma 9.17 gives

Theorem 9.18. *Let u be a viscosity solution to f.b.p. in C_1. If all the hypotheses of Theorem 9.1 hold and, in particular, $F(u)$ is given by the Lipschitz graph $x_n = f(x', t)$, then for each t, $|t| < 1$, $\nabla_{x'} f$ is continuous in $B'_{1/2}$. That is, $F_t = F(u) \cap \{t\}$ is a C^1-surface in $B'_{1/2} \times (-1, 1)$.*

Proof. We iterate Lemma 9.17. Set $u_1(x,t) = u(x, \bar{c}\delta_0^2 t)$ where $\delta_0 = \delta$ and $\bar{c}, \delta$ are as in Lemma 9.17. By this lemma u_0 is monotone increasing along any direction $\tau \in \Gamma(\nu_1, \theta_1, \theta_1^t)$ with $\theta_1^t \geq \theta_0$ and $\delta_1 = \frac{\pi}{2} - \theta_1 \leq \delta_0 - c\delta_0^3$ in $B'_{1/2} \times (-1,1)$.

Suppose now that $u_k = u_k(x,t)$, $k \geq 1$, satisfies the hypotheses of Lemma 9.17 in a cone $\Gamma(\nu_k, \theta_k, \theta_k^t)$ with $\theta_k^t \geq 0$,

$$\delta_k = \frac{\pi}{2} - \theta_k \leq \delta_{k-1} - b_1 \delta_{k-1}^3 , \qquad |\nu_k - \nu_{k-1}| \leq b_2 \delta_{k-1}$$

in $B_{1/2} \times (-1,1)$. Set

$$u_{k+1}(x,t) = u_k(2^{-r_k} x, \bar{c} 2^{-r_k} \delta_k^2 t) \cdot 2^{r_k}$$

where r_k is the smallest integer such that $2^{-r_k} < \bar{c}\delta_k^2$. Then by Lemma 9.17, u_{k+1} is monotone in a cone $\Gamma(\nu_{k+1}, \theta_{k+1}, \theta_{k+1}^t)$ with $\theta_{k+1}^t \geq \theta_0$ and

$$|\nu_{k+1} - \nu_k| \leq b_2 \delta_k , \qquad \delta_{k+1} \leq \delta_k - b_1 \delta_k^2$$

in $B'_{1/2} \times (-1,1)$.

The recurrence relation implies $\delta_k \to 0$ as $k \to +\infty$ and the conclusion follows. □

9.9. Free boundary regularity in space and time

The fact that the defect angle in space can be made as small as we prefer, allows us to use the results in Sections 9.4 and 9.5.

In the final iteration process the delicate balance between the defect angles in space and time gives a modulus of continuity in time and an improved one in space.

As we already noticed in Section 9.5, for some A, B, $0 < B - A \leq c\mu$, u is monotone increasing along the directions $e_t + Be_n$ and $-e_t - Ae_n$. To enlarge the cone in time, we have to lower B or to increase A.

The main lemma, which shows the initial step in the iteration process, follows.

Lemma 9.19 (Basic iteration). *Let u be a viscosity solution to f.b.p. in C_1. Suppose the hypotheses of Theorem 9.1 hold. In particular, assume that the following hold.*

(i) *u is monotone increasing along any direction of a space cone*

$$\Gamma_x(e_n, \theta), \qquad 0 \leq \theta_0 \leq \theta < \frac{1}{2}\pi.$$

(ii) *There exist constant $\bar{c}_1$ (positive) and A, B such that u is monotone increasing along the directions $e_t + Be_n$ and $-e_t - Ae_n$, with*

$$0 < B - A \leq \bar{c}_1 \mu .$$

Then if $\delta = \frac{\pi}{2} - \theta \ll \mu^3$, *there exist constants* c_1, c_2, c_0 *(positive) and* A_1, B_1, *depending only on* n, θ_0, *and a spatial unit vector* ν_1 *such that, in* $B_{1/2} \times (-\frac{c_0\delta}{2\mu}, \frac{c_0\delta}{2\mu})$,

(a) u *is monotone increasing along every* $\tau \in \Gamma(\nu_1, \theta_1)$ *with*

$$\delta_1 = \frac{\pi}{2} - \theta_1 \le \delta - c_1\frac{\delta^2}{\mu}, \qquad |\nu_1 - e_n| \le c_1\delta, \tag{9.47}$$

(b) u *is monotone increasing along the directions* $e_t + B_1\nu_1$ *and* $-e_t - A_1\nu_1$ *with*

$$0 < B_1 - A_1 \le \bar{c}_1\mu_1 \quad \text{and} \quad \mu_1 \le \mu - c_1\delta\ . \tag{9.48}$$

Proof. We first perform a dilation in time setting

$$w(x,t) = u\left(x, \frac{\delta t}{\mu}\right)\ .$$

After this dilation, the coefficient of the t-derivative in the heat equations as well as in the free boundary conditions is multiplied by μ/δ. Also, the regions $B_1' \times (-1,1)$ and $B_{1/8}'(\pm\frac{3}{4}e_n) \times (-\frac{c_0\delta}{\mu}, \frac{c_0\delta}{\mu})$ are mapped to $B_1' \times (-\frac{\mu}{\delta}, \frac{\mu}{\delta})$ and $B_{1/8}'(\pm\frac{3}{4}e_n) \times (-c_0, c_0)$, respectively.

Condition (ii) becomes: w is monotone increasing along the directions

$$e_t + \frac{\delta}{\mu}Be_n \quad \text{and} \quad -e_t - \frac{\delta}{\mu}Ae_n$$

with

$$0 < B - A \le \bar{c}_1\delta\ .$$

To enlarge the cone in space and to prove (a), consider the spatial vectors $\tau \in \Gamma_x(e_n, \theta - \delta)$, with $|\tau| \le \delta$ and let

$$\varepsilon = |\tau| \sin\delta\ .$$

If we define $w_1(x,t) = w(x - \tau, t)$, clearly we have

$$\sup_{y \in B_\varepsilon(x)} w_1(y,t) \le w(x,t) \qquad \forall\ (x,t) \in B_{1-\varepsilon}' \times \left(-\frac{\mu}{\delta}, \frac{\mu}{\delta}\right)\ .$$

We proceed now as in Section 9.4, by deleting, say, $\frac{1}{10}$ of the original cone, containing a neighborhood of both the generatrices of $\Gamma_x(e_n, \theta)$ opposite to $\nabla u(\pm\frac{3}{4}e_n, -\frac{1}{2}a^2)$. This is possible thanks to Lemma 9.6 and Corollary 9.7. We conclude that for any vector τ in the remaining cone $\Gamma_x^1(e_n, \theta)$, we have

$$D_\tau w \ge c\delta D_{e_n} w \ \text{ in } B_{1/8}'\left(\pm\frac{3}{4}e_n\right) \times (-c_0, c_0)\ . \tag{9.49}$$

This inequality extends to derivatives along directions having a t-component of order δ. Indeed, if for the spatial unit vector τ (9.49) holds, then for $\lambda_2^2+\lambda_1^2=1$, since

$$|D_t w| \le \tilde{c}\frac{\delta}{\mu} D_n w,$$

we have

$$\lambda_1 D_\tau w + \lambda_2 D_t w \ge \left(c\lambda_1\delta - \tilde{c}\lambda_2\frac{\delta}{\mu}\right) De_n w \ge \tilde{c}\delta D_{e_n} w$$

as long as $|\lambda_1| \le \delta$, since $\delta/\mu \ll 1$. As a consequence, if ρ is a unit vector in $\mathbb{R}^{n+1}$ and $\bar{\tau} = \tau + \varepsilon\rho$, we can write, for every $(x,t) \in B'_{1/8}(\pm\frac{3}{4}e_n)\times(-c_0,c_0)$, respectively,

$$\begin{aligned} w((x,t)-\bar{\tau}) - w(x,t) &= -D_{\bar{\tau}} w(\bar{x},\bar{t}) \le -c\delta\varepsilon D_{e_n} w(x,t) \\ &\le -c\varepsilon\delta w^{\pm}\left(\pm\frac{3}{4}e_n, 0\right) \end{aligned}$$

(since $|\bar{\tau}| \ge c\varepsilon$ and by Corollary 13.14). Therefore

$$v_\varepsilon(x,t) \equiv \sup_{B_\varepsilon(x,t)} w_1 \le w(x,t) - c\varepsilon\delta w^{\pm}\left(\pm\frac{3}{4}e_n, 0\right),$$

respectively, in $B'_{1/8}(\pm\frac{3}{4}e_n)\times(-c_0,c_0)$. From Harnack's principle (Lemma 9.2), for a small positive $\bar{h}$, in the same set we obtain

$$v_{(1+\bar{h}\delta)\varepsilon} \le w - c\varepsilon\delta w^{\pm}\left(\pm\frac{3}{4}e_n, 0\right).$$

Hence, w satisfies the hypotheses of the propagation lemma with $T = C$ and $h = \bar{h}$. By taking account of the effect of the dilation in time, we conclude that, for a small $\gamma > 0$, in $B'_{1/2}\times(-\frac{c_0}{2},\frac{c_0}{2})$,

$$v_{(1+\gamma\bar{h}\delta^2/\mu)\varepsilon} \le w\ .$$

This amounts to saying that w is monotone increasing along all the directions of the form

$$\bar{\tau} = \tau + \left(1 + \gamma\bar{h}\frac{\delta^2}{\mu}\right)\varepsilon\rho \tag{9.50}$$

in $B'_{1/2}\times(-\frac{c_0}{2},\frac{c_0}{2})$. The convex envelope of the old cone $\Gamma'_x(e_n,\theta)$ and the set of directions (9.50) contains a new cone (in space) $\Gamma_x(\nu_1,\theta_1)$ such that

$$|\nu_1 - \nu| \le c\delta \quad \text{and} \quad \theta_1 - \theta = \bar{c}\bar{h}\frac{\delta^2}{\mu}$$

which, after rescaling back in time, corresponds to (a) with $c_1 = \bar{c}\bar{h}$.

To prove (b), notice first that the new axis ν_1 is shifted with respect to e_n of order δ in a spatial direction orthogonal to e_n. Since $\delta \ll \mu^3$, we can use the results of Lemma 9.11 which, in the present situation, read

$$D_t w + \frac{\delta}{\mu} B D_{e_n} w \geq c\delta D_{e_n} w \qquad \text{or} \qquad - D_t w - A\frac{\delta}{\mu} D_{e_n} w \geq c\delta D_{e_n} w \ .$$

Suppose that the left inequality holds and call $\bar{\rho}, \bar{\rho}_1$, respectively, the unit vectors in the directions

$$e_t + \frac{\delta}{\mu} B e_n \qquad \text{and} \qquad e_t + \frac{\delta}{\mu} B \nu_1 .$$

Then it is easy to check that if $|e^1| = 1$, $e^1 \cdot e_n = 0$ and $\lambda_1^2 + \lambda_2^2 = 1$,

$$\lambda_1 D_{\bar{\rho}_1} w + \lambda_2 D_{e^1} w \geq \left(c\delta - \tilde{c}\frac{\delta^2}{\mu} B - \tilde{c}_1 \delta^2 \right) D_{e_n} w \geq \tilde{c}\delta D_{\nu_1} w$$

in $B'_{1/8}(\pm\frac{3}{4}e_n) \times (-c_0, c_0)$, as long as $|\lambda_2| \leq 2\delta$, since

$$D_{\bar{\rho}_1} w \geq c D_{\bar{\rho}} w - \tilde{c}\frac{\delta^2}{\mu} B D_{e_n} w \ .$$

Let now ρ^* denote the direction below $\bar{\rho}_1$ (with respect to time) in the (e_t, ν_1)-plane, which makes an angle δ with $\bar{\rho}_1$.

For any small vector τ in the ρ^* direction, again set

$$\varepsilon = |\tau| \sin\delta \ , \qquad w_1(x,t) = w((x,t) - \tau) \ .$$

Then

$$v_\varepsilon(x,t) \equiv \sup_{B_\varepsilon(x,t)} w_1 \leq w(x,t) \quad \text{in} \quad B'_{1-\varepsilon} \times \left(-\frac{\mu}{\delta} + \varepsilon, \frac{\mu}{\delta} - \varepsilon \right) \ .$$

Proceeding exactly as before, we conclude that

$$v_{(1+\bar{c}\bar{h}\delta^2/\mu)\varepsilon} \leq w \quad \text{in} \quad B'_{1/2} \times \left(-\frac{c_0}{2}, \frac{c_0}{2} \right) \ .$$

This implies that in the same set, w is monotone increasing along the directions

$$e_t + \left(\frac{\delta}{\mu} B - \bar{c}\bar{h}\frac{\delta^2}{\mu} \right) \nu_1;$$

that is, rescaling back in time, u is monotone increasing along the directions

$$e_t + (B - \bar{c}\bar{h}\delta)\nu_1 \ .$$

Therefore (b) holds with $B_1 = B - \bar{c}\bar{h}\delta$, $A_1 = A$ and

$$B_1 - A_1 = B - A - \bar{c}\bar{h}\delta \leq \bar{c}_1(\mu - c_2\delta),$$

so that the proof is complete. □

We are now ready to complete the proof of Theorem 9.1.

Proof of Theorem 9.1. By the results in the previous sections, for λ large enough, the function $u_\lambda(x,t) = \frac{1}{\lambda}u(\lambda x, \lambda t)$, which we call u again, falls under the hypotheses of the propagation lemma. Now we proceed inductively, by applying Lemma 9.19 to

$$u_k(x,t) = 2^k u(2^{-k}, 2^{-k}t), \qquad k \geq 1 .$$

In this way we define a sequence of space cones $\Gamma_x(\nu_k, \theta_k)$ and sequences $\{A_k\}$, $\{B_k\}$, $\{\delta_k\}$, $\{\mu_k\}$ with the following properties: in

$$B'_{2^{-k}} \times \left(-\frac{c_0\delta_k}{2^k\mu_k}, \frac{c_0\delta}{2^k\mu^k} \right) ,$$

(a) u is monotone increasing along every $\tau \in \Gamma_x(\nu_k, \theta_k)$,

(b) u is monotone increasing along the directions

$$e_t + B_k\nu_k \quad \text{and} \quad -e_t - A_k\nu_k,$$

(c) the sequences $\{\delta_k\}$ and $\{\mu_k\}$ satisfy the recurrent relations

$$\delta_{k+1} \leq \delta_k - c_1\frac{\delta_k^2}{\mu_k},$$
$$\mu_{k+1} \leq \mu_k - c_2\delta_k$$

as long as $\delta_k \ll \mu_k^3$,

(d) $|\nu_{k+1} - \nu_k|, |A_{k+1} - A_k|, |B_{k+1} - B_k| \leq \bar{c}h\delta_k$ and

$$0 < B_k - A_k \leq c_2\mu_k .$$

From (c) we obtain the asymptotic behavior

$$\delta_k \sim \frac{c_1(\eta)}{k^{3/2-\eta}}, \qquad \mu_k \sim \frac{c_2(\eta)}{k^{1/2-\eta}}$$

for any small $\eta > 0$. Then using (a), (b) and (d), the assertion (9.1) of the main theorem follows. To prove (9.2), notice that at each time level $t_0 \in (-\frac{1}{2}, \frac{1}{2})$, $\Omega^\pm(u) \cap \{t = t_0\}$ is a Liapunov-Dini domain. Since u_t is bounded, the results of [W] apply and therefore $\nabla u^\pm$ are continuous up to the free boundary at each time level t_0.

On the other hand $\nabla u^\pm$ have nontangential limits everywhere on $F(u)$; therefore $\nabla u^\pm$ are continuous in space and time. From the free boundary condition, which now holds in a pointwise sense everywhere, we deduce that $u_t^\pm$ are continuous in (x,t) on $F(u)$. From this and the existence of nontangential limits of $u_t^\pm$ everywhere on $F(u)$ we conclude that $u_t^\pm$ is continuous in (x,t) up to $\Omega^\pm(u)$, respectively. □

Chapter 10

Flat Free Boundaries Are Smooth

10.1. Main result and strategy

According to the results in the previous chapter, under a nondegeneracy condition preventing simultaneous vanishing of the heat flow from both phases, Lipschitz free boundaries enjoy instantaneous regularization and viscosity solutions are classical.

Coming back to the counterexamples of Section 9.3, we realized that in the absence of nondegeneracy (in the above sense) the achievement of further regularity of the free boundary depends on its Lipschitz constant in space. Our main result, indeed, implies that if this Lipschitz constant is small enough (depending on dimension and the oscillation in time of the solution), then the above regularity results hold; i.e., roughly speaking, we are actually in a "nondegenerate" situation.

It turns out that to be in a situation with enough "nondegeneracy" to inply regularity, it is not necessary for the free boundary to be a Lipschitz graph or even a graph at all. It is enough to have a suitable flatness condition that we express, as in the elliptic case, in the following flexible version of ε-monotonicity.

Definition 10.1. We say that u is *ε-monotone* along the directions of a cone $\Gamma(e,\theta)$ if for a small $\delta > 0$ and every $\tau \in \Gamma(e, \theta - \delta), \varepsilon' \geq \varepsilon$,

$$\sup_{q \in B_{\varepsilon' \sin \delta}(p)} u(q - \varepsilon'\tau) \leq u(p) .$$

In the case of our free boundary problem, the results in Section 13.4 can be summarized as follows.

Corollary 10.1. *Let u be a viscosity solution to our free boundary problem in $\mathcal{C}_1 = B_1' \times (-1,1)$, ε-monotone along every $\tau \in \Gamma_x(e_n, \theta^*) \cup \Gamma_t(\nu, \theta^t)$, $\nu \in \operatorname{span}\{e_n, e_t\}$.*

Then there exist positive C, c such that at a distance greater than $C\sqrt{\varepsilon}$ from $F(u)$, u is fully monotone along any

$$\tau \in \Gamma_x(e_n, \theta^* - c\varepsilon) \cup \Gamma_t(\nu, \theta^t - c\sqrt{\varepsilon}) \ .$$

We can now state the main result of this chapter.

Theorem 10.2. *Let u be a viscosity solution of an f.b.p. in $\mathcal{C}_2 = B_2' \times (-2,2)$, ε-monotone along all directions $\tau \in \Gamma_x(e_n, \theta_0^*) \cup \Gamma_t(e_n, \theta^*)$. Set $M_0 = \sup_{\mathcal{C}_2} u$ and assume $u(e_n, -\frac{3}{2}) = 1$, $(0,0) \in F(u)$.*

Moreover, let $G = G(a,b)$ be a Lipschitz function with Lipschitz constant L_G such that

$$D_a G \geq c^* > 0 \ , \quad D_b G \leq -c^* \ .$$

Then if ε and $\delta_0 = \frac{\pi}{2} - \theta_0^*$ are small enough, depending only on n and θ^*, the following conclusions hold.

(1) *In $\mathcal{C}_1$ the free boundary is a C^1-graph, say $x_n = f(x', t)$, in space and time. Moreover, there exists a positive constant C_1 depending only on $n, M_0, L_G, c^*, a, \theta^*$ such that for every $(x', x_n, t), (y' y_n, s) \in F(u)$,*
   (a) $|\nabla_{x'} f(x',t) - \nabla_{x'} f(y',t)| \leq c_1(-\log|x'-y'|)^{-4/3}$,
   (b) $|D_t f(x',t) - D_t f(x',s)| \leq c_1(-\log|s-t|)^{-1/3}$.

(2) *$u \in C^1(\bar{\Omega}^+(u)) \cap C^1(\bar{\Omega}^-(u))$ and on $F(u) \cap \mathcal{C}_1$*

$$u_\nu^+ \geq c_2 > 0$$

with $c_2 = c_2(n, M_0, L_G, c^*, a, \theta^*)$.

In particular, Theorem 10.2 holds for 0-monotone (i.e., fully monotone) functions. In fact, as an immediate consequence we have the following corollary.

Corollary 10.3. *In Theorem 10.2 replace the ε-monotonicity hypothesis with the following one:*

> *The free boundary $F(u)$ is given by the graph of a Lipschitz function $x_n = f(x', t)$ with Lipschitz constant L_1 in space and L_2 in time. Then there exists $L_0 = L_0(n, L_2)$ such that, if $L_1 \leq L_0$, then conclusions (1) and (2) in Theorem 10.2 hold.*

Strategy of the proof. The main problems arise of course from the lack of *a priori* control of the normal derivatives (in the viscosity sense) of u at the free boundary. Let us see what relevant ideas are needed to overcome this difficulty.

Step 1. Interior enlargement of the cone in space. Saying that u is monotone along any direction τ belonging to a spatial cone $\Gamma_x(e_n, \theta^*)$ amounts to asking that for any $\varepsilon' > 0$

$$u_{\varepsilon'}(p) = \sup_{q \in B'_{\varepsilon' \sin\theta}(p)} u(q - \varepsilon' e_n) \leq u(p) \ . \tag{10.1}$$

The function $u_{\varepsilon'}$ measures the opening of the monotonicity cone, and one tries to get inequality (10.1) in a smaller region for τ belonging to a larger cone.

Thanks to Corollary 10.1, this can be done $\sqrt{\varepsilon'}$-away from the free boundary, for any $\varepsilon' > 0$, using the same techniques of Sections 9.2–9.4.

Step 2. More delicate is the improvement of the space-time cone $\Gamma_t(e_n, \theta^*)$, since in Section 9.5 we used full monotonicity up to the free boundary. The idea is to approximate u by suitable monotone supersolutions and subsolutions $\bar{v}, \underline{v}$ which fall under the hypotheses of Lemma 9.11.

In this way we obtain an improvement in monotonicity for both $\bar{v}$ and $\underline{v}$ that can be transferred to u via an estimate of the approximation error.

Step 3. We want to carry to the free boundary the gain obtained in Steps 1 and 2. What we can do is to carry only ε-monotonicity in a larger cone.

We need however to counterbalance the lack of nondegeneracy through a rather delicate control from below of u_ν^+ in the viscosity sense, at regular points of the free boundary. This sort of Hopf maximum principle is to be expected if in the iteration of the various steps the opening speed of the cones of ε-monotonicity is enough to reconstruct in a neighborhood of $(0,0)$ a Lyapunov-Dini domain.

This is exactly what happens, but at the same time it is necessary, at each iteration step, to decrease at a suitable rate the ε in the ε-monotonicity of u. This is done in the next step.

Step 4. Since ε-monotonicity implies full monotonicity $\sqrt{\varepsilon}$-away from the free boundary, it is possible to improve the ε-monotonicity itself, i.e., to decrease ε, at the price of giving up a small portion of the enlarged cone. This improvement of ε-monotonicity up to the free boundary can be obtained via a family of perturbations in the style of Section 9.6. Also at this stage, the key point is the control of u_ν^+ at regular points of $F(u)$.

Step 5. We perform a double iteration procedure that consists, at each step, of a cone enlargement and of an ε-monotonicity improvement, in a sequence of contracting cylinders. Again, due to the underlying double homogeneity of the problem, the sequence of contracting domains is neither hyperbolically nor parabolically scaling and the opening speed of the monotonicity cones is logarithmic. This produces the logarithmic modulus of continuity of $D_t\varphi, \nabla f$ appearing in the theorem.

The next section deals with Steps 1 and 2.

10.2. Interior enlargement of the monotonicity cone

The first step in the strategy of the proof of Theorem 10.2 is the enlargement of the monotonicity cone away from the free boundary. According to Corollary 10.1 our ε-monotone solution is fully monotone $\sqrt{\varepsilon}$-away from the free boundary and therefore the technique in Sections 9.2–9.4 can be used to get an enlargement of the spatial section of the cone. Precisely, we have

Lemma 10.4. *Let u be a viscosity solution to our f.b.p. in $\mathcal{C}_1 = B_1' \times (-1,1)$. Suppose that u is fully monotone along every $\tau \in \Gamma_x(e_n, \theta^x) \cup \Gamma_t(\nu, \theta^t)$, with $\nu \in \operatorname{span}\{e_n, e_t\}$, in the domain $\mathcal{C}_1 \cap \{d(p, F(u)) > c\varepsilon^\gamma\}$, for $0 < \gamma \leq \frac{1}{2}$ and $\varepsilon > 0$ small.*

Then there exist $c, h, 0 < h < 1$, $\bar{\theta}^x$ and $\bar{e}$, all dependent only on θ^x, θ^t, a, n, such that

(i) *$\Gamma_x(e_n, \theta^x) \subset \Gamma_x(\bar{e}, \bar{\theta}^x)$ and $\frac{\pi}{2} - \bar{\theta}^x \leq h(\frac{\pi}{2} - \theta^x)$,*
(ii) *u is fully monotone in $\Gamma_x(\bar{e}, \bar{\theta}^x)$ in the cylinder $B_{1/8}'(\frac{3}{4}e_n) \times (-\frac{c\delta}{\mu}, \frac{c\delta}{\mu})$ provided $\delta \leq \mu$, where $\delta = \frac{\pi}{2} - \theta^x$, $\mu = \frac{\pi}{2} - \theta^t$.*

The enlargement of the cone of monotonicity in time away from the free boundary, done in Section 9.5, requires full monotonicity up to the free boundary. This is not our case here and therefore we have to argue differently. The idea is to use suitable monotone replacements of u, constructed by taking supremum or infimum of u over balls of radius ε^γ. First of all let us state a lemma analogous to Lemma 5.6.

Lemma 10.5. *Let u be a viscosity solution of our f.b.p., ε^γ-monotone along every direction $\tau \in \Gamma_x(e_n, \theta^x) \cup \Gamma_t(\nu, \theta^t)$, $\nu \in \operatorname{span}\{e_n, e_t\}$, and fully monotone outside an $M\varepsilon^\gamma$-neighborhood of $F(u)$. Then there exist constants c, C and C_0, depending only on n, θ^x and θ^t, such that*

$$c d_p D_{e_n} u(p) \leq u(p) \leq C d_p D_{e_n} u(p)$$

where $d_p = d(p, F(u))$, for every p with $d_p \geq c_0 \varepsilon^\gamma$.

The proof follows closely the proof of Lemma 5.6 mentioned above. Now let u as in Lemma 10.5 and, for $0 < \gamma' < \gamma < \frac{1}{2}$, define

$$\bar{v}(p) = \inf_{B_{\varepsilon^{\gamma'}}(p)} u, \qquad \underline{v}(p) = \sup_{B_{\varepsilon^{\gamma'}}(p)} u$$

and

$$\bar{u}_\varepsilon(p) = u(p + 4\varepsilon^{\gamma'} e_n)\ , \qquad \underline{u}_\varepsilon(p) = u(p - 4\varepsilon^{\gamma'} e_n)\ .$$

Lemma 10.6. *The following hold.*

(i) *In* $\Omega^+(\bar{v}) \cup \Omega^+(\underline{v})$

$$|\bar{v} - \bar{u}_\varepsilon| \le c\varepsilon^{\gamma'} D_{e_n} \bar{u}_\varepsilon$$

and in $\Omega^-(\bar{v}) \cup \Omega^-(\underline{v})$

$$|\underline{v} - \underline{u}_\varepsilon| \le c\varepsilon^{\gamma'} D_{e_n} \underline{u}_\varepsilon.$$

(ii) $\bar{v}$ *and* $\underline{v}$ *are monotone increasing along every*

$$\tau \in \Gamma_x(e_n, \theta^x - c\varepsilon^{1-\gamma'}) \cup \Gamma_t(\theta^t - c\varepsilon^{1-\gamma'}, \nu)\ .$$

(iii) $\bar{v}$ *is supercaloric in* $\Omega^+(\bar{v}) \cup \Omega^-(\bar{v})$ *and* $\underline{v}$ *is subcaloric in* $\Omega^+(\underline{v}) \cup \Omega^-(\underline{v})$.

(iv) *Each point of* $F(\bar{v})$ *is regular from the left and if* $B(y_0, s_0) \subset \Omega^-(\bar{v})$ *touches* $F(\bar{v})$ *at* (x_0, t_0) *with slope* $-\frac{\beta_+}{\alpha_+}$, *then* $\alpha_+ > 0$ *and near* x_0

$$\bar{v}(x, t_0) \le \alpha_+ \langle x - x_0, \nu\rangle^+ - \alpha_- \langle x - x_0, \nu\rangle^- + o(|x - x_0|)$$

where $\nu = \frac{y_0 - x_0}{|y_0 - x_0|}$, *with*

$$\frac{\beta_+}{\alpha_+} \ge G(\alpha_+, \alpha_-)\ .$$

(v) *Each point of* $F(\underline{v})$ *is regular from the right and if* $B(y_0, s_0) \subset \Omega^+(\underline{v})$ *touches* $F(\underline{v})$ *at* (x_0, t_0) *with slope* $-\frac{\beta_-}{\alpha_-}$, *then* $\alpha_- > 0$ *and near* x_0

$$\underline{v}(x, t_0) \ge \alpha_+ \langle x - x_0, \nu\rangle^+ - \alpha_- \langle x - x_0, \nu\rangle^- + o(|x - x_0|)$$

with

$$\frac{\beta_-}{\alpha_-} \le G(\alpha_+, \alpha_-)\ .$$

Proof. (i) is an easy consequence of Lemma 10.5; (ii) follows from Lemma 5.4; (iii) is obvious since $\bar{v}$ and $\underline{v}$ are infimum and supremum of translations of caloric functions. Finally, (iv) and (v) follow from Lemma 9.13. □

Remark. Properties (iii) and (iv), (v) say that $\bar{v}$ and $\underline{v}$ are R-supersolutions and R-subsolutions, respectively, in the terminology of Section 7.4.

Now we replace $\bar{v}$ and $\underline{v}$ by their caloric counterparts in their positive and negative regions, in strips centered at free boundary points and of height $a = \min(a_1, a_2)$. We shall treat only the $\bar{v}$ case, since that of $\underline{v}$ can be done in an analogous way.

Let $(0,0) \in F(\bar{v})$ be the center of the strip

$$R_\eta = \left\{ |x'| < \frac{2\eta}{3},\ |x_n| < \frac{2\eta}{3},\ |t| < a\eta \right\}$$

and let

$$R_\eta^\pm = \Omega^\pm(\bar{v}) \cap R_\eta .$$

Denote by v the function that satisfies

$$\Delta v - a_1 v_t = 0 \ \text{ in } \ R_1^+ , \qquad \Delta v - a_2 v_t = 0 \ \text{ in } \ R_1^-$$

and

$$v = \bar{v} \ \text{ on } \ \partial_p R_1^\pm .$$

By the maximum principle and Lemma 10.6

$$|\bar{u}_\varepsilon - v| \le c\varepsilon^{\gamma'} D_{e_n}\bar{u}_\varepsilon \ \text{ and } \ |\underline{u}_\varepsilon - v| \le c\varepsilon^{\gamma'} D_{e_n}\bar{u}_\varepsilon$$

in $\bar{R}_1^\pm$, respectively.

Also, since $\bar{v}$ is supercaloric in $R_1^+ \cup R_1^-$, $v^+ \le \bar{v}^+$ and $v^- \ge \bar{v}^-$. Therefore at each point $(x_0, t_0) \in F(v) = F(\bar{v})$, we have the asymptotic development

$$v(x, t_0) \le \alpha_+^* \langle x - x_0, \bar{\nu}\rangle^+ - \alpha_-^* \langle x - x_0, \bar{\nu}\rangle^- + o(|x - x_0|)$$

with $\alpha_+^* < \alpha_+$ and $\alpha_-^* \ge \alpha_-$, and by the monotonicity properties of G

$$\frac{\beta_+^*}{\alpha_+^*} = \frac{\beta_+}{\alpha_+} \ge G(\alpha_+, \alpha_-) \ge G(\alpha_+^*, \alpha_-^*) .$$

Let $M_\pm := \max_{R_1^\pm} v^\pm$ and $m_\pm := \min_{\Sigma^\pm} v^\pm$ where

$$\Sigma^\pm = \left\{ x_n = \pm\frac{2}{3} \right\} \cap \partial R_1^\pm .$$

Since on $\partial R_1^\pm, v = \bar{v}$, then $\frac{M_+}{m_+}$ and $\frac{M_-}{m_-}$ are controlled above by a constant depending only on n, M_0, θ^x, and θ^t. This means that after an appropriate rescaling, v^+ and v^- satisfy the hypotheses of Theorem 13.13. In particular, in a δ-neighborhood of $F(v)$ in $R_{7/8}$ we have that v is monotonically increasing along the directions of a space-time cone $\Gamma(\bar{\theta}, e_n)$ where $\bar{\theta} = \bar{\theta}(n, \theta^x, \theta^t)$. Actually v is monotone increasing in all of $R_{4/5}$ for some cone of directions. This is precisely the content of the following lemma.

Lemma 10.7. *Let $\gamma'' < \gamma'/4$. Then, in $R_{4/5}$, v is monotone increasing along any direction $\tau \in \Gamma_x(\theta^x - c\varepsilon^{\gamma''}, e_n)$ and $\tau \in \Gamma_x(\theta^t - c\varepsilon^{\gamma''}, \nu) \subset \mathrm{Sp}\{e_n, e_t\}$, with $c = c(n, L_1, L_2)$.*

Proof. By Lemma 10.6, on $F(v)$, $D_\tau v^+ \geq 0$ for any

$$\tau \in \Gamma_x(\theta^x - c\varepsilon^{1-\gamma'}, e_n) \cup \Gamma_t(\theta^t - c\varepsilon^{1-\gamma'}, \nu) .$$

Therefore for any $p \in R^+_{7/8}$

$$D_\tau v^+(p) \geq \int_{\partial_p R^+_{7/8} \setminus F(v)} D_\tau v^+ \, d\omega^p \geq - \int_{\partial_p R^+_{7/8} \setminus F(v)} |D_\tau v^+| \, d\omega^p \equiv -z(p)$$

where ω denotes the caloric measure of $\Delta - a_1 D_t$ in $R^+_{7/8}$. By Lemma 13.11,

$$\|D_t v^+\|_{L^2(R^+_{7/8})}, \|\nabla v^+\|_{L^2(R^+_{7/8})} \leq c(n, \bar{\theta}) \|v^+\|_{L^2(R^+_{8/9})} .$$

Replacing, if necessary, $\partial_p R^+_{7/8}$ by a nearby parallel surface, we may assume that $D_\tau v^+ \in L^2(\partial_p R^+_{7/8})$ with

$$\|D_\tau v^+\|_{L^2(\partial_p R^+_{7/8})} \leq c M_+ ,$$

which implies that $z(\frac{4}{5} e_n) \leq \bar{c} M_+$. Since $z = 0$ on $F(v)$, by Corollary 13.8 in $R^+_{4/5}$, we have

$$z(p) \leq c \frac{z(\frac{4}{5} e_n)}{v^+(\frac{4}{5} e_n)} v^+(p) \leq c v^+(p) .$$

Therefore, by Corollary13.14, if $d_p = \operatorname{dist}(p, F(v)) \leq \delta$, then

$$v^+(p) \leq c d_p D_{e_n} v^+(p) .$$

Hence, for $\gamma'' < \frac{\gamma'}{4}$,

$$D_\tau v^+(p) + \varepsilon^{\gamma''} D_{e_n} v^+(p) \geq (\varepsilon^{\gamma''} - c d_p) D_{e_n} v^+(p) \geq 0$$

if $d_p \leq c^{-1} \varepsilon^{\gamma''}$ ($\varepsilon^{\gamma''} < \delta$).

Now consider the points p with $d_p \geq c^{-1} \varepsilon^{\gamma''}$. For the function $\bar{u}_\varepsilon(p) = u(p + 4\varepsilon^{\gamma'} e_n)$, the free boundary $F(u_\varepsilon)$ is at a distance greater than $3\varepsilon^{\gamma'}$ from R^+_1.

Recall that on $\partial_p R^+_1$ we have

$$|\bar{v} - \bar{u}_\varepsilon| \leq c \varepsilon^{\gamma'} D_{e_n} \bar{u}_\varepsilon;$$

therefore by the maximum principle,

$$|v^+ - \bar{u}_\varepsilon| \leq c \varepsilon^{\gamma'} D_{e_n} \bar{u}_\varepsilon .$$

By Schauder estimates, whenever $d_p \geq \bar{c} \varepsilon^{\gamma''}$, $p \in R^+_{4/5}$,

$$|D_t v^+ - D_t \bar{u}_\varepsilon|, |\nabla v^+ - \nabla \bar{u}_\varepsilon| \leq c(\varepsilon^{\gamma' - 2\gamma''}) D_{e_n} \bar{u}_\varepsilon \leq c \varepsilon^{\gamma''} D_{e_n} \bar{u}_\varepsilon$$

since $\gamma'' < \gamma'/4$; in particular,

$$D_\tau v^+ - D_\tau \bar{u}_\varepsilon \geq c \varepsilon^{\gamma''} D_{e_n} \bar{u}_\varepsilon .$$

Therefore, if $\tau \in \Gamma_x(\theta^x - c\varepsilon^{\gamma''}, e_n) \cup \Gamma_t(\theta^t - c\varepsilon^{\gamma''}, \nu)$,

$$D_\tau v^+ \geq 0 \, .$$

For v^- we can proceed in an analogous manner. The proof is complete. □

Finally, we are in a position to apply the method of Section 9.5. Let $e_t + Be_n$ and $-e_t - Ae_n$ be the vectors of the generatrix of the cone $\Gamma_t(\theta^t - c\varepsilon^{\gamma''}, \nu)$ ($\nu \in \mathrm{Sp}\{e_n, e_t\}$). Notice that if $\mu = \frac{\pi}{2} - \theta^t$,

$$B - A \leq c(\mu + \varepsilon^{\gamma''}) \leq \bar{c}\mu$$

provided that $\varepsilon^{\gamma''} \ll \mu$.

Lemma 10.8. *Let* $\alpha_\pm = D_{e_n} u^\pm(\frac{3}{4}e_n)$. *If*

$$G(\alpha_+, \alpha_-) \geq -b \equiv -\frac{A+B}{2} \qquad (\textit{resp.},\ \ G(\alpha_+, \alpha_-) \leq -b) \, ,$$

then there exist $c, \bar{c} > 0$ *such that if* δ *is small and* $\delta \leq c\mu^3$

$$D_t u^+ + (B - c\mu) D_{e_n} u^+ \geq 0 \qquad (\textit{resp.},\ D_t u^+ + (A + c\mu) D_{e_n} u^+ \leq 0)$$

in $B_{1/30}(\frac{3}{4}e_n) \times (-\bar{c}\frac{\delta}{\mu}, \bar{c}\frac{\delta}{\mu})$.

Proof. Suppose $G(\alpha_+, \alpha_-) \geq -b$. Set $\alpha^*_\pm = D_{e_n} v^\pm(\frac{3}{4}e_n)$. In every strip $R_{4/5}(p_0)$ centered at $p_0 \in F(v)$ we apply Lemma 9.11 to v. Therefore, since $G(\alpha^*_+, \alpha^*_-) \geq -b - c\varepsilon^{\gamma''}$, in $B_{1/30}(\frac{3}{4}e_n) \times (-\bar{c}\frac{\delta}{\mu}, \bar{c}\frac{\delta}{\mu})$ we have

$$D_t v^+ + B D_{e_n} v^+ \geq (C\mu - c\varepsilon^{\gamma''}) D_{e_n} v^+ \geq \bar{c}\mu D_{e_n} v^+ \, .$$

Hence the conclusion follows, since

$$|D_t v^+ - D_t u^+| \leq c\varepsilon^{\gamma''} D_{e_n} u^+ \, , \quad |D_{e_n} v^+ - D_{e_n} u^+| \leq c\varepsilon^{\gamma''} D_{e_n} u^+ \, .$$

If $G(\alpha_+, \alpha_-, e_n) \leq -b$, we use the caloric replacement of $\underline{v}$ instead of $\bar{v}$. □

10.3. Control of u_ν at a "contact point"

In order to be able to propagate the gain in Section 10.2 from the interior to the free boundary, we need to have a control of u_ν at regular points of $F(u)$. We are dealing with the following situation:

Let $\Gamma_k := \Gamma_x(\theta_k, \nu_k)$ be a sequence of spatial cones,

$$D = \{x = (x', x_n) \in \mathbb{R}^n : |x'| < 2 \, , \ g(x') < x_n < 2\}$$

with g a Lipschitz function and $0 \in F' = \{x : x_n = g(x')\}$. Suppose that the following hold.

(a) $\delta_k = \frac{\pi}{2} - \theta_k \leq \frac{C_0}{(k+N)^{1+\alpha}}$, $a > 0$, $N \gg 1$, and $|\nu_k - \nu_{k+1}| \leq \bar{c}(\delta_k - \delta_{k+1})$.

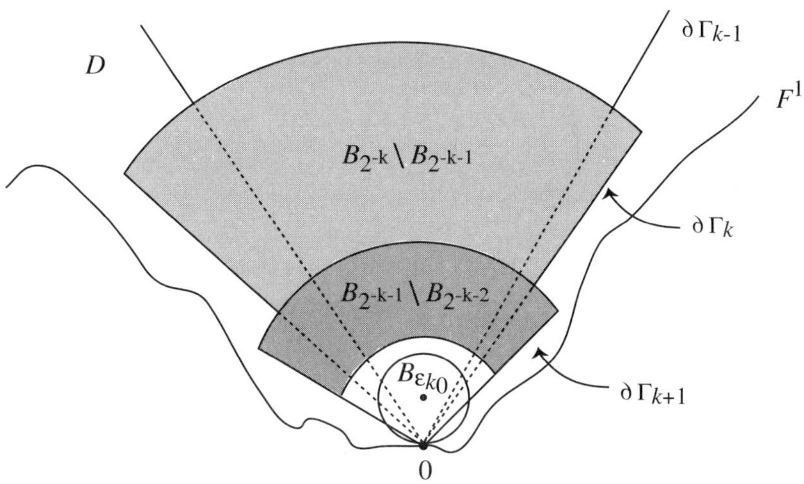

Figure 10.1. The geometry in the contact point lemma

(b) There exists k_0 such that, for $k \leq k_0$,

$$\Gamma_k \cap [B_{2^{-k}}(0) \setminus B_{2^{-k-1}}(0)] \subset D \cap [B_{2^{-k}}(0) \setminus B_{2^{-k-1}}(0)]$$

and a ball $B_{\varepsilon_{k_0}}, \varepsilon_k = 2^{-4k}$ tangent from inside to F' at 0, with ν_{k_0} as inward normal.

Then the following lemma holds.

Lemma 10.9 (Contact point lemma). *Let $\{\Gamma_k\}_{k\geq 1}$ and D be as above and let w be a positive superharmonic function, continuous in $\bar{D}$ and vanishing on F'. Then there exists a constant $c = c(n, a, N)$ such that, near 0,*

$$w(x) \geq c w_0 \langle x, \nu_{k_0} \rangle$$

where $w_0 = \min\{w(x) : x \in \bar{D} \cap \{x_n = 2\}\}$.

Proof. We will construct a Lyapunov-Dini domain

$$D' = \{(x', x_n) \in \mathbb{R}^n : x_n > \varphi(x'), |x'| < 2, x_n < 2\}$$

such that

(i) $\varphi(x') > g(x')$ if $|x'| < 2$ and $\varphi(0) \sim \varepsilon_{k_0}$,

(ii) $\varphi \in C^1$ and $\omega_{\nabla\varphi}(r) \leq C_1[\log(\frac{8}{r})]^{-1-a'} (r \leq 4)$ where $a' = a/2$ and $\omega_{\nabla\varphi}$ denotes the modulus of continuity of $\nabla\varphi$.

Suppose that $\nu_{k_0} = e_n$ and set

$$\psi(s) = \int_0^s c_1 \left[\log\left(\frac{8}{r}\right)\right]^{-1-a'} dr\ .$$

Then $\psi(s)$ is increasing and $\psi(s) \geq c_2 s(\log \frac{8}{s})^{-1-a'}$. Therefore the graph $x_n = \psi(|x'|)$ intersects $\partial\Gamma_{k_0}$ at points $\bar{x}'$ such that

$$|\bar{x}'| \leq 8 \exp\left(-\left(\frac{\delta_{k_0}}{c_2}\right)^{-\frac{1}{1+a'}} \right) = 8 \exp\left(-c_3 \delta_{k_0}^{-\frac{1}{1+a'}} \right)$$

with $\psi(|\bar{x}'|) \leq c_3 \delta_{k_0} \exp(-c_3 \delta_{k_0}^{-1/1+a'})$. Since

$$\delta_{k_0}^{-\frac{1}{1+a'}} \geq c_0^{-\frac{1}{1+a'}} (k_0 + N)^{\frac{1+a}{1+a'}}$$

for some N large and $\frac{1+a}{1+a'} > 1$, we have

$$\psi(|\bar{x}'|) \leq \frac{1}{10} \varepsilon_{k_0} \, .$$

If we set $\varphi(x') = \psi(|x'|) + \varepsilon_{k_0}$, choosing c_1 large enough and using (a), (b), it is easily seen that the domain $D' = \{(x', x_n) : \varphi(x') < x_n < 2\}$ has the desired properties.

To prove that w satisfies the inequality above, denote by $Z(x)$ the harmonic measure in D' of the set $\partial D' \cap \{x_n = 2\}$ and by $\bar{w}$ the Poisson integral of w in D. Using the results of [W], we infer that there exists a constant $c = c(n)$ such that

$$Z_\nu(0) \geq c \, .$$

On the other hand, using the Hopf principle and the maximum principle, if x_n is small, we can write

$$\frac{\bar{w}(0, x_n)}{x_n} \geq c \frac{\bar{w}(p_{k_0})}{\varepsilon_{k_0}} \geq c w_0 \frac{Z(p_{k_0})}{\varepsilon_{k_0}} \geq c w_0$$

where $w_0 =$ minimum of $\bar{w}$ on $D \cap \{x_n = 2\}$ and $p_{k_0} = (0, \varepsilon_{k_0})$. Since $w \geq \bar{w}$, the proof is complete. □

10.4. A continuous family of perturbations

The full monotonicity away from the free boundary yields a gain in the ε-monotonicity on the free boundary. This can be achieved by means of a continuous family of perturbations, as in Section 5.2. Although the construction of this family follows a different approach here, the underlying ideas are exactly those expressed in Section 5.1. We introduce the following notation:

$N_{b\varepsilon^\gamma} = \{p = (x', x_n, t) : d(p, F) < b\varepsilon^\gamma\}$ for $b > 0 \, , 0 < \gamma < 1$.

$\mathcal{C}_{b,R,T} = N_{b\varepsilon^\gamma} \cap \{|x'| < R\} \cap \{|t| < T\}$.

$\Omega_{\varepsilon,R,T}$ is a smooth domain such that $\mathcal{C}_{\frac{b}{2},R,T} \subset \Omega_{\varepsilon,R,T} \subset \mathcal{C}_{b,R,T}$.

Here we assume that u is ε^γ-monotone along all directions $\tau \in \Gamma_x(\theta^x, e_n) \cup \Gamma_t(\theta^t, \nu)$ and, as in the main theorem, that δ is very small ($\delta = \frac{\pi}{2} - \theta^x$, $\mu = \frac{\pi}{2} - \theta^t$).

Lemma 10.10. *Let b_1, b_2, b_3, and D be positive constants; then there exists a family of C^2-functions $\psi_\eta = \psi_\eta(x,t)$, $\eta \in [0,1]$, in $\Omega_{\varepsilon,R,T}$ such that*

(i) $0 < \bar{b} \le \psi_\eta \le 1 + \eta$ *for some constant* $\bar{b}$,

(ii) $\Delta\psi_\eta - b_1 D_t\psi_\eta - b_2|\nabla\psi_\eta| - b_3\frac{|\nabla\psi_\eta|^2}{\psi_\eta} \ge 0$,

(iii) $D_t\psi_\eta, |\nabla\psi_\eta| \le \frac{C}{D}\varepsilon^{\beta-\gamma}, \quad 0 < \beta < \gamma < 1$,

(iv) $D_t\psi_\eta \ge 0$,

(v) $\psi_\eta \le 1$ *in* $\Omega_{\varepsilon,R,T} \cap \Big(\{-T < t < -T + \varepsilon^\alpha\} \cup \{R - \frac{\varepsilon^{\alpha/4}}{2} < |x'| < R\}\Big)$,

(vi) $\psi_\eta \ge 1 + \eta(1 - C\varepsilon^\beta)$ *in* $\Omega_{\varepsilon,R,T} \cap \Big(\{t > -T + 2\varepsilon^\alpha\} \cap \{|x'| < R - \frac{\varepsilon^{\alpha/4}}{2}\}\Big)$ *where* $0 < \alpha < \gamma - \beta$.

Proof. Let $g_1 = g_1(t) \in C^\infty([-T,T])$ and $g_2 = g_2(x) \in C^\infty(B_R)$ such that

$$g_1(t) = \left(2 - \frac{T+t}{2\varepsilon^\alpha}\right) \qquad \text{for } -T \le t \le -T + 2\varepsilon^\alpha + 2\varepsilon^{\alpha+\beta},$$

$g_1(t)$ strictly decreasing from $1 + \varepsilon^\beta$ to $1 + \frac{1}{2}\varepsilon^\beta$ for $-T + 2\varepsilon^\alpha + 2\varepsilon^{\alpha+\beta} < t \le T$ with $g_1'(t) < -c\varepsilon^\beta$,

$$g_1 = 1 \text{ if } R - \frac{\varepsilon^{\alpha/4}}{2} < |x| < R,$$
$$g_2 = 0 \text{ if } |x| < R - \varepsilon^{\alpha/4}.$$

Let F_ε^+ be the lateral part of $\partial\Omega_{b,2R,2T} \cap \{u > 0\}$. Under the dilation $x \to \varepsilon^\gamma\tilde{x}$, $t \to \varepsilon^\gamma\tilde{t}$, F_ε^+ becomes a uniformly smooth surface $\tilde{F}_\varepsilon^+$ at a distance of order 1 from the dilated free boundary. Due to the flatness of the free boundary the space curvature of $\tilde{F}_\varepsilon^+$ is bounded by $c\delta$. Therefore, for each t, the spatial distance function $d_\varepsilon = d(x,t)$ from $\tilde{F}_\varepsilon^+$ is well defined up to a distance of order $1/\delta$, and we have $|\Delta_{\tilde{x}} d_\varepsilon| \le c\delta$, $|D_t d_\varepsilon| \le CL_1$, where L_1 is the Lipschitz constant in time of F.

Define

$$G(x,t) = g_1(t) + g_2(x) + A\varepsilon^{2\gamma-\alpha}\left(d_\varepsilon\left(\frac{x}{\varepsilon^\gamma}\right) - \sigma d_\varepsilon^2\left(\frac{x}{\varepsilon^\gamma}\right)\right).$$

If $\sigma = \sigma(L_0) > 0$ and $A > 0$ are chosen properly, G has the following properties:

(a) $D_t G \le 0$, $|D_t G| \le c\varepsilon^{-\alpha}$,

(b) $|\nabla G| \le c\varepsilon^{-\alpha/4}$, and

(c) $\Delta G \le -cA\varepsilon^{-\alpha}$.

Now, for $C > 1$ large enough, set $F(x,t) = (\frac{1+G(x,t)}{3})^{\frac{1}{1-2C}}$; then, it is easy to see that the functions

$$\psi_\eta(x,t) = 1 + \eta \left(\frac{F(x,t) - 1}{3^{\frac{1}{2C-1}} - 1} \right) \qquad \forall\, \eta \in [0,1]\,,$$

have the required properties. □

Lemma 10.11. *Let u be a viscosity solution of a free boundary problem satisfying the hypothesis of the main theorem. For $0 < \varepsilon,\ \sigma \ll \lambda < 1$, let $\bar{u}(q) = u(q - \lambda\varepsilon\tau)$ and let*

$$v_\eta(p) = \sup_{q \in B_{\sigma\psi_\eta}(p)} \bar{u}(q)\,, \qquad \tau \in \mathbb{R}^{n+1}\,,\ |\tau| = 1\,, \tag{10.2}$$

where ψ_η are the auxiliary functions constructed in the previous lemma. Suppose the supremum in (10.2) occurs uniformly away from the top and the bottom points of the ball, at a distance not smaller than ρ. Then the following hold.

(i) *v_η is subcaloric in $\Omega^+(v_\eta)$ and $\Omega^-(v_\eta)$.*

(ii) *$\partial\Omega^+(v_\eta)$ is uniformly Lipschitz in space-time with Lipschitz constant in space $L \le \tan(\frac{\pi}{2} - \theta_0^x) + c\varepsilon^{1-\alpha}$.*

(iii) *If $(x_0, t_0) \in F(v_\eta)$ and $(y_0, s_0) \in F(\bar{u})$ with*

$$(y_0, s_0) \in \partial B_{\sigma\psi(x_0,t_0)}(x_0, t_0)\,,$$

then (x_0,t_0) is a regular point from the right. Moreover, if near (y_0, s_0) along the paraboloid $s = s_0 - \gamma\langle y - y_0, \nu\rangle^2$ ($\gamma > 0$), u has the asymptotic expansion

$$\bar{u}(y,s) = \alpha_+\langle y - y_0, \nu\rangle^+ - \alpha_-\langle y - y_0, \nu\rangle^- + o(|y - y_0|)$$

where $\nu = \frac{y_0 - x_0}{|y_0 - x_0|}$, then near (x_0, t_0) along the paraboloid $t = t_0 - \gamma\langle x - x_0, \nu\rangle^2$

$$\begin{aligned} v_\eta(x,t) \ge{} & \alpha_+ \left\langle x - x_0, \nu + \frac{\sigma\psi_\eta(x_0,t_0)}{|y_0 - x_0|}\nabla(\sigma\psi_\eta) \right\rangle^+ \\ & - \alpha_- \left\langle x - x_0, \nu + \frac{\sigma\psi_\eta(x_0,t_0)}{|y_0 - x_0|}\nabla(\sigma\psi_\eta) \right\rangle^- + o(|x - x_0|) \end{aligned}$$

with $\frac{s_0 - t_0}{|y_0 - x_0|} \le G(\alpha_+, \alpha_-, \nu)$.

Proof. (i) At the point where the supremum occurs, we have the estimate $|u_t| \le c|\nabla_x u|$, with c depending on ρ. The proof then is similar to that of Lemma 9.12.

Except for minor changes, the proofs of (ii) and (iii) follow those of Lemma 4.9 and Lemma 9.13, respectively. □

10.5. Improvement of ε-monotonicity

In this section, using the family of subsolutions of the previous section, we show how to obtain a gain in ε-monotonicity on the free boundary. Also, repeating the process for this gain a finite number of times, ε can be decreased enough so that the hypotheses of our main inductive argument are fulfilled.

Let u be a viscosity solution to our free boundary problem in $\mathcal{C}_{R,T} = B'_R(0) \times (-T, T)$, ε-monotone along every direction $\tau \in \Gamma_x(\theta^x, e_n) \cup \Gamma_t(\theta^t, \nu)$ where $\nu \in \mathrm{Sp}\{e_n, e_t\}$. If $\delta = \frac{\pi}{2} - \theta^x$ and $\delta \ll \mu = \frac{\pi}{2} - \theta^t$, $\varepsilon \ll \delta$, we know that in $\mathcal{C}_{R-\varepsilon,T-\varepsilon}$, after a dilation in time $t \to \frac{\delta}{\mu}t$,

$$\sup_{q \in B_{\varepsilon \sin \delta}(p)} u(q - \varepsilon\tau) \le u(p)$$

for any $\tau \in \Gamma_x(\theta^x - \delta, e_n) \cup \Gamma_t(\theta^t_* - \mu, \nu_*)$, $|\tau| = 1$, where $\Gamma_t(\theta^t_*, \nu_*)$ is the dilated cone in space-time. Since for any $\lambda < 1$ close to 1

$$B_\sigma(p - \lambda\varepsilon\tau) \subset B_{\varepsilon \sin \delta}(p - \varepsilon\tau) \qquad (\sigma = \varepsilon[\sin\delta - (1-\lambda)]) \ ,$$

it follows that

$$\sup_{B_\sigma(p)} u(q - \lambda\varepsilon\tau) \le u(p) \ .$$

On the other hand, by Corollary 10.1, outside an ε^γ-neighborhood of $F(u)$, for $0 < \gamma \le \frac{1}{2}$, u is fully monotone in $\Gamma_x(\theta^x, e_n) \cup \Gamma_t(\theta^t, \nu)$, except for a correction of order ε for θ^x and $\sqrt{\varepsilon}$ for θ^t, which can be neglected since $\varepsilon \ll \delta$. This implies that for any $0 < \lambda \le 1$

$$\sup_{B_{\lambda\varepsilon \sin \delta}(p)} u(q - \lambda\varepsilon\tau) \le u(p) \ .$$

Now, in order to gain in ε-monotonicity near $F(u)$, we find an intermediate radius $\sigma\psi_\eta$, using the auxiliary function ψ_η constructed in the previous section. First, a technical lemma.

Lemma 10.12. *Let there be α, β, γ $(\le \frac{1}{2})$, and let $\Omega_{\varepsilon,R,T}$ be as in Section 10.4. Suppose that in $\Omega_{\varepsilon,R-c_1\varepsilon^\alpha,T-c_2\varepsilon^\alpha}$*

$$w_1(p) = \sup_{B_{\ell_1}(p)} u(q - \lambda\varepsilon\tau) \le u(p)$$

where $\ell_1 = \varepsilon(\lambda \sin\delta - c\varepsilon^\beta)$, for any $\tau \in \Gamma_x(\theta^x - \delta, e_n)$ or $\tau \in \Gamma_t(\theta^t_* - \mu, \nu_*)$ $(\nu_* \in \mathrm{Sp}\{e_n, e_t\})$. Then on $\partial\Omega_{\varepsilon,R-c_1\varepsilon^\alpha,T-c_2\varepsilon^\alpha}$ at a distance $M\varepsilon^\gamma$ (M large) from $F(u)$

$$w_1(p) \le (1 - \bar{c}\varepsilon^{1+\beta-\gamma})u(p) \ .$$

Proof. Set $w_2(p) = \sup_{B_{\ell_2}(p)} u(q - \lambda\varepsilon\tau)$ where $\ell_2 = \varepsilon\lambda\sin\delta$. Then using Lemma 10.5, we deduce

$$w_1(p) \le w_2(p) - c\varepsilon^{\beta+1}|\nabla u(p)| \le u(p) - \bar{c}\varepsilon^{\beta+1-\gamma}u(p)$$

for $p \in \partial\Omega_{\varepsilon,R-c_1\varepsilon^\alpha,T-c_2\varepsilon^\alpha}$ and $\operatorname{dist}(p, F(u)) \geq M\varepsilon^\gamma$. □

Lemma 10.13 (Basic Iteration). *Let u be a viscosity solution to our free boundary problem in $\mathcal{C}_{R,T} = B'_R(0)\times(-T,T)$ that is ε-monotone along every direction $\tau \in \Gamma_x(\theta^x, e_n) \cup \Gamma_t(\theta^t, \nu)$, $\nu \in \operatorname{Sp}\{e_n, e_t\}$. Suppose that $\delta \ll \mu^3$. Then there exists an $\varepsilon_0 > 0$ ($\varepsilon_0 \ll \delta$) and $0 < \lambda = \lambda(\varepsilon_0) < 1$ such that if $\varepsilon < \varepsilon_0$, u is $\lambda\varepsilon$-monotone in the cones $\Gamma_x(\theta^x - \bar{c}\varepsilon^\beta, e_n)$ and $\Gamma_t(\theta^t - \bar{c}\varepsilon^\beta, \nu)$, in the domain $\mathcal{C}_{R-\varepsilon^\alpha,T-\varepsilon^\alpha} \cap \Omega_{\varepsilon,R,T}$ where $0 < \alpha < \beta < \frac{1}{2}$.*

Proof. Choose $\bar{\eta}$ such that

$$\sigma(1-\bar{\eta}) = \lambda\varepsilon \sin\delta - c\varepsilon^{1+\beta}$$

where $\sigma = \varepsilon(\sin\delta - (1-\lambda))$. Take $1-\lambda = \frac{1}{2}\sin\delta$; then since $0 < \varepsilon \ll \delta \ll 1$, we have $\frac{1}{3} < \bar{\eta} < 1$. Perform the dilation in time $t \to \frac{\delta}{\mu}t$. This assures that the supremum of the new u in balls is taken uniformly away from the top and the bottom. Now set

$$\bar{v}_\eta = v_\eta + \bar{c}\varepsilon^{1+\beta-\gamma} w$$

where v_η is the function defined in Lemma 10.11 and w has the properties

$$\begin{cases} \Delta w - a_1 w_t = 0 & \text{in } \Omega^+(v_\eta) \cap \Omega_{\varepsilon,R,T}, \\ w = \tilde{u} & \partial_p\Omega_{\varepsilon,R,T} \cap \{p : d(p, F(v_\eta)) > M\varepsilon^\gamma\}, \\ w = 0 & \text{everywhere else in } \Omega_{\varepsilon,R,T} \end{cases}$$

where $\tilde{u}$ is the caloric function in $\mathcal{C}_{R,T} \cap \Omega^+(v_\eta)$ with boundary values zero on $F(v_\eta)$ and equal to u everywhere else on the parabolic boundary of $\mathcal{C}_{R,T} \cap \Omega^+(v_\eta)$. Notice that by the maximum principle $\tilde{u} \leq u$ in $\mathcal{C}_{R,T} \cap \Omega^+(v_\eta)$.

Following our discussion above, we want to show that for every $\eta \in [0, \bar{\eta}]$

$$\bar{v}_\eta \leq u$$

in $\Omega_{\varepsilon,R,T} \cap \mathcal{C}_{R-c\varepsilon^\alpha,T-c\varepsilon^\alpha}$.

The set $\{\eta \in [0,\bar{\eta}] : \bar{v}_\eta \leq u\}$ is nonempty since by the maximum principle and Lemma 10.12, $\eta = 0$ belongs to this set. Also, it is easy to see that this set is closed. To show it is open, it is enough to show that $\Omega_{\varepsilon,R,T} \cap \{\bar{v}_{\eta_0} > 0\}$ is compactly contained in $\Omega_{\varepsilon,R,T} \cap \{u > 0\}$, if $\bar{v}_{\eta_0} \leq u$, for $\eta_0 \in [0, \bar{\eta})$. Suppose not; then there exists a point $(x_0, t_0) \in F(\bar{v}_{\eta_0}) \cap F(u)$ which is, due to the construction of ψ_η, a regular point and $d((x_0,t_0), \partial_p\Omega_{\varepsilon,R,T}) \geq c\varepsilon^\gamma$. Since $F(v_\eta)$ is Lipschitz, by Corollary 13.8, we have $w \geq c\tilde{u}$ in a ε^γ-neighborhood of (x_0, t_0) in $\{v_\eta > 0\}$.

On the other hand, by Lemma 13.15, $w_- = \tilde{u}(1-\tilde{u}^r)$, for a suitable small $r > 0$, is a positive superharmonic function on the hyperplane $t = t_0$ and therefore, by Lemma 10.9, $(w_-)_\nu \geq \tilde{c} > 0$ at x_0, where $\nu = \nu^*/|\nu^*|$ with $\nu^* = (y_0 - x_0 + \sigma^2\psi_\eta|\nabla\psi_\eta|)/|y_0 - x_0|$. Notice that at (x_0, t_0) we have $\tilde{u}_\nu = (w_-)_\nu$.

Now, $\bar{v}_{\eta_0}$ has the following asymptotic development at the contact point (x_0, t_0) along $t = t_0 - \gamma\langle x - x_0, \nu\rangle^2$:

$$\bar{v}_{\eta_0}(x,t) \geq \bar{\alpha}_+\langle x - x_0, \nu\rangle^+ - \alpha_-\langle x - x_0, \nu\rangle^- + o(|x - x_0|)$$

where

$$\bar{\alpha}_+ \geq \alpha_+(1 - \sigma|\nabla\psi_{\eta_0}|) + c\varepsilon^{1-\gamma+\beta}$$

with α_+ and α_- as in Lemma 10.11.

Also, along the same paraboloid, near (x_0, t_0),

$$u(x,t) = \alpha_+^{(2)}\langle x - x_0, \nu\rangle^+ - \alpha_-^{(2)}\langle x - x_0, \nu\rangle^- + o(|x - x_0|)$$

and

$$\begin{aligned} G(\alpha_+^{(2)}, \alpha_-^{(2)}) \leq \frac{\beta_+^{(2)}}{\alpha_+^{(2)}} &\equiv \left(\frac{s_0 - t_0 + \sigma^2\psi_{\eta_0}(x_0,t_0)D_t\psi_{\eta_0}(x_0,t_0)}{|y_0 - x_0|}\right)|\nu^*|^{-1} \\ &\leq \left(\frac{s_0 - t_0}{|y_0 - x_0|} + C\sigma|D_t\psi_{\eta_0}|\right)(1 + C\sigma|\nabla\psi_{\eta_0}|) \\ &\leq (\alpha_+, \alpha_-) + C\delta\varepsilon^{1-\alpha} . \end{aligned}$$

Since G is Lipschitz, $\alpha < \gamma - \beta$ and $\frac{\partial G}{\partial \alpha_+} \geq c^* > 0$, we have

$$G(\alpha_+^{(2)}, \alpha_-^{(2)}) \leq G(\bar{\alpha}_+, \alpha_-) - C\varepsilon^{1-\gamma+\beta}(c - \delta) \leq G(\bar{\alpha}_+, \alpha_-) .$$

By the Hopf maximum principle, $\alpha_+^{(2)} > \bar{\alpha}_+$ and $\alpha_-^{(2)} < \alpha_-$, so we arrive to a contradiction.

Since $\lambda\sin\delta - c\varepsilon^\beta > \lambda\sin(\delta - \bar{c}\varepsilon^\beta)$, the conclusion follows easily. $\square$

Corollary 10.14. *Let u be as in Lemma 10.13 with $\varepsilon = \varepsilon_k = 1/16^k$ (k large), $\delta = \delta_k = c/k^{1+\alpha}$ (some $a > 0$), and $\lambda_k = 1 - \frac{1}{2}\sin\delta_k$. Then there exist α' ($< \alpha$) and β' ($< \beta$) such that u is ε_{k+1}-monotone for every direction $\tau \in \Gamma_x(\theta^x - c\varepsilon_{k+1}^{\beta'}, e_n) \cup \Gamma_t(\theta^t - c\varepsilon_{k+1}^{\beta'}, \nu)$, in the domain $\mathcal{C}_{R-\varepsilon_{k+1}^{\alpha'}, T-\varepsilon_{k+1}^{\alpha'}}$.*

Proof. It is enough to iterate Lemma 10.13 a number of times such that

$$\lambda_k^s \leq \frac{1}{16}$$

and to choose α' and β' so that

$$\sum_s c\varepsilon_k^\beta \lambda_k^{\beta s} \leq c\varepsilon_k^\beta \frac{1}{1 - \lambda_k^\beta} \leq c\varepsilon_{k+1}^{\beta'}$$

and

$$\sum_s c\varepsilon_k^\alpha \lambda_k^{\alpha s} \leq c\varepsilon_k^\alpha \frac{1}{1 - \lambda_k^\gamma} \leq c\varepsilon_{k+1}^{\alpha'} . \qquad \square$$

10.6. Propagation of cone enlargement to the free boundary

The enlarged cone of full monotonicity of Section 10.2 yields only an enlarged cone of ε-monotonicity on the free boundary. This is precisely what we prove in this section, using the continuous family of perturbations constructed in Lemma 9.15. For the reader's convenience we repeat here the statement of that lemma.

Lemma 9.15. *Let $0 < T_0 \le T$ and $C > 1$. There exist positive constants $\bar{C}, k$ and h_0, depending only on C and T_0, such that, for any h, $0 < h < h_0$, there is a family of C^2-continuous functions φ_η, $0 \le \eta \le 1$, defined in the closure of*

$$D = \left[B' \setminus \left\{\bar{B}'_{1/8}\Big(\frac{3}{4}e_n\Big) \cup \bar{B}'_{1/8}\Big(-\frac{3}{4}e_n\Big)\right\}\right] \times (-T, T)$$

such that

(1) $1 \le \varphi_\eta \le 1 + \eta h$ *in* $\bar{D}$,
(2) $\Delta\varphi_\eta - c_1 D_t\varphi_\eta - C\frac{|\nabla\varphi_\eta|^2}{\varphi_\eta} - c_2|\nabla\varphi| \ge 0$ *in* D,
(3) $\varphi_\eta \equiv 1$ *outside* $B'_{8/3} \times (-\frac{7}{8}T, T)$,
(4) $\varphi_\eta \ge 1 + \kappa\eta h$ *in* $B'_{1/2} \times (-\frac{T}{2}, \frac{T}{2})$,
(5) $D_t\varphi_\eta, |\nabla\varphi| \le \bar{C}\eta h$ *in* $\bar{D}$,
(6) $D_t\varphi_\eta \ge 0$ *in* D,

provided c_1 and c_2 are small enough.

The next lemma is perfectly analogous in both statement and proof to Lemma 10.11.

Lemma 10.15. *Let u_1 be a viscosity solution of our free boundary problem in $\mathcal{C}_2$ satisfying the hypotheses of the main theorem. For $0 < \varepsilon$, $\sigma < 1$ define*

$$V_\eta(p) = \sup_{q \in B_{\varepsilon\varphi_{\sigma\eta}}(p)} u_1(q) \tag{10.3}$$

where $\varphi_{\sigma\eta}$ are the functions constructed in the preceding lemma. Suppose the supremum in (10.3) occurs uniformly away from the top and the bottom points of the ball at a distance not smaller than $p > 0$. Then the following hold.

(1) *V_η is subcaloric in $\Omega^+(V_\eta)$ and $\Omega^-(V_\eta)$.*
(2) *$\partial\Omega^+(V_\eta)$ is uniformly Lipschitz with Lipschitz constant $L' \le \tan(\frac{\pi}{2} - \theta_0^x) + C\varepsilon|\nabla\varphi_{\sigma\eta}|$.*
(3) *If $(x_0, t_0) \in F(V_\eta)$ and $(y_0, s_0) \in F(u_1)$ with*

$$(y_0, s_0) \in \partial B_{\varepsilon\varphi_{\sigma\eta}(x_0,t_0)}(x_0, t_0)\ ,$$

then (x_0, t_0) *is a regular point from the right and if, near* (y_0, s_0) *along the paraboloid* $s = s_0 - \gamma\langle y - y_0, \nu\rangle^2$ $(\gamma > 0)$, u_1 *has the asymptotic expansion*

$$u_1(y, s) = \alpha_+^{(1)}\langle y - y_0, \nu\rangle^+ - \alpha_-^{(1)}\langle y - y_0, \nu\rangle^- + o(|y - y_0|)$$

where $\nu = \frac{y_0 - x_0}{|y_0 - x_0|}$, *then, near* (x_0, t_0), *along the paraboloid* $t = t_0 - \gamma\langle x - x_0, \nu\rangle^2$,

$$\begin{aligned} V_\eta(x, t_0) \geqq & \alpha_+^{(1)} \left\langle x - x_0, \nu + \frac{\varepsilon^2 \varphi_{\sigma\eta}(x_0, t_0)}{|y_0 - x_0|} \nabla\varphi_{\sigma\eta}(x_0, t_0) \right\rangle^+ \\ & - \alpha_-^{(1)} \left\langle x - x_0, \nu + \frac{\varepsilon^2 \varphi_{\sigma\eta}(x_0, t_0)}{|y_0 - x_0|} \nabla\varphi_{\sigma\eta}(x_0, t_0) \right\rangle^- \\ & + o(|x - x_0|) \end{aligned}$$

with $\frac{s_0 - t_0}{|y_0 - x_0|)} \leq G(\alpha_+, \alpha_-, \nu)$.

The next lemma is a propagation lemma similar to Lemma 9.16. The key point here is the use of the estimate of Section 10.3 in order to control the degeneracy of the solution at free boundary points (see Lemma 10.9).

Lemma 10.16. *Let* $u_1 \leq u_2$ *be two viscosity solutions of our free boundary problem in* $\mathcal{C}_2$ *satisfying the hypotheses of the main theorem, with* u_1 *as in Lemma 10.15. Assume further that*

$$v_\varepsilon(x, t) = \sup_{B_\varepsilon(x,t)} u_1 \leq u_2(x, t)$$

for $(x, t) \in \mathcal{C}_{1,T}$ *and for some* h *small*

$$u_2(x, t) - v_{(1+h\sigma)\varepsilon}(x, t) \geq C\sigma\varepsilon u_2\left(\frac{3}{4}\varepsilon, 0\right)$$

for every $(x, t) \in B'_{1/8}(\frac{3}{4}e_n) \times (-T, T) \subset \{u_2 > 0\}$. *Then if* $\varepsilon > 0$ *and* $h > 0$ *are small enough, there exists* $0 < c < 1$ *such that in* $\mathcal{C}_{1/2, T/2}$

$$v_{(1+ch\sigma)\varepsilon}(x, t) \leq u_2(x, t) .$$

Proof. We construct a continuous family of functions $\bar{V}_\eta \leq u_2$ for $0 \leq \eta \leq 1$ such that $\bar{V}_1 \geq v_{(1+ch\sigma)\varepsilon}$ in $\mathcal{C}_{1/2,T/2}$. Set

$$\bar{V}_\eta(x, t) = V_\eta(x, t) + C\sigma\varepsilon W(x, t)$$

where V_η are defined in Lemma 10.15 and W is a continuous function in

$$D = \left[B'_{9/10}(0) - B'_{1/8}\left(\frac{3}{4}e_n\right)\right] \times \left(-\frac{9}{10}T, \frac{9}{10}T\right)$$

defined as

$$\Delta W - a_1 D_t W = 0 \quad \text{in } D \cap \{V_\eta > 0\},$$

$$W = \begin{cases} \tilde{u}\left(\frac{3}{4}e_n, 0\right) & \text{on } \partial B'_{1/8}\left(\frac{3}{4}e_n\right) \times \left(-\frac{9}{10}T, \frac{9}{10}T\right), \\ 0 & \text{in } \left(D \cap \overline{\{V_\eta \le 0\}}\right) \\ & \cup \left(\partial_p D - \left\{\partial B'_{1/8}\left(\frac{3}{4}e_n\right) \times \left(-\frac{9}{10}T, \frac{9}{10}T\right)\right\}\right) \end{cases}$$

where $\tilde{u}$ is the caloric function in $C^+ = \Omega^+(V_\eta) \cap [\bar{B}'_{1/8}(\frac{3}{4}e_n) \times (-T, T)]$ with boundary values zero on $F(V_\eta)$ and equal to u_2 everywhere else on the parabolic boundary of C^+. Notice that $\tilde{u} \le u_2$ in C^+ and $\tilde{u}(\frac{3}{4}e_n) \ge cu_2(\frac{3}{4}e_n)$.

We want to show that the set $S = \{\eta \in [0,1] : \bar{V}_\eta \le u_2\}$ is open and closed. $S \neq \varnothing$ and closed follow by the maximum principle. To show it is open, assume that $\bar{V}_{\eta_0} \le u_2$ for some $\eta_0 \in [0,1]$; then it is enough to prove that $D \cap \{\bar{V}_{\eta_0} > 0\}$ is compactly contained in $D \cap \{u_2 > 0\}$. If not, there exists $(x_0, t_0) \in F(\bar{V}_{\eta_0}) \cap F(u_0) \cap D$, which is by construction a regular point. Since $F(V_\eta)$ is Lipschitz, by Corollary 13.8, $W \ge c\tilde{u}$ in $\{V_\eta > 0\}$ strictly away from the parabolic boundary of D.

By Lemma 13.15, at time level $t = t_0$, $w_- = \tilde{u}(1 - \tilde{u}^r)$, for some small $r > 0$, is superharmonic and therefore by Lemma 10.9, $(w_-)_\nu \ge \tilde{c} > 0$ at x_0, where $\nu = \frac{\nu^*}{|\nu^*|}$ with

$$\nu^* = (y_0 - x_0 + \varepsilon^2 \varphi_{\sigma\eta} \nabla_x \varphi_{\sigma\eta})/|y_0 - x_0| \, .$$

Hence at (x_0, t_0), $(u_2)_\nu = (w_-)_\nu \ge \tilde{c} > 0$.

Near (x_0, t_0), we have, along the paraboloid $t = t_0 - \gamma\langle x - x_0, \nu\rangle^2$, $\gamma > 0$, the asymptotic development

$$\bar{V}_{\eta_0}(x,t) \ge \bar{\alpha}_+ \langle x - x_0, \nu\rangle^+ - \alpha_-^{(1)}\langle x - x_0, \nu\rangle^- + o(|x - x_0|)$$

where

$$\bar{\alpha}_+ \ge \alpha_+^{(1)}(1 - C\sigma\varepsilon h) + C\sigma\varepsilon\alpha_+^{(2)}$$

and

$$u_2(x, t_0) = \alpha_+^{(2)}\langle x - x_0, \nu\rangle^+ - \alpha_-^{(2)}\langle x - x_0, \nu\rangle^- + o(|x - x_0|) \, .$$

Now,

$$G(\alpha_+^{(2)}, \alpha_-^{(2)}) \le \frac{\beta_+^{(2)}}{\alpha_+^{(2)}} \equiv \frac{s_0 - t_0 + \varepsilon^2 \varphi_{\sigma\eta}(x_0, t_0) D_t \varphi_{\sigma\eta_0}(x_0 t_0)}{|y_0 - x_0|} |\nu^*|^{-1}$$

$$\le \left(\frac{s_0 - t_0}{|y_0 - x_0|} + C\sigma\varepsilon h \right) (1 + C\sigma\varepsilon h)$$

$$\le G(\alpha_+^{(1)}, \alpha_-^{(1)}) + \bar{C}\sigma\varepsilon h \ .$$

Since G is Lipschitz and $\frac{\partial G}{\partial \alpha_+} \ge c^* > 0$, $\frac{\partial G}{\partial \alpha_-} \le -c^* < 0$, we have

$$G(\alpha_+^{(2)}, \alpha_-^{(2)}) \le G(\bar{\alpha}_+, \alpha_-^{(1)}) - C\sigma\varepsilon(c\tilde{c} - h) \le G(\bar{\alpha}_+, \alpha_-^{(1)}) \ .$$

On the other hand, the Hopf maximum principle gives $\alpha_+^{(2)} > \bar{\alpha}_+$ and $\alpha_-^{(2)} < \alpha_-^{(1)}$, so we arrive at a contradiction. □

Lemma 10.17 (Cone Basic Iteration). *Let u be a viscosity solution of our free boundary problem in $\mathcal{C}_{R,S}$ (with R and S less than 1 and R close to 1), ε-monotone along any direction in $\Gamma_x(\theta^x, e_n) \cup \Gamma_t(\theta^*, e_n)$. Suppose that $\delta = \frac{\pi}{2} - \theta^*$, $\delta \ll \mu = \frac{\pi}{2} - \theta^*$, and $\varepsilon \ll \delta$. Then there exist positive constants c_1, c_2, and C and unit vectors e and ν_1 where e is spatial and $\nu_1 \in \mathrm{Sp}\{e, e_t\}$, such that, in $B_{1/2}(0) \times (-\frac{C\delta}{2\mu}, \frac{C\delta}{2\mu})$, u is ε-monotone in $\Gamma_x(\theta_1^x, e_n) \cup \Gamma_t(\theta_1^t, \nu_1)$ with*

$$\frac{\pi}{2} - \theta_1^x = \delta_1 \le \delta - c_1 \frac{\delta^2}{\mu}$$

and

$$\frac{\pi}{2} - \theta_1^t = \mu_1 \le \mu - c_2 \delta \ .$$

Proof. In order to enlarge the space cone, take any vector

$$\tau \in \Gamma_x(\theta^x - \delta, e_n)$$

with $|\tau| \ll \delta$ and let $\varepsilon = |\tau| \sin \delta$. Define

$$u_1(x,t) = u(x - \tau, t)$$

and argue as in Lemma 9.19, so that the hypotheses of the preceding lemma are fulfilled (with $u_2 = u$). Hence the result. To enlarge the time cone, take $\tau \in \Gamma_t(\theta^t - \delta, \nu_1)$ with $|\tau| \ll \delta$ and $\varepsilon = |\tau| \sin \delta$ and proceed likewise. □

10.7. Proof of the main theorem

We are now ready to finish the proof of Theorem 10.2. We shall apply Corollary 10.14 and Lemma 10.16 inductively in the following manner. For simplicity, let

$$\delta_k \sim \frac{1}{(k+\bar{k})^{4/3}}\,, \quad \bar{k} \gg 1\,, \quad \mu_k \sim \frac{1}{k^{1/3}}\,, \quad \text{and} \quad \varepsilon_k = \frac{1}{(32)^{k+\bar{k}}}\,.$$

At the kth step we have, in $B'_{2^{-k}}(0) \times (-\frac{C\delta_k}{2^k\mu_k}, \frac{C\delta_k}{2^k\mu_k})$, u ε_k-monotone along every $\tau \in \Gamma_x(\theta^x_k, e^{(k)}) \cup \Gamma_t(\theta^t_k, \nu_k)$, where $\theta^x_k = \frac{\pi}{2} - \delta_k$, $\theta^t_k = \frac{\pi}{2} - \mu_k$, $e^{(k)}$ a spatial unit vector, and $\nu_k \in \mathrm{Sp}\{e^{(k)}, e_t\}$.

Set

$$u_k(x,t) = 2^k u(2^{-k}x, 2^{-k}t)$$

whose domain of definition is $B'_1 \times (-\frac{C\delta_k}{\mu_k}, \frac{C\delta_k}{\mu_k})$ and which is $\bar{\varepsilon}_k$-monotone ($\bar{\varepsilon}_k = 2^k\varepsilon_k$) in $\Gamma_x(\theta^x_k, e^{(k)}) \cup \Gamma_t(\theta^t_k, \nu_k)$. Notice that by Corollary 10.1 u_k is fully monotone away from the free boundary at a distance ε_k^γ, with $\gamma = \frac{3}{8}$. Apply Corollary 10.14 to u_k with $R = 1$ and $T = \frac{C\delta_k}{\mu_k}$ (with $\alpha = \frac{1}{8}$ and $\beta = \frac{1}{6}$, for instance) in order to have in

$$B'_{1-\bar{\varepsilon}^{\alpha'}_{k+1}} \times \left(-\frac{C\delta_k}{\mu_k} + \bar{\varepsilon}^{\alpha'}_{k+1}, \frac{C\delta_k}{\mu_k} - \bar{\varepsilon}^{\alpha'}_{k+1}\right)$$

u_k $\bar{\varepsilon}_{k+1}$-monotone along any direction in

$$\Gamma_x(\theta^x_k - c\bar{\varepsilon}^{\beta'}_{k+1}, e^{(k)}) \cup \Gamma_t(\theta^t_k - c\bar{\varepsilon}^{\beta'}_{k+1}, \nu_k)\,.$$

Now, apply Lemma 10.16 to u_k with $R = 1 = \bar{\varepsilon}^{\alpha'}_{k+1}$ and $S = \frac{C\delta_k}{\mu_k} - \bar{\varepsilon}^{\alpha'}_{k+1}$ to have in

$$B'_{1/2} \times \left(-\frac{C\delta_k}{2\mu_k}, \frac{C\delta_k}{2\mu_k}\right)$$

u_k $\bar{\varepsilon}_{k+1}$-monotone in $\Gamma_x(\theta^x_{k+1}, e^{(k+1)}) \cup \Gamma_t(\theta^t_{k+1}, \nu_{k+1})$ where

$$\frac{\pi}{2} - \theta^x_{k+1} = \delta_{k+1} \le \delta_k - c_1\frac{\delta_k^2}{\mu_k} - c\bar{\varepsilon}^{\beta'}_{k+1}\,,$$

$$\frac{\pi}{2} - \theta^t_{k+1} = \mu_{k+1} \le \mu_k - c_2\delta_k - c\bar{\varepsilon}^{\beta'}_{k+1}\,.$$

Taking $\beta' < \frac{1}{4}$, we see that the above choices of δ_k, μ_k, and ε_k are compatible with these inequalities. Hence the proof of the first assertion of the theorem follows easily.

To prove the second assertion, we observe that at each time level $t_0, \Omega^\pm \cap \{t = t_0\}$ is a Lyapunov-Dini domain. By the results in [W], $\nabla_x u^\pm$ are continuous up to the free boundary, at each level of time. This, in turn, implies that u_t is bounded. We can now argue as in the end of the proof of Theorem 9.1 to complete the proof.

10.8. Finite time regularization

In this section we show two applications of our theory regarding "small perturbations" of nice solutions.

The interesting fact is that they are small L^∞ perturbations, a bound that is easy to obtain just from the uniform continuity of the temperature, a well-known property of the Stefan problem. In fact, our first application is to solutions of the classical Stefan problem (here $a_1 = a_2$) defined in the infinite cylinder $\bar{B}_1 \times [0, +\infty)$ and having a nice asymptotic behavior as $t \to +\infty$. The main result is that no matter how "bad" the initial data is, the solution regularizes after a finite time. It is well known that finite waiting time is a common feature of solutions to degenerate diffusions.

Note that harmonic functions are stationary solutions of the Stefan problem:

Theorem 10.18. *Suppose u is a solution of the two-phase Stefan problem in $\bar{B}'_1 \times [0, +\infty)$ converging for $t \to +\infty$ to a harmonic function $u_\infty = u_\infty(x)$, $x \in B_1$, uniformly in any compact subset of B_1. Suppose that at $x_0 \in F(u_\infty)$, $|\nabla u_\infty(x_0)| \neq 0$. Then there exist $T^* > 0$ and a neighborhood V of x_0 such that in $V \times [T^*, +\infty)$, $F(u)$ is a C^1-graph and u is a classical solution.*

Remark. T^* depends on a bound from below of $|\nabla u_\infty|$ in V.

Proof. In a neighborhood V of x_0, u_∞ is monotone along the directions in a cone $\Gamma(\theta^\infty, \nu)$. Therefore, choosing V such that $|\nabla u_\infty(x)| \geq c > 0$ in V, if T^* is large enough, $u(x,t)$, $t \geq T^*$, is ε-monotone along the directions of $\Gamma(\theta^\infty, \nu)$ and the directions of a space-time cone $\Gamma_t(\theta^t, \nu)$ (with the same ν and θ^t large). The conclusions follow now from Theorem 10.2. □

The second application pertains to being close to a traveling wave.

Traveling waves for the Stefan problem are of the form

$$u_0(x,t) = (A+1)(e^{t-x_n} - 1)^+ - A(e^{t-x_n} - 1)^- .$$

Lemma 10.19. *Let u be a solution of the two-phase Stefan problem in $\mathbb{R}^N \times (0, \infty)$ with initial values*

$$u(x,0) = (A+1)(e^{-x_n} - 1 + \varphi(x))^+ - A(e^{-x_n} - 1 + \varphi(x))^-$$

where A is a positive constant and φ is a continuous function, compactly supported in $\mathbb{R}^n$. Then as $t \to \infty$, uniformly in x,

$$u(x,t) - (A+1)(e^{t-x_n} - 1)^+ - A(e^{t-x_n} - 1)^- \to 0 .$$

Proof. If $M > 0$ is chosen large enough, we claim that the function

$$v(x,t) = (A+1)\left(e^{t-x_n} - 1 + Mf\left(\frac{x_n}{\sqrt{t+1}}\right)\right)^+$$
$$- A\left(e^{t-x_n} - 1 + Mf\left(\frac{x_n}{\sqrt{t+1}}\right)\right)^-,$$

where $f(s) = \frac{\pi}{2} - \arctan(s)$, is a supersolution to the two-phase Stefan problem. It is enough to check that

$$v_t^+ \geq (v_{x_n}^+)^2 - v_{x_n}^+ v_{x_n}^- \tag{10.4}$$

along the free boundary of v, i.e., at the zero level set of v. Now

$$v_t^+(x,t) = (A+1)\left(e^{t-x_n} + Mf'\left(\frac{x_n}{\sqrt{t+1}}\right)\left(-\frac{1}{2}\frac{x_n}{(t+1)^{3/2}}\right)\right),$$

$$v_{x_n}^+ = (A+1)\left(-e^{t-x_n} + Mf'\left(\frac{x_n}{\sqrt{t+1}}\right)\frac{1}{\sqrt{t+1}}\right),$$

$$v_{x_n}^- = A\left(-e^{t-x_n} + Mf'\left(\frac{x_n}{\sqrt{t+1}}\right)\frac{1}{\sqrt{t+1}}\right).$$

Substituting in (10.4), we have to check that

$$e^{t-x_n} + Mf'\left(\frac{x_n}{\sqrt{t+1}}\right)\left(-\frac{1}{2}\frac{x_n}{(t+1)^{3/2}}\right)$$
$$\geq \left(-e^{t-x_n} + Mf'\left(\frac{x_n}{\sqrt{t+1}}\right)\frac{1}{\sqrt{t+1}}\right)^2$$

on $v(x,t) = 0$, or

$$(1 - Mf(s)) + Mf'(s)\left(-\frac{1}{2}\frac{s}{(t+1)}\right)$$
$$\geq (1 - Mf(s))^2 - 2(1 - Mf(s))Mf'(s)\frac{1}{\sqrt{t+1}} + M^2(f'(s))^2\frac{1}{t+1}$$

where $s = \frac{x_n}{\sqrt{t+1}}$. Equivalently,

$$f(s)(1 - Mf(s)) + f'(s)\left(-\frac{s}{2(t+1)}\right)$$
$$\geq -2(1 - Mf(s))f'(s)\frac{1}{\sqrt{t+1}} + M(f'(s))^2\frac{1}{t+1}.$$

Now, since $1 - Mf(s) > 0$ on $v = 0$, we have $s \geq M - \frac{c}{M}$ for some c. Since also $f'(s) = \frac{-1}{1+s^2}$, the second term on the left is larger than the second on the right. Therefore the inequality is true if

$$f(x) \geq -2f'(s)\frac{1}{\sqrt{t+1}}$$

or

$$\arctan\frac{1}{s} \geq \frac{2}{1+s^2}\frac{1}{\sqrt{t+1}} .$$

Since $s \geq M - \frac{c}{M}$, the inequality is true if M is large enough.

In an analogous way, the function

$$\begin{aligned} w(x,t) = (A+1)\left(e^{t-x_n} - 1 - Mf\left(\frac{x_n}{\sqrt{t+1}}\right)\right)^+ \\ - A\left(e^{t-x_n} - 1 - Mf\left(\frac{x_n}{\sqrt{t+1}}\right)\right)^- \end{aligned}$$

is a subsolution to the two-phase Stefan problem in $\mathbb{R}^n \times (0, +\infty)$, if M is large enough.

Now, given φ, we can always find M large such that v and w are a supersolution and subsolution, respectively, and

$$w(x,0) \leq u(x,0) \leq v(x,0) .$$

By the maximum principle, this implies

$$w(x,t) \leq u(x,t) \leq v(x,t)$$

for every $t \geq 0$. Since both v and w converge to the traveling wave u_0 for $t \to +\infty$, uniformly in x, the proof is complete. □

The above lemma shows that a compactly supported perturbation of an initial traveling wave has a two-plane asymptotic free boundary as $t \to +\infty$, uniformly in x.

Therefore, with the same proof as Theorem 10.18, we have

Theorem 10.20. *Let $u(x,0)$ be a compactly supported perturbation of a traveling wave initial data, i.e., let*

$$u_0(x,0) = (A+1)(e^{-x_n} - 1)^+ - A(e^{-x_n} - 1)^- \qquad (A > 0)$$

and let φ_0 be a compactly supported continuous function in $\mathbb{R}^n$ such that

$$u_0(x,0) - \varphi_0(x) \leq u(x,0) \leq u_0(x,0) + \varphi_0(x) .$$

Then after a finite time $T^* = T^*(u_0, \varphi_0)$, the free boundary of the solution $u(x,t)$ to the Stefan problem with initial data $u(x,0)$ is a smooth graph

$$x_n = g(x', t) .$$

Part 3

Complementary Chapters: Main Tools

Chapter 11

Boundary Behavior of Harmonic Functions

11.1. Harmonic functions in Lipschitz domains

In this section we study the properties of harmonic functions in Lipschitz domains, since it is a theory we encounter often in proving regularity results for both elliptic and parabolic problems.

The importance of rescaling is by now very well recognized in a mathematical problem, and Lipschitz domains are precisely a class of rescaling invariant domains.

A bounded domain $\Omega \in \mathbb{R}^n$ is a Lipschitz domain if it is locally given by the domain above the graph of a Lipschitz function. This means that for any $x \in \partial\Omega$ there is a ball B centered at x such that, in a suitable coordinate system $(x', x_n) = (x_1, \dots, x_{n-1}, x_n)$ with origin at x,

$$\Omega \cap B = \{(x', x_n) : x_n > f(x')\} \cap B$$

where f is a Lipschitz function with Lipschitz constant less than or equal to L (independent of x) and $f(0) = 0$.

Such domains satisfy both an interior and exterior cone condition and therefore they are regular for the Dirichlet problem.

If $x \in \partial\Omega$, a nontangential region at x is a truncated cone of the type

$$\Gamma(x) = \{y \in \Omega;\ \operatorname{dist}(y, \partial\Omega) \geq \gamma|x - y|\} \cap B_\rho(x)$$

with some positive γ and ρ.

We say that *a property holds nontangentially* near $x \in \partial\Omega$ if it holds in every nontangential region at x, with $\rho \leq \rho_0$, small.

The main points of the theory we are going to develop are the following. Suppose our Lipschitz domain is

$$\Omega = \{|x'| < 1,\ f(x') < x_n < 4L\} \qquad (L > 1) \tag{11.1}$$

with $f(0) = 0$. Let u, v be two positive harmonic functions in Ω, continuously vanishing on $\{x_n = f(x')\}$. We will prove the following.

(a) $u(\frac{1}{2}e_n)$ controls u in $B_{1/2} \cap \Omega$, i.e.,

$$\sup_{B_{1/2}\cap\Omega} u \le c(n, L) u\left(\frac{1}{2}e_n\right).$$

(b) The ratio $u(\frac{1}{2}e_n)/v(\frac{1}{2}e_n)$ controls u/v in $B_{1/2} \cap \Omega$, i.e.,

$$\sup_{B_{1/2}\cap\Omega} \frac{u}{v} \le c(n, L)\frac{u(\frac{1}{2}e_n)}{v(\frac{1}{2}e_n)}.$$

This implies that u/v is Hölder continuous up to the boundary $\partial\Omega$, in $B_{1/2}$.

(c) u and u_{x_n} are comparable in the natural way, i.e., there exists $\delta = \delta(n, L) > 0$ such that $u_{x_n} > 0$ in $B_\delta \cap \Omega$ and

$$u(x) \sim u_{x_n}(x) \cdot \operatorname{dist}(x, \partial\Omega).$$

The points (a) and (b) are true not just for harmonic functions in Lipschitz domains but also for solutions of uniformly elliptic equations with bounded measurable coefficients. Indeed, this class is invariant under dilations and bilipschitz transformations.

More precisely, let u be a solution of

$$\mathcal{L}u = \operatorname{div}(A(x)\nabla u) = 0$$

with A bounded, measurable and uniformly elliptic:

$$\lambda^{-1} I \le A(x) \le \lambda I, \tag{11.2}$$

in a Lipschitz domain D.

Let $y = T(x)$ be a bilipschitz transformation from Ω onto the Lipschitz domain Ω'. Then $v(y) = u(T^{-1}(y))$ is also a solution of an equation in the same class. In fact, for any test function $\varphi \in C_0^\infty(\Omega)$, one has

$$\int_\Omega (\nabla\varphi)^T A \nabla u \, dx = 0$$

which, after the change of variables $y = T(x)$, becomes

$$\int_{\Omega'} (\nabla\psi)^T (Dy)^T A (Dy) \nabla v |\det(Dy)|^{-1} \, dy$$

where $\psi = \varphi(T^{-1}(y))$ and Dy is the Jacobian matrix of the transformation.

The new matrix

$$M = |\det Dy|^{-1}(Dy)^T A(Dy) \qquad (x = T^{-1}(y))$$

is again uniformly elliptic.

Therefore, the study of solutions of uniformly elliptic equations with bounded measurable coefficients in our basic Lipschitz domain (11.1) can be reduced to the study in the half ball

$$B_1^+ = B_1(0) \cap \{x_n > 0\}$$

by the local transformation

$$y' = x', \qquad y_n = x_n - f(x')$$

and a dilation.

Remark 11.1. Let u be a solution of $\mathcal{L}u = 0$ in B_1^+, continuously vanishing on $\{x_n = 0\}$. If we extend u to $\{x_n < 0\}$ in an odd fashion, i.e.,

$$u(x_1, \dots, x_n) = -u(x_1, \dots, -x_n)$$

when $x_n < 0$, then the extended u is a solution in B_1 (therefore across $x_n = 0$) of the elliptic equation

$$\operatorname{div}(A(x)\nabla u) = 0$$

where the elements a_{ij} of A are defined for $x_n < 0$ as follows:

$$a_{in}(x_1, \dots, x_n) = -a_{in}(x_1, \dots, -x_n), \qquad i \neq n,$$
$$a_{nj}(x_1, \dots, x_n) = -a_{nj}(x_1, \dots, -x_n), \qquad j \neq n,$$

and

$$a_{ij}(x_1, \dots, x_n) = a_{ij}(x_1, \dots, -x_n)$$

in the other cases.

This remark is very useful since it allows one to use interior results, such as the following classical ones.

Theorem 11.2 (DeGiorgi oscillation lemma). *Let v be a subsolution of $\mathcal{L}v = 0$ in B_1 satisfying*

(a) $v \leq 1$,

(b) $|\{v \leq 0\}| = a > 0$.

Then

$$\sup_{B_{1/2}} v \leq \mu(a) < 1 \ .$$

Theorem 11.3 (DeGiorgi-Nash-Moser Interior Harnack inequality). *Let v be a nonnegative solution in B_1. Then for $r < 1$*

$$\sup_{B_r} u \leq c(1-r)^{-p} \inf_{B_r} u$$

with c, p depending only on n and λ. (For r close to one, we may choose the constant c equal to one by making p large.)

Theorem 11.4 (Littman, Stampacchia, Weinberger—Behavior of the fundamental solution). *The fundamental solution of $\mathcal{L}$ behaves like that of the Laplacian, more precisely: Let $B_1 \subset \Omega$, and let G satisfy*

(a) $\mathcal{L}(G) = -\delta_0$ *(Dirac's),*

(b) $G|_{\partial G} = 0$.

Then on $B_{1/2}$

$$\frac{C_1}{r^{n-2}} \leq G \leq \frac{C_2}{r^{n-2}} \qquad (r = |x|)$$

with C_1, C_2 depending only on n and λ.

The function G in Theorem 11.4 is the $\mathcal{L}$-Green function for Ω with pole at $y = 0$. We will denote by $G_{\mathcal{L}}(x, y)$ the $\mathcal{L}$-Green function with pole y.

Notice that with respect to the variable y, $G_{\mathcal{L}}(x, \cdot)$ satisfies the adjoint equation

$$\mathcal{L}^* G_{\mathcal{L}}(x, \cdot) = \operatorname{div}(A^T(y) \nabla_y G_{\mathcal{L}}(x, \cdot)) = -\delta_x \ .$$

Clearly, if $A = A^T$, then $G_{\mathcal{L}}(x, y) = G_{\mathcal{L}}(y, x) = G_{\mathcal{L}^*}(x, y)$. Here is a proof of Theorem 11.4 in this case. Notice that we can assume $A \in C^\infty$, so that $G_{\mathcal{L}} \in C^\infty$ off the diagonal $x = y$.

On any sphere $|x| = h$, $h < \frac{1}{2}$, all values of $G_{\mathcal{L}}$ are comparable from Harnack's inequality, since in the ring

$$R_h = \left\{ \frac{1}{2} h < |x| < 2h \right\}$$

$G_{\mathcal{L}}$ is a nonnegative solution of $\mathcal{L}v = 0$.

We note the following.

(a) The theorem is true for $\mathcal{L} = \Delta$.

(b) $\mathcal{L}$-capacity and Δ-capacity are comparable. Precisely, take a closed set $E \subset B_1$ and minimize the Dirichlet integral $\int_{B_1} |Dv|^2$ among all functions in $H_0^1(B_1)$, such that $u \equiv 1$ on E (in the H^1 sense). The minimizer u_0 is harmonic in $B_1 \setminus E$ and equal to 1 on E. Moreover

$$\int_{B_1} |Du_0|^2 = \operatorname{cap}_\Delta(E) \ .$$

On the other hand, if we minimize

$$\int_{B_1} (\nabla v)^T A(\nabla v),$$

the minimizer u_1 is a solution of $\mathcal{L}v = 0$ in $B_1 \setminus E$, is equal to 1 on E and

$$\int_{B_1} (\nabla u_1)^T A \nabla u_1 = \operatorname{cap}_{\mathcal{L}}(E) \ .$$

Clearly

$$\lambda^{-2} \operatorname{cap}_{\Delta}(E) \leq \operatorname{cap}_{\mathcal{L}}(E) \leq \lambda^2 \operatorname{cap}_{\Delta}(E) \ . \tag{11.3}$$

If $E_h = \{G_{\mathcal{L}}(x,0) > h\}$, then

$$\operatorname{cap}_{\mathcal{L}}(E_h) = \frac{1}{h}$$

independently of the elliptic operator $\mathcal{L}$. In fact, the minimizer is $g_h = \frac{1}{h} \min(G_{\mathcal{L}}, h)$ so that $(G_{\mathcal{L}} = G_{\mathcal{L}}(\cdot, 0))$

$$1 = -\int_{B_1} \mathcal{L}G_{\mathcal{L}} \cdot g_h = \int_{B_1} (\nabla g_h)^T A \nabla G_{\mathcal{L}} = h \operatorname{cap}_{\mathcal{L}}(E_\lambda) \ .$$

(c) Let $\Sigma_h = \{|x| < h\}$, $h \leq \frac{1}{2}$. Set

$$a = \max_{|x|=h} G_{\mathcal{L}}(x,0) \ , \quad b = \min_{|x|=h} G_{\mathcal{L}}(x,0) \ .$$

Then from (b),

$$\frac{1}{a} = \operatorname{cap}_{\mathcal{L}}(E_a) \leq \operatorname{cap}_{\mathcal{L}}(\Sigma_h) \leq \operatorname{cap}_{\mathcal{L}}(E_b) = \frac{1}{b} \ .$$

From Harnack's inequality $a \sim b$, and therefore

$$\operatorname{cap}_{\mathcal{L}}(\Sigma_h) \sim G_{\mathcal{L}}(x,0)^{-1} \ . \tag{11.4}$$

From (11.4) and (11.3) we easily conclude the proof. □

11.2. Boundary Harnack principles

The first two main results we are going to prove are expressed in normalized form in the following two theorems ([AthC], [CFMS], [JK]).

Theorem 11.5 (Boundary Harnack inequality or Carleson estimate). *Let u be a positive solution to $\mathcal{L}u = 0$ in B_1^+, continuously vanishing on $\{x_n = 0\}$.*

Normalize u so that

$$u\left(\frac{1}{2}e_n\right) = 1 \ .$$

Then in $B_{1/2}^+$,

$$u \leq M$$

with $M = M(n, \lambda)$.

Theorem 11.6 (Comparison principle). *Let u, v be a positive solution to $\mathcal{L}w = 0$ in B_1^+, continuously vanishing on $\{x_n = 0\}$, with*

$$u\left(\frac{1}{2}e_n\right) = v\left(\frac{1}{2}e_n\right) = 1\ .$$

Then in $\bar{B}_{1/2}^+$,

$$\frac{v}{u} \quad \textit{is of class} \quad C^\alpha$$

and

$$\left\|\frac{v}{u}\right\|_{\mathcal{L}^\infty(B_{1/2}^+)}, \quad \left\|\frac{v}{u}\right\|_{C^\alpha(\bar{B}_{1/2}^+)} \le C(n, \lambda)\ .$$

With the results of the previous section we are ready for the proof of Theorem 11.5.

Proof of Theorem 11.5. We start with three remarks.

(a) If $Y_0 \in \{x_n = 0\}$, then $\sup_{B_r(Y_0)} u$ decreases polynomially, i.e., for $r < R$

$$\sup_{B_r(Y_0)} u \le \left(\frac{r}{R}\right)^\alpha \sup_{B_R(Y_0)} u\ .$$

Indeed, when extended by $u \equiv 0$ for $x_n < 0$, u is a subsolution of $\mathcal{L}u = 0$ and

$$|\{u = 0\} \cap B_r(Y_0)| = \frac{1}{2}|B_r(Y_0)|.$$

From the DeGiorgi oscillation lemma, with $\mu = \mu(1/2)$, we get

$$\sup_{B_{r/2}(Y_0)} u \le \mu \sup_{B_r(Y_0)} u\ .$$

(b) From the interior Harnack inequality

$$\sup_{B_{3/4} \cap \{x_n \ge s\}} u \le s^{-p} u\left(\frac{1}{2}e_n\right) = s^{-p}\ .$$

(c) Since u takes continuously the value zero at $\{x_n = 0\}$, the sup of u in $B_{1/2}^+$ is attained, i.e.,

$$\sup_{B_{1/2}^+} u = u(X_0) = M\ .$$

We will now show that if $M \ge M_0$ large, we can construct a sequence of points X_k, all contained in $B_{3/4}^+$, $X_k \to \{x_n = 0\}$, and such that $u(X_k)$ goes to $+\infty$.

Construction. We will denote by Y_k the projection of X_k onto the $\{x_n = 0\}$ axis.

From the interior Harnack inequality,

$$M = u(X_0) \le |X_0 - Y_0|^{-p} .$$

Thus, with $\varepsilon = 1/p$

$$d_0 = |X_0 - Y_0| \le M^{-\varepsilon},$$

that is, X_0 is very close to the $\{x_n = 0\}$-plane.

Now we use the oscillation lemma backwards: since

$$\sup_{B_{d_0}(Y_0)} u \ge u(X_0) \ge M ,$$

this implies that

$$\sup_{B_{2d_0}(Y_0)} u = u(X_1) \ge TM$$

(for $T = \frac{1}{\lambda(1/2)}$, a universal constant bigger than one).

Again, by Harnack's inequality, as with d_0, we obtain that

$$d_1 = |X_1 - Y_1| \le (TM)^{-\varepsilon}$$

and, by using the oscillation lemma backwards again, as with $u(x_1)$, we obtain that

$$u(X_2) = \sup_{B_{2d_1}} u \ge Tu(X_1) \ge T^2 M.$$

Once more, by Harnack's inequality,

$$d_2 = |X_2 - Y_2| \le (T^2 M)^{-\varepsilon} .$$

We repeat the process inductively, and we get a sequence of points X_k satisfying

(a) $u(X_k) \ge T^k M$,

(b) $|X_k - Y_k| \le (T^k M)^{-\varepsilon}$,

(c) $|X_k - X_{k-1}| \le 4(T^{k-1} M)^{-\varepsilon}$.

All we have to make sure of is that in this construction we always stayed inside, say $B_{9/16}$. But T is universal, ε is universal and we can choose M as large as we please, so we can make $\sum |X_k - X_{k-1}| \le 1/16$, and get, for $M \ge M_0$, a contradiction. □

The proof of Theorem 11.6 is divided into two main steps.

Step 1. Show that $\frac{v}{u}$ remains bounded in $B^+_{1/2}$ up to $\{x_n = 0\}$.

Step 2. Rescaling and iteration of Step 1.

Proof of Step 1. We want to show that $\frac{v}{u}$ remains bounded in $B^+_{1/2}$.

Since $u(\frac{1}{2}e_n) = v(\frac{1}{2}e_n) = 1$, we have from Theorem 11.5 that

$$v \le M \text{ in } B^+_{2/3}$$

and from the interior Harnack inequality that (M large)

$$u \ge \frac{1}{M} > 0 \text{ in } B^+_{2/3} \cap \left\{x_n \ge \frac{1}{8}\right\} .$$

Let v_1 and v_2 be such that $\mathcal{L}v_1 = \mathcal{L}v_2 = 0$ in $B^+_{2/3}$ with

$$v_1 = 1 \text{ on } \partial B^+_{2/3} \cap \{x_n > 0\} , \quad v_1 = 0 \text{ on } \{x_n = 0\},$$

$$v_2 = 1 \text{ on } \partial B^+_{2/3} \cap \left\{x_n > \frac{1}{8}\right\} , \quad v_2 = 0 \text{ on the rest of } \partial B^+_{2/3} .$$

Then by the maximum principle, in $B^+_{2/3}$ we have

$$v \le M v_1 \quad \text{and} \quad u \ge \frac{1}{M} v_2 .$$

Step 1 follows if we can prove that $v_1 \le C v_2$ in $B^+_{1/2}$.

This is a consequence of the following lemma, where we set

$$Q_{2r} = Q_{2r}(re_n) = \{0 < x_n < 2r, \ |x_j| < r, \ j = 1, \dots, n-1\} .$$

Lemma 11.7. *Let F_1, F_2 be two faces of Q_2, different from $\{x_n = 0\}$. Let v_i $(i = 1, 2)$ be the function satisfying*

(a) $\mathcal{L}v_i = 0$ *in* Q_2,

(b) $v_i|_{\partial Q_2} = \chi_{F_i}$.

Then in $Q_{1/2}$,

$$v_1 \le C v_2 .$$

The function v_i is called the $\mathcal{L}$-harmonic measure of F_i in $Q_2 = Q_2(e_n)$ and the lemma expresses a *doubling property of $\mathcal{L}$-harmonic measure.* It states that two adjacent "balls" along the boundary of a Lipschitz domain have comparable $\mathcal{L}$-harmonic measure. $\mathcal{L}$-harmonic measure may be absolutely singular with respect to Lebesgue measure so this is a nontrivial result. When $\mathcal{L} = \Delta$ though, harmonic and surface measures are mutually absolutely continuous. We will come back to these questions in the next section.

Proof. From the oscillation lemma, by extending $(1 - v_i)$ identically equal to zero across F_i, we get that $(1 - v_i) \le \mu < 1$, near F_i, say on the cube Q_{F_i} of sides one with one face lying on F_i (see Figure 11.1). Thus $v_i(e_n) \ge (1 - \mu) > 0$ and v_i is strictly positive inside Q_2, say in $\tilde{Q}_1(e_n)$, the cube of sides one centered at e_n.

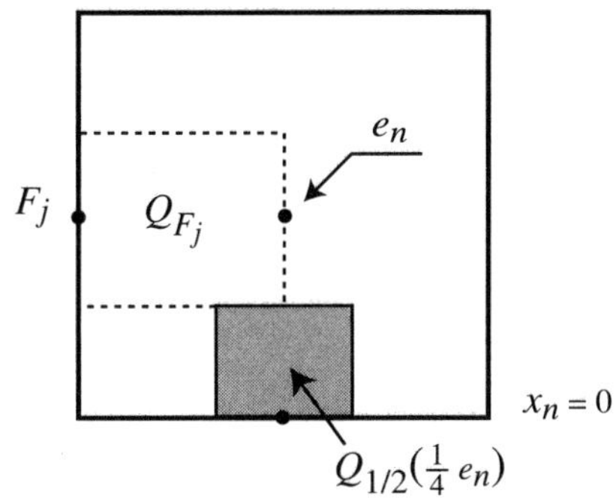

Figure 11.1

Let $G(X, Y)$ denote the $\mathcal{L}$-Green function in the cube Q_2. From Theorem 11.4, $G(X, e_n)$ is bounded for X on the boundary of $\tilde{Q}_1(e_n)$, vanishes on ∂Q_2 and hence

$$G(X, e_n) \leq c(\mu) v_2(X) \tag{11.5}$$

in $Q_2(e_n) \setminus \tilde{Q}_1(e_n)$.

We now show that for X in $Q_{1/2} = Q_{1/2}(\frac{1}{4}e_n)$, it also holds that $v_1(X) \leq CG(X, e_n)$. For that, we "freeze" X in $Q_{1/2}$ and recall that $\mathcal{L}_Y G(X, Y) = 0$, i.e., G is a solution in Y for X fixed, as long as $X \neq Y$, in particular for $Y \notin Q_{3/4}(\frac{3}{8}e_n)$.

Therefore, Theorem 11.5 applies and with X always frozen in $Q_{1/2}$,

$$G(X, Y) \leq CG(X, e_n)$$

for, say, $Y \notin Q_{3/4}(\frac{3}{8}e_n)$.

Since G vanishes in ∂Q_2, the standard energy estimate says that

$$\begin{aligned}\int_{Q_2(e_n) \setminus Q_1(\frac{1}{2}e_n)} & |\nabla_Y G(X, Y)|^2 \, dy \\ & \leq C \int_{Q_2(e_n) \setminus Q_{3/4}(\frac{3}{8}e_n)} G^2(x, y) \, dy \leq CG^2(X, e_n) \ .\end{aligned}$$

We now take a C^∞-function η, vanishing in a $\frac{1}{4}$ neighborhood of F_1, and $\eta \equiv 1$ on $Q_1(\frac{1}{2}e_n)$ and represent $v_1(X)$ for X in $Q_{1/4}(\frac{1}{8}e_n)$ by the formulas (no boundary terms left)

$$\int_{Q_2} (\eta v_1)(Y) L_Y G(X, Y) \, dy + \int_{Q_2} \nabla^T(\eta v_1) A \nabla_Y G(X, Y) \, dy = 0$$

and

$$\int_{Q_2} \eta(Y) G(X, Y) L v_1(Y) \, dy + \int_{Q_2} \nabla_Y^T(\eta(G) A \nabla v_1 \, dy = 0 \ .$$

After subtracting, this gives us

$$v_1(X) = \int_{Q_2} \nabla^T \eta A[v_1 \nabla_Y G - G \nabla v_1]\, dy\ .$$

But on the support of $\nabla\eta$,

$$\|v_1\|_{H^1} \le C \quad \text{and} \quad \|G\|_{H^1} \le CG(X, e_n)\ . \tag{11.6}$$

So $v_1(X) \le CG(X, e_n)$.

(Note that $\int(\nabla\eta v_1)^2$ is bounded from the standard energy inequality since η vanishes near F_1.)

This completes the proof of Lemma 11.7 and therefore of Step 1 in the proof of Theorem 11.6. □

Proof of Step 2. The proof of Step 2 consists of showing a C^α estimate by iteration, in the following way.

Lemma 11.8. *There are constants a_k, b_k, $\frac{1}{M} \le a_k \le b_k \le M$, and a constant $\Lambda < 1$, such that on $B^+_{2^{-k}}$,*

$$a_k u \le v \le b_k u \quad \textit{and} \quad (b_k - a_k) \le \Lambda(b_{k-1} - a_{k-1})\ .$$

Proof. By induction: renormalize $B^+_{2^{-k}}$ to B^+_1 by the transformation $\bar{u}(X) = u(2^{-k}X)$, define the positive functions

$$w_1(X) = \frac{(\bar{v} - a_k\bar{u})(X)}{b_k - a_k},$$
$$w_2(X) = \frac{(b_k u - v)(X)}{b_k - a_k}$$

and look at the positive numbers $w_1(\frac{1}{2}e_n)$, $w_2(\frac{1}{2}e_n)$. One of them is bigger than $\frac{1}{2}\bar{u}(\frac{1}{2}e_n)$ since $w_1(\frac{1}{2}e_n) + w_2(\frac{1}{2}e_n) = \bar{u}(\frac{1}{2}e_n)$.

Say it is $w_1(\frac{1}{2}e_n)$. Then by the inductive hypothesis, $2w_1(X)$ is a non-negative solution of $\mathcal{L} = 0$, vanishes on $\{x_n = 0\}$ and $2w_1(\frac{1}{2}e_n) \ge \bar{u}(\frac{1}{2}e_n)$. Hence

$$\frac{2w_1}{\bar{u}}\bigg|_{B^+_{1/2}} \ge \frac{1}{M}$$

or, renormalizing back, in $B^+_{2^{-(k+1)}}$,

$$\frac{v - a_k u}{(b_k - a_k)u} \ge \frac{1}{2M},$$

that is, in $B_{2^{-(k+1)}}$

$$\left[a_k + \frac{1}{2M}(b_k - a_k)\right] u \le v \le b_k u\ .$$

So $b_{k+1} = b_k$ and $a_{k+1} = a_k + \frac{1}{2M}(b_k - a_k)$. □

11.3. An excursion on harmonic measure

Let $\mathcal{L} = \operatorname{div}(A(x)\nabla)$ be a uniformly elliptic operator. Given a smooth function g on ∂B_1, we can solve the Dirichlet problem

$$\begin{cases} \mathcal{L}u = 0 & \text{in } B_1, \\ u = g & \text{on } \partial B_1 . \end{cases}$$

For x fixed in B_1, the map

$$g \mapsto u(x)$$

is continuous from $C(\partial B_1)$ to $\mathbb{R}$ and it is monotone, i.e., by the maximum principle

$$g_1 \le g_2 \Longrightarrow u_1(x) \le u_2(x) .$$

Thus there exists a nonnegative Borel measure ω^x on ∂B_1 such that

$$u(x) = \int_{\partial B_1} g(\sigma)\, d\omega^x(\sigma) .$$

The measure ω^x is called $\mathcal{L}$-harmonic measure (harmonic measure if $\mathcal{L} = \Delta$) in B_1, evaluated at x. If A is smooth and $G(x,y)$ is the $\mathcal{L}$-Green function in B_1, then from the divergence theorem,

$$u(x) = \int_{\partial B_1} g(\sigma)\partial_{\nu_{\mathcal{L}}} G(x,\sigma)\, d\sigma$$

where $\partial_{\nu_{\mathcal{L}}} G(x,\cdot)$ is the conormal derivative of G with respect to $\mathcal{L}$, that is,

$$\partial_{\nu_{\mathcal{L}}} G(x,\sigma) = A(\sigma)\nabla_y G(x,\sigma)\nu(\sigma)$$

with ν the interior unit normal to ∂B_1 at σ.

Therefore, in this case

$$d\omega^x(\sigma) = \partial_{\nu_{\mathcal{L}}} G(x,\sigma)\, d\sigma = K(x,\sigma)\, d\sigma \tag{11.7}$$

(K is the Poisson kernel).

For general uniformly elliptic A, the $\mathcal{L}$-harmonic measure could be completely singular with respect to surface measure (see [MM], [CFK]).

Nevertheless, Lemma 11.7 shows that ω^x is a *doubling measure*, a property that we can express in nonrescaled version as follows: for $P \in \partial B_1$, let $\Delta_r(P) = B_r(P) \cap \partial B_1$ be a *surface disc* of radius r and $x \in B_{1/2}$. Then

$$\omega^x(\Delta_{2r}(P)) \le c(n,\lambda)\omega^x(\Delta_r(P)) .$$

Moreover, (11.5) and (11.6) (in nonrescaled form) can be written as

$$\omega^x(\Delta_r(P)) \sim r^{n-2} G(x,(1-r)P) \tag{11.8}$$

which may be considered a weak form of (11.7).

In the case of the Laplacian in Lipschitz domains, harmonic measure enjoys many more properties, as the next lemma states (see [D]).

Lemma 11.9. *Let Ω be a bounded Lipschitz domain in $\mathbb{R}^n$. Then if $x \in \Omega$ is fixed and σ denotes surface measure on $\partial\Omega$, the following hold.*

(a) *$\omega^x \ll \sigma$ and $\sigma \ll \omega^x$.*

(b) *If $K = \frac{d\omega^x}{d\sigma}$, then $K(x, \cdot) \in L^2(\partial\Omega)$ and*

$$\left(\frac{1}{|\Delta|}\int_\Delta K^2\, d\sigma\right)^{\frac{1}{2}} \leq c\left(\frac{1}{|\Delta|}\int_\Delta K\, d\sigma\right) \tag{11.9}$$

for every surface disc $\Delta \subset \partial\Omega$, where the constant c depends only on n, the Lipschitz character of Ω and $\operatorname{dist}(x, \partial\Omega)$.

(c) *There exist positive constants c and δ, depending only on n, the Lipschitz character of Ω and* $\operatorname{dist}(x, \partial\Omega)$ *such that, for every surface disc Δ and every Borel set $E \subseteq \Delta$,*

$$\frac{\omega^x(E)}{\omega^x(\Delta)} \leq c\left(\frac{|E|}{|\Delta|}\right)^\delta . \tag{11.10}$$

Inequality (11.9) is a *reverse Schwarz* inequality for K while (11.10), in the theory of weights (see [CF]), says that ω^x is an A_∞ weight with respect to surface measure.

A quick proof of (a) and (b) can be obtained as follows. Note first that, by approximation, it is enough to prove (a) and (b) assuming Ω is smooth, as long as the estimates under consideration depend only on n, $\operatorname{dist}(x, \partial\Omega)$ and the Lipschitz character of Ω. Let $g(\cdot) = G(x, \cdot)$ be the Green function in Ω, $\Delta_r = \Delta_r(P)$ a surface disc centered at $P \in \partial\Omega$, and φ a smooth function vanishing outside $B_{2r}(P)$, such that $0 \leq \varphi \leq 1$ and $\varphi = 1$ in $B_r(P)$.

We can assume $r \leq r_0$ small so that, in $B_{4r}(P)$, $\partial\Omega$ is described by the graph $y_n = f(y')$, and $\operatorname{dist}(x, \partial\Omega) \geq 10r_0$.

Observe that, from (11.8), since $\partial\Omega$ is smooth,

$$g_\nu(P) \sim K(x, P) \sim g_{y_n}(P) . \tag{11.11}$$

We have

$$\int_{B_{2r}(P)\cap\Omega} \{-\Delta(\varphi^2 g) g_{y_n} + \varphi^2 g \Delta g_{y_n}\}\, dy = \int_{\Delta_{2r}} \varphi^2 g_\nu g_{x_n}\, d\sigma . \tag{11.12}$$

From (11.11)

$$\int_{\Delta_{2r}} \varphi^2 g_\nu g_{x_n}\, d\sigma \geq c \int_{\Delta_r} K^2\, d\sigma . \tag{11.13}$$

On the other hand $\Delta g_{y_n} = 0$ in $B_{2r}(P) \cap \Omega$ and

$$\int_{B_{2r}(P)\cap\Omega} \Delta(\varphi^2 g) g_{y_n}\, dy = \int_{B_{2r}(P)\cap\Omega} \{g\Delta\varphi^2 + 2\nabla\varphi^2\nabla g\}\, g_{y_n}\, dy .$$

Using standard energy estimates and $|\nabla\varphi| \le \frac{c}{r}$, $|\Delta\varphi| \le \frac{c}{r^2}$, we get

$$\begin{aligned}\Big|\int_{B_{2r}(P)\cap\Omega} \Delta(\varphi^2 g) g_{y_n}\,dy\Big| &\le \frac{c}{r^3}\int_{B_{2r}(P)\cap\Omega} g^2\,dy\\ &\le c\,r^{n-3} g^2((1-r)P) \quad \text{(from Theorem 11.5)}\\ &\le c\,r^{1-n}[\omega^x(\Delta_{2r})]^2 \quad \text{from (11.8)}\\ &\le c\,r^{1-n}[\omega^x(\Delta_r)]^2 = c\,r^{n-1}\left(\frac{1}{|\Delta_r|}\int_{\Delta_r} K\,d\sigma\right)^2 \quad \text{(doubling)}\,.\end{aligned}$$

This inequality and (11.12), (11.13) give (11.9). □

11.4. Monotonicity properties

We are now ready to prove the third of our main results:

Theorem 11.10. *Let u be a nonnegative harmonic function in the domain*

$$\Omega = \{|x'| < 1,\ f(x') < x_n < 4L\}$$

where f is Lipschitz with constant L, $f(0) = 0$.

Assume u vanishes continuously on $\{x_n = f(x')\}$. Then there exists $\delta > 0$, $\delta = \delta(n, L)$, such that

$$u_{x_n}(x) \sim \operatorname{dist}(x, \partial\Omega) u(x) \tag{11.14}$$

in

$$\Omega_\delta = \left\{|x'| < \frac{1}{2},\ f(x') < x_n < 4\delta L\right\}\,.$$

Theorem 11.10 is a consequence of the following two lemmas.

Lemma 11.11. *Let u as in Theorem 11.5. Assume $u_{x_n} \ge 0$ in Ω. Then (11.14) holds in $\Omega_{1/2}$.*

Proof. Let $x = \delta e_n \in \Omega_{1/2}$. From Theorem 11.5, we have

$$u(\eta\delta e_n) \le c\eta^\alpha u(\delta e_n)$$

so that if $\eta = \eta(n, L)$ is small enough, we have $u(\eta\delta e_n) \le \frac{1}{2}u(\delta e_n)$. Therefore

$$\frac{1}{2}u(\delta e_n) \le \int_{\eta\delta}^{\delta} u_{x_n}(s e_n)\,ds \le u(\delta e_n)\,. \tag{11.15}$$

Since u_{x_n} is positive, along the segment $\eta\delta e_n, \delta e_n$ all the values of u_{x_n} are comparable, by Harnack's inequality, and (11.14) follows from (11.15).

Lemma 11.12. *Let u as in Theorem 11.5. Then for some $\delta = \delta(n, L) > 0$, $u_{x_n} \ge 0$ in Ω_δ.*

Proof. Let h be the harmonic measure in Ω of the set $\partial\Omega \setminus \operatorname{graph}(f)$. By comparing h with its translations along the direction e_n in their common domain of definition, we obtain

$$h_{x_n} \geq 0 \quad \text{in} \quad \Omega \;.$$

Set $v = \gamma h$ and adjust γ so that

$$u(Le_n) = v(Le_n) \;.$$

Then from Theorem 11.6

$$u \sim v \tag{11.16}$$

in $\Omega_{1/2}$ and

$$\left| \frac{u(x)}{v(x)} - \frac{u(y)}{v(y)} \right| \leq c|x-y|^{\alpha} \;.$$

Let us freeze y at distance d from the graph of f. Then $\frac{u(y)}{v(y)} \approx c\frac{u(Le_1)}{v(Le_1)} = c$.

In $B_{d/2}(y)$ we get

$$\begin{aligned} \left| u(x) - \frac{u(y)}{v(y)} v(x) \right| &\leq c|x-y|^{\alpha} v(x) \leq c d^{\alpha} v(x) \\ &\leq c d^{\alpha+1} v_{x_n}(x) \quad \text{(from Lemma 11.11)} \;. \end{aligned}$$

We now take the x_n derivative on the unfrozen variable and evaluate at y. From standard interior estimates for $w = u - \kappa v$, $\kappa = u(y)/v(y)$, we get

$$|u_{x_n}(y) - \kappa v_{x_n}(y)| \leq c d^{\alpha} v_{x_n}(y),$$

that is,

$$u_{x_n}(y) \geq (\kappa - c d^{\alpha}) v_{x_n}(y) \;.$$

The last term is positive if $d \leq \delta(n, L)$ is small enough. □

Lemma 11.12 holds if instead of u_{x_n} we consider the derivative of u along a direction τ entering the domain Ω. As a consequence, in a neighborhood of the graph of f, there exists an entire cone of directions along which u is nondecreasing. Precisely we have

Corollary 11.13. *Let u as in Theorem 11.5. There exists $\delta = \delta(n, L)$ and $\theta = \theta(n, L)$ such that for every $\tau \in \Gamma(e_n, \theta)$*

$$D_{\tau} u \geq 0 \quad \textit{in} \quad \Omega_{\delta} \;.$$

We will call $\Gamma(e_n, \theta)$ a *cone of monotonicity*. Observe that the existence of a monotonicity cone in Ω_{δ} implies that the level surfaces of u are all Lipschitz surfaces, with a common Lipschitz constant, with respect to a common direction.

11.5. ε-monotonicity and full monotonicity

In the regularization of "flat" free boundaries, the key notion is "ε-monotonicity".

Definition 11.1. We say that a function u is ε-monotone in the direction τ if

$$u(x + \lambda\tau) \geq u(x)$$

for every $\lambda \geq \varepsilon$.

The following proposition establishes the connection between ε-monotonicity and flatness of level surfaces.

Proposition 11.14. *Let u be ε-monotone in every direction $\tau \in \Gamma(\theta, e)$. Then the level surfaces of u, that is, the sets $\partial\{u > t\}$, are contained in a $(1 - \sin\theta)\varepsilon$-neighborhood of the graph of a Lipschitz function, with Lipschitz constant* $\operatorname{cotg}\theta$.

Proof. Let V be the union of the cones $x + \varepsilon e + \Gamma(\theta, e)$, for $x \in \partial\{u > t\}$. Then $V \subset \{u > t\}$ and $A = \partial V$ is the graph of a Lipschitz function with Lipschitz constant $\operatorname{cotg}\theta$. Moreover, if $y \in A$, then $\operatorname{dist}(y, \partial\{u > t\}) \leq (1 - \sin\theta)\varepsilon$. □

When we particularize to harmonic functions, ε-monotonicity in a ball of radius $M\varepsilon$, M large, implies full monotonicity in a smaller ball. This is a consequence of the following lemma (note that $u(x + e_1) \geq u(x)$ is satisfied by the harmonic function $e^{\pi y}\cos\pi x$).

Lemma 11.15. *Let $\delta > 0$, let u be harmonic in $B_{M/\delta} = B_{M/\delta}(0)$ $(M \gg \delta)$ and in $B_{M/2\delta}$ let*

$$u(x + \lambda\tau) \geq u(x) \tag{11.17}$$

for every λ with $1 \leq \lambda \leq 1 + \delta$. Then if $M = M(n)$ is large enough,

$$D_\tau u(0) \geq c(n, \delta)[u(\tau) - u(0)]\ . \tag{11.18}$$

Proof. Iterating (11.17) $\frac{1}{\delta}$ times, we deduce that

$$u(x + \lambda\tau) \geq u(x)$$

for every $\lambda \geq \frac{1}{\delta}$.

Rescaling $(u(x) \mapsto \delta u(x/\delta))$, we may assume

$$u(x + \lambda\tau) \geq u(x)$$

for every $\lambda \geq 1$; that is, u is 1-monotone along τ. Thus, for $1 < \lambda < M/2$,

$$h_\lambda(x) = u(x + \lambda\tau) - u(x)$$

is harmonic, nonnegative in $B_{M/2}$. Harnack's inequality gives

$$0 \le c_1 \le \frac{h_\lambda(x)}{h_\lambda(y)} \le c_2$$

for every $x, y \in B_{M/4}$.

If, in particular, λ is an integer, $\lambda < \frac{1}{8}M$ and $y \in B_{M/8}$, we have

$$h_\lambda(y) = \sum_{j=1}^{\lambda} h_1(y + (j-1)\tau) \sim \lambda h_1(y) \; .$$

Therefore, if $x, y \in B_{M/8}$

$$0 \le c_1 \le \frac{h_\lambda(x)}{\lambda h_1(y)} \le c_2 \; . \tag{11.19}$$

Since for λ real, $\lambda \ge 2$,

$$h_{[\lambda]-1} \le h_\lambda \le h_{[\lambda]+2},$$

the inequality (11.19) holds for any λ, $2 \le \lambda \le \frac{M}{8}$ in $B_{M/8}$. Moreover, in $B_{M/8}$ we also have

$$|D_\tau h_\lambda| \le \frac{c(n)}{M} h_\lambda(0) \; .$$

Now,

$$D_\tau h_\lambda(0) = D_\tau u(\lambda\tau) - D_\tau u(0)$$

and therefore we can write

$$\begin{aligned} c_1 h_1(0) \le h_1(2\tau) &= u(3\tau) - u(2\tau) \\ &= \int_2^3 D_\tau u(\lambda\tau)\, d\lambda = \int_2^3 D_\tau h_\lambda(0)\, d\lambda + D_\tau u(0) \\ &\le \frac{c}{M} \int_2^3 h_\lambda(0)\, d\lambda + D_\tau u(0) \\ &\le \frac{c}{M} h_1(0) + D_\tau u(0) \end{aligned}$$

and if $M = M(n)$ is large enough, the lemma follows. □

Here is a rescaled corollary.

Corollary 11.16. *Let u be harmonic in B_1. Then there is a critical value $\varepsilon_0 = \varepsilon_0(n) > 0$ such that if*

$$u(x + \lambda\tau) \ge u(x)$$

for every $\lambda > \varepsilon_0$, then

$$D_\tau u(0) \ge \frac{c(n)}{\varepsilon_0} [u(\varepsilon_0\tau) - u(0)] \; .$$

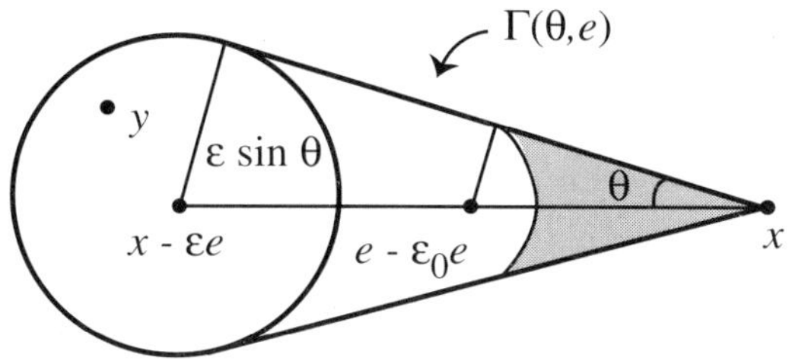

Figure 11.2. For an ε_0-monotone function in $\Gamma(\theta, e)$, $u(y) \leq u(x)$ when y is not in the shaded region

We call the statements in Lemma 11.15 and Corollary 11.16 *ε_0-monotonicity implies full monotonicity.*

We conclude this section with an equivalent definition of ε-monotonicity, which turns out to be more flexible when applied to free boundary problems:

Definition 11.2. We say that u is ε_0-monotone in the cone $\Gamma(\theta, e)$ if for any $\varepsilon > \varepsilon_0$

$$\sup_{B_{\varepsilon \sin\theta}(x)} u(y - \varepsilon e) \leq u(x).$$

Clearly Definition 11.2 implies Definition 11.1. On the other hand Definition 11.1 implies Definition 11.2 with $\varepsilon_0 = \varepsilon(1 + \sin\theta)$.

11.6. Linear behavior at regular boundary points

In this section we examine the behavior of a positive harmonic function in a domain Ω near a regular point of a part of $\partial\Omega$, where it vanishes.

If $x_0 \in \partial\Omega$, we say that x_0 is a *regular point from the right* (*left*) if there exists a ball B such that $B \subset \Omega$ ($B \subset \mathcal{C}\Omega$) and $B \cap \partial\Omega = \{x_0\}$, that is, if $\partial\Omega$ at x_0 satisfies an interior (exterior) ball condition. In this case we say that the ball B *touches* $\partial\Omega$ at x_0 from the right or from the left, respectively. The first result follows.

Lemma 11.17. *Let u be a positive harmonic function in a domain Ω. Assume that $x_0 \in \partial\Omega$ and that u vanishes in $B_1(x_0) \cap \partial\Omega$. Then the following hold.*

(a) *If x_0 is regular from the right, with touching ball B, either near x_0, in B, u grows more than any linear function or it has the asymptotic development*

$$u(x) \geq \alpha\langle x - x_0, \nu\rangle + o(|x - x_0|) \tag{11.20}$$

with $\alpha > 0$, where ν is the unit normal to ∂B at x_0, inward to Ω. Moreover, equality holds in (11.20) in every nontangential region.

(b) *If x_0 is regular from the left, near x_0,*

$$u(x) \le \beta\langle x - x_0, \nu\rangle^+ + o(|x - x_0|) \tag{11.21}$$

with $\beta \ge 0$ and equality in every nontangential region. Moreover, if $\beta > 0$, then B is tangent to $\partial\Omega$ at x_0.

Proof. (a) Suppose the ball B has radius R and center $y_0 : B = B_R(y_0)$. Without loss of generality we may assume that $x_0 = 0$ and $\nu = e_n$. Let h be the harmonic function in the ring $B_R(y_0) \setminus \bar{B}_{R/2}(y_0)$, vanishing on $\partial B_R(y_0)$ and equal to 1 on $\partial B_{R/2}(y_0)$. Then near 0,

$$h(x) = \frac{c(n)}{R} x_n + o(|x|) \ . \tag{11.22}$$

Let

$$\alpha_0 = \sup\{m : u(x) \ge m h(x) \text{ in } B_1 \cap B_R(y_0)\}$$

and $(k \ge 1)$

$$\alpha_k = \sup\{m : u(x) \ge m h(x) \text{ in } B_{2^{-k}} \cap B_R(y_0)\} \ .$$

The sequence α_k is increasing. Let $\tilde{\alpha} = \sup \alpha_k$. If $\tilde{\alpha} = \infty$, u grows more than any linear function near x_0. If $\tilde{\alpha} < \infty$, set $\alpha = \frac{c(n)}{R}\tilde{\alpha}$. Then from the definition of $\tilde{\alpha}$,

$$u(x) \ge \tilde{\alpha} h(x) + o(|x|) = \alpha x_n + o(|x|) \ . \tag{11.23}$$

Now assume there is a sequence of points x^k, with $|x^k| = r_k \sim d(x^k, \partial B_R(y_0))$ such that, for some $\delta_0 > 0$,

$$u(x^k) - \tilde{\alpha} h(x^k) > \delta_0 |x^k| \ .$$

Then also

$$u(x^k) - \alpha_{k_j} h(x^k) > \delta_0 |x^k| \tag{11.24}$$

where k_j is such that $2^{-6k_j} < r_k \le 2^{-4k_j}$.

Now the function

$$w(x) = u(x) - \alpha_{k_j} h(x)$$

is harmonic and positive in $B_{2^{-k_j}} \cap B_R(y_0)$. From (11.24) and Harnack's inequality, we have

$$u(x) - \alpha_{k_j} h(x) \ge c\delta_0 |x^k|$$

in a fixed portion of ∂B_{r_k}. From the Poisson representation formula, we have in $B_{r_{k/2}} \cap B_R(y_0)$

$$u(x) - \alpha_{k_j} h(x) \ge c_0 \delta_0 |x^k| \ . \tag{11.25}$$

For k large and x near zero (11.25) implies

$$u(x) - \alpha x_n \ge c_0 \delta_0 x_n + o(|x|),$$

which contradicts the definition of α.

To prove (b), extend u to zero outside Ω. Now u is subharmonic and let $\tilde{h}$ be the Kelvin transform of h with respect to ∂B. Let

$$\beta_0 = \inf\{m : m\tilde{h}(x) \geq u(x) \ \text{ in } \ \mathcal{C}B_R \cap B_1\},$$
$$\beta_k = \inf\{m : m\tilde{h}(x) \geq u(x) \ \text{ in } \ \mathcal{C}B_R \cap B_{2^{-k}}\} \ .$$

The sequence β_k is decreasing. Let $\tilde{\beta} = \inf \beta_k$. Then $\tilde{\beta} \geq 0$ and if $\beta = \frac{c(n)}{R}\tilde{\beta}$, we have, near 0,

$$u(x) \leq \tilde{\beta}\tilde{h}(x) + o(|x|) = \beta x_n^+ + o(|x|) \ . \tag{11.26}$$

For the proof that equality holds in (11.26) inside every nontangential region, one can proceed as for the equality in (11.20).

Finally, if $\beta > 0$ and B_R is not tangent to $\partial\Omega$ at 0, there exists a sequence of points $x^k \in \mathcal{C}\Omega$ approaching the origin with $|x^k| \sim d(x^k, \partial\Omega)$ and $u(x^k) = 0$. Since u is subharmonic, this is incompatible with a nontrivial linear behavior of u at 0, along any nontangential direction. □

Remark 11.18. If in Lemma 11.17 we know that u is Lipschitz up to the boundary, then clearly α and β in (11.20) and (11.21) are bounded by the Lipschitz constant of u.

Moreover, equality in (11.20) holds in B near x_0, not only along nontangential domains.

The same conclusion holds if we know that u is monotonically increasing along any direction of a cone $\Gamma(\theta, \bar{\nu})$, with axis $\bar{\nu}$ such that $\langle \bar{\nu}, \nu \rangle \geq \eta_0 > 0$.

In the asymptotic developments (11.20) and (11.21), α and β represent natural substitutes for normal derivatives at points x_0 where u is not smooth. In Remark 11.18 we have seen that if u is Lipschitz, then α and β are uniformly bounded. We conclude this chapter with a lemma that somehow reverses the situation: boundedness of α and β implies Lipschitz continuity.

Lemma 11.19 (Lipschitz regularity of harmonic functions with control on u_ν). *Let u be a nonnegative harmonic function in $\Omega \cap B_1$. Assume that $0 \in \partial\Omega$, that u vanishes on $\partial\Omega_1 \cap B_1$ and that whenever $x_0 \in \partial\Omega \cap B_1$ is a regular point from the right,*

$$\liminf_{\varepsilon \to 0^+} \frac{u(x_0 + \varepsilon\nu)}{\varepsilon} \leq \alpha_0$$

where ν is the inner unit normal to the touching ball at x_0. Then u is Lipschitz in $\Omega \cap B_{1/2}$ and

$$\|u\|_{\mathrm{Lip}(B_{1/2})} \leq c\alpha_0 \ .$$

Proof. It is enough to show that $u(x) \le c d(x, \partial\Omega)$. Let $d(x, \partial\Omega) = h$ $(< \frac{1}{2})$. Then $B_h(x)$ touches $\partial\Omega$ at some point x_0. By Harnack's inequality,

$$\inf_{B_{h/2}(x)} u \ge c_0 u(x) \ .$$

If v is the harmonic function in the ring $B_h(x) \backslash \bar{B}_{h/2}(x)$ vanishing on $\partial B_h(x)$, equal to $c_0 u(x)$ on $B_{h/2}(x)$, then $v \le u$ in the ring. Since

$$\lim_{\varepsilon \to 0^+} \frac{v(x_0 + \varepsilon\nu)}{\varepsilon} = c_1 \frac{u(x)}{h},$$

we get

$$c_1 \frac{u(x)}{h} \le \alpha_0$$

and the proof is complete. □

Chapter 12

Monotonicity Formulas and Applications

12.1. A 2-dimensional formula

In one-phase problems, the Lipschitz continuity of the solution u in, say, $B_{1/2} = B_{1/2}(0)$ comes simply from the fact that 0 belongs to the free boundary and that u is defined in B_1.

For instance, let u be the viscosity solution of the free boundary problem in Section 6.1 and let $u^- \equiv 0$. If $x \in B_{1/2}$ and $u(x) = \lambda > 0$, let $h = \operatorname{dist}(x, F(u))$ and let $x_0 \in \partial B_h(x) \cap F(u)$. Let us show that $\lambda \leq ch$.

By Harnack's inequality, $u \geq \bar{c}\lambda$ in $B_{h/2}(x)$ and hence, if w is the radially symmetric harmonic function in the ring $B_h(x) \setminus \bar{B}_{h/2}(x)$ with $w_{|\partial B_{h/2}(x)} = \bar{c}\lambda$ and $w_{|\partial B_h(x)} = 0$, then

$$u \geq w$$

in the ring.

If ν is the unit inner normal to $\partial B_h(x)$ at x_0, we have

$$w_\nu = c\frac{\lambda}{h}$$

while, near x_0, nontangentially

$$\begin{aligned} u(x) &= \alpha\langle x - x_0, \nu\rangle^+ + o(|x - x_0|) \\ &= G(0)\langle x - x_0, \nu\rangle^+ + o(|x - x_0|) . \end{aligned}$$

Therefore we get

$$c\frac{\lambda}{h} \leq G(0)$$

or

$$\lambda \leq Ch\ .$$

Clearly the above calculation is not enough for two-phase problems, where the solution may have both positive and negative part with large slope.

One way to solve this problem is through some monotonicity formulas. The simplest one follows.

Theorem 12.1. *Let $B_1 \subset \mathbb{R}^2$ and let $u_1, u_2 \in H^1(B_2)$, continuous and nonnegative in B_2, supported and harmonic in disjoint domains Ω_1, Ω_2, respectively, with $0 \in \partial\Omega_i$ and*

$$u_i = 0 \quad \text{along} \quad \partial\Omega_i \cap B_1 = \Gamma_i \qquad (i = 1, 2).$$

Then the quantity

$$J(R) = \frac{1}{R^4} \int_{B_R} |\nabla u_1|^2\, dx \cdot \int_{B_R} |\nabla u_2|^2\, dx$$

is monotone increasing in R, $R \leq 3/2$.

Proof. We want to prove that $J'(R) \geq 0$ a.e. in $0 < R < \frac{3}{2}$. Observe that $\varphi_i(r) = \frac{d}{dr} \int_{B_r} |\nabla u_i|^2\, dx = \int_{\partial B_r} |\nabla u_i|^2\, d\sigma$ is in $L^1(0, 2)$. By rescaling, it is enough to prove that $J'(1) \geq 0$. We have

$$\begin{aligned} J'(1) = \int_{\partial B_1} |\nabla u_1|^2\, dx \cdot \int_{B_1} |\nabla u_2|^2\, dx + \int_{B_1} |\nabla u_1|^2\, dx \cdot \int_{\partial B_1} |\nabla u_2|^2\, dx \\ - 4 \int_{B_1} |\nabla u_1|^2\, dx \cdot \int_{B_1} |\nabla u_2|^2\, dx\ . \end{aligned}$$

Dividing by $J(1)$, we get

$$\frac{J'(1)}{J(1)} = \frac{\displaystyle\int_{\partial B_1} |\nabla u_1|^2\, dx}{\displaystyle\int_{B_1} |\nabla u_1|^2\, dx} + \frac{\displaystyle\int_{\partial B_1} |\nabla u_2|^2\, dx}{\displaystyle\int_{B_1} |\nabla u_2|^2\, dx} - 4\ .$$

Now $2|\nabla u_i|^2 \leq \Delta u_i^2$ so that

$$\int_{B_1} |\nabla u_i|^2\, dx \leq \int_{\partial B_1} u_i (u_i)_r\, d\sigma \leq \left(\int_{\partial B_1} u_i^2\, d\sigma \right)^{1/2} \left(\int_{\partial B_1} (u_i)_r^2\, d\sigma \right)^{1/2}$$

where u_r denotes the exterior radial derivative of u along ∂B_1.

Moreover, if u_θ denotes the tangential derivative of u along ∂B_1,

$$\int_{\partial B_1} |\nabla u_i|^2\, dx \geq 2 \left(\int_{\partial B_1} (u_i)_r^2\, d\sigma \right)^{1/2} \cdot \left(\int_{\partial B_1} (u_i)_\theta^2\, d\sigma \right)^{1/2}. \tag{12.1}$$

Hence, it is enough to show that

$$\frac{\left(\int_{\partial B_1} (u_1)_\theta^2\, d\sigma\right)^{1/2}}{\left(\int_{\partial B_1} u_1^2\, d\sigma\right)^{1/2}} + \frac{\left(\int_{\partial B_1} (u_2)_\theta^2\, d\sigma\right)^{1/2}}{\left(\int_{\partial B_1} u_2^2\, d\sigma\right)^{1/2}} - 2 \geq 0\ .$$

Now the quotients

$$\int_{\Gamma_i} v_\theta^2\, d\sigma \Big/ \int_{\Gamma_i} v^2\, d\sigma$$

are minimized by the first eigenfunction of the domains $\partial B_1 \cap \bar{\Omega}_i$, respectively. Since $\Omega_1 \cap \Omega_2 = \emptyset$, we can optimize further and ask: given any domain Γ in the unit circle of prescribed measure μ, which is the one with the smallest eigenvalue, i.e.,

$$\inf_{v \in H_0^1(\Gamma)} \frac{\int_\Gamma v_\theta^2\, d\sigma}{\int_\Gamma v^2\, d\sigma}\ ?$$

A symmetrization argument gives that the optimal domain is a connected arc. Moreover, the larger the arc, the smaller the quotient; thus the sum

$$\frac{\left(\int_{\Gamma_1} (u_1)_\theta^2\, d\sigma\right)^{1/2}}{\left(\int_{\Gamma_1} (u_1)^2\, d\sigma\right)^{1/2}} + \frac{\left(\int_{\Gamma_2} (u_2)_\theta^2\, d\sigma\right)^{1/2}}{\left(\int_{\Gamma_2} (u_2)^2\, d\sigma\right)^{2}} = I_1 + I_2$$

takes its minimum for two adjacent, complementary arcs with u_1 and u_2 the corresponding eigenfunctions.

If the arcs have length $\alpha 2\pi$ and $(1-\alpha)2\pi$, the corresponding eigenfunctions are

$$\sin\frac{\theta}{2\alpha} \quad \text{and} \quad \sin\frac{\theta}{2(1-\alpha)}$$

and thus

$$I_1 + I_2 \geq \frac{1}{2\alpha} + \frac{1}{2(1-\alpha)} = \frac{1}{2\alpha(1-\alpha)} \geq 2 \qquad (0 \leq \alpha \leq 1),$$

which completes the proof. □

Remark 12.2. The above computations become exact for linear functions. More precisely, if $v(x) = \lambda x^+$ $(\lambda > 0)$, equality holds in (12.1) since both the normal and the tangential gradient have the same L^2-norm on ∂B_1.

12.2. The n-dimensional formula

In dimension $n > 2$ it seems to be natural to consider the quantity

$$\frac{1}{R^{2n}} \int_{B_R} |\nabla u_1|^2 \, dx \int_{B_R} |\nabla u_2|^2 \, dx. \tag{12.2}$$

However, this time, the above computation fails because the estimate

$$\begin{aligned} \int_{\partial B_1} |\nabla u|^2 \, d\sigma &= \int_{\partial B_1} [(u_r)^2 + |\nabla_\theta u|^2] \, d\sigma \\ &\geq 2 \left(\int_{\partial B_1} (u_r)^2 \right)^{1/2} \left(\int_{\partial B_1} |\nabla_\theta u|^2 \, d\sigma \right)^{1/2} \end{aligned}$$

is not exact, the L^2-norms of the normal and tangential gradient being different, even for linear functions.

To understand the higher-dimensional situation, we first relate the eigenfunctions of the Laplace-Beltrami operator on the sphere to the homogeneity of harmonic functions. In $\mathbb{R}^n$, the Laplacian in polar coordinates reads

$$u_{rr} + \frac{n-1}{r} u_r + \frac{1}{r^2} \Delta_\theta u$$

where Δ_θ is the Laplace-Beltrami operator that one obtains by minimizing the Dirichlet integral $\int_{\partial B_1} |\nabla_\theta u|^2 \, d\sigma$.

Given a domain $\Gamma \subset \partial B_1$, an eigenfunction $g \in H_0^1(\Gamma)$ of $-\Delta_\theta$ can be extended to a homogeneous harmonic function $u = u(x)$ in the cone generated by Γ by writing

$$u(x) = r^\alpha g(\theta)$$

and choosing $\alpha = \alpha(\Gamma)$ according to

$$\Delta_\theta g + \lambda g = 0 \quad \text{in } \Gamma$$

or

$$\int_\Gamma |\nabla_\theta g|^2 \, d\sigma = \lambda \int_\Omega g^2 \, d\sigma$$

and

$$\lambda = \inf_{v \in H_0^1(\Gamma)} \frac{\displaystyle\int_\Gamma |\nabla_\theta v|^2}{\displaystyle\int_\Gamma v^2}.$$

Since

$$\Delta u = r^{\alpha-2} \Big[\alpha(\alpha - 1) + \alpha(n-1) \Big] g(\theta) + r^{\alpha-2} \Delta_\theta g = 0,$$

$\alpha = \alpha(\Gamma)$ must satisfy

$$\alpha(\alpha - 1) + \alpha(n-1) - \lambda = 0$$

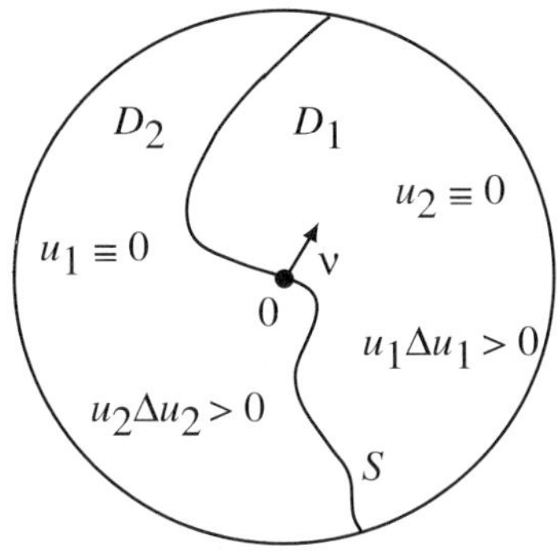

Figure 12.1

or

$$\alpha^2 + (n-2)\alpha - \lambda = 0 \, . \tag{12.3}$$

A particular case corresponds to $\alpha = 1$, $\lambda = n - 1$ and $g(\theta) = \cos \theta_{n-1} =$ the restriction to the upper unit sphere of $u(x) = x_n^+$, where $0 < \theta_{n-1} < \frac{\pi}{2}$.

In this case, on ∂B_1,

$$u_r^2 = u^2$$

while

$$\int_{\partial B_1} |\nabla_\theta u|^2 \, d\sigma = (n-1) \int_{\partial B_1} u^2 \, d\sigma$$

and thus, inequality (12.1) is not exact.

The computations become exact if instead of looking at (12.2) we look at

$$J(R) = \frac{1}{R^2} \int_{B_R} \frac{|\nabla u_1|^2}{|x|^{n-2}} \, dx \cdot \frac{1}{R^2} \int_{B_R} \frac{|\nabla u_2|^2}{|x|^{n-2}} \, dx \equiv J_1(R) \cdot J_2(R). \tag{12.4}$$

Let us make a few remarks.

- Each one of the factors J_1, J_2 can be understood as an average of $|\nabla u_i|^2$, i.e., we are dividing the volume integral in a domain of size R^n by a factor $R^2 \cdot r^{n-2}$.
  In fact, if u_i is the positive part of a linear function, $u_i(x) = \alpha_i x_n^+$, then
  $$J_i(R) = c(n) \alpha_i^2$$
  where $c(n) = \frac{\omega_n}{4}$ and ω_n is the surface of the unit ball in $\mathbb{R}^n$.
- J_i has a linear scaling. That is, if $u_\lambda(x) = \frac{1}{\lambda} u(\lambda x)$,
  $$J_i\left(\frac{R}{\lambda}, u_\lambda\right) = J_i(R, u).$$

- If the supports of u_1 and u_2 are two domains D_1 and D_2 separated by a nice surface S through the origin and $D_\nu u_i(0)$ exist (ν normal unit vector to S at 0), then
$$\lim_{R\to 0} J_i(R) = c(n)(D_\nu u_i)^2.$$
- As we shall see in the proof of Theorem 12.3, each one of the J_i is controlled from above by $\frac{1}{r^2} \fint_{B_r} u_i^2$ so that we have a good control of $J(R)$.

As before we end up with checking that the quantity

$$\text{(12.5)} \qquad \frac{J'(1)}{J(1)} = \frac{\displaystyle\int_{\partial B_1} |\nabla u_1|^2\, d\sigma}{\displaystyle\int_{B_1} \frac{|\nabla u_1|^2}{|x|^{n-2}}\, dx} + \frac{\displaystyle\int_{\partial B_1} |\nabla u_2|^2\, d\sigma}{\displaystyle\int_{B_1} \frac{|\nabla u_2|^2}{|x|^{n-2}}\, dx} - 4$$

is nonnegative.

But now, transforming the volume integrals into boundary integrals, we get

$$\begin{aligned}
\int_{B_1} \frac{|\nabla u_i|^2}{|x|^{n-2}}\, dx &= \frac{1}{2}\int_{B_1} \frac{\Delta(u_i)^2}{|x|^{n-2}}\, dx \\
&= \int_{\Gamma_i} u_i (u_i)_r \, d\sigma + \frac{n-2}{2}\int_{B_1} \frac{\nabla(u_i)^2\cdot x}{|x|^n}\, dx \\
&= \int_{\Gamma_i} u_i (u_i)_r \, d\sigma + \frac{n-2}{2}\int_{\Gamma_i} u^2\, d\sigma
\end{aligned}$$

since $u_i(0) = 0$ and therefore

$$\int_{B_1} (u_i)^2 \Delta\left(\frac{1}{|x|^{n-2}}\right) dx = 0\ .$$

This means we must examine quotients of the type

$$\text{(12.6)} \qquad \frac{\displaystyle\int_\Gamma [(u_r)^2 + |\nabla_\theta u|^2]\, d\sigma}{\displaystyle\int_\Gamma \left[u u_r + \frac{n-2}{2} u^2\right] d\sigma} \qquad \Big(u \in H_0^1(\Gamma)\Big).$$

We look for the optimal partition of u_r^2 and $|\nabla_\theta u|^2$ for a given domain Γ.

If $\lambda = \lambda(\Gamma)$ is the corresponding eigenvalue, then we can write for any $0 < \beta < 1$

$$\int_\Gamma |\nabla_\theta u|^2\, d\sigma \geq \lambda \int_\Gamma u^2\, d\sigma = \beta\lambda \int_\Gamma \cdots + (1-\beta)\lambda \int_\Gamma \cdots\ .$$

We let the first piece go with $\int_\Gamma u_r^2$ and we let the other one control $\frac{n-2}{2}\int_\Gamma u^2$ directly. Precisely, we write

(12.7)

$$\int_\Gamma [u_r^2 + |\nabla_\theta u|^2]\, d\sigma \geq 2\Big(\int_\Gamma u_r^2\, d\sigma\Big)^{1/2}\Big(\beta\lambda\int_\Gamma u^2\, d\sigma\Big)^{1/2} + (1-\beta)\lambda\int_\Gamma u^2\, d\sigma\ ,$$

and we want to control

$$\Big(\int_\Gamma u_r^2\, d\sigma\Big)^{1/2}\Big(\int_\Gamma u^2\, d\sigma\Big)^{1/2} + \frac{n-2}{2}\int_\Gamma u^2\, d\sigma\ .$$

Therefore we balance the two terms in (12.7), choosing β such that

$$\frac{2}{n-2}(1-\beta)\lambda = 2(\beta\lambda)^{1/2}$$

or

$$\beta\lambda + (n-2)(\beta\lambda)^{1/2} - \lambda = 0,$$

which is nothing but the equation for the positive homogeneity constant $\alpha = \alpha(\Gamma)$ in (12.3). Thus if we choose

$$\beta\lambda = \alpha^2,$$

the quotient in (12.6) is greater than or equal to 2α. Going back to (12.5), we are left to prove that

$$\frac{J'(1)}{J(1)} \geq 2[\alpha_1 + \alpha_2 - 2] \geq 0 \qquad \big(\alpha_i = \alpha_i(\Gamma_i) > 0\big) \tag{12.8}$$

for any couple of disjoint domains $\Gamma_1, \Gamma_2 \subset \partial B_1$.

The positive set function $\alpha = \alpha(\Gamma)$, $\Gamma \subset \partial B_1$, is called the *characteristic constant of* Γ (see [FH]).

Thus, our optimization problem is reduced to showing that given two adjacent domains Γ_1 and Γ_2 on the unit sphere, the sum of their characteristic constants is greater than or equal to 2. Let us list a few facts.

(a) ([S]) By symmetrization, among all domains Γ on ∂B_1 with prescribed $(n-1)$-dimensional surface area, a spherical cap, i.e., a set of the form

$$\Gamma^* = \partial B_1 \cap \{x_n > s\} \qquad (-1 < s < 1)$$

has the smallest characteristic constant α, and thus, the smallest eigenvalue λ, with

$$\lambda = \inf_{u(s)=0} \frac{\displaystyle\int_s^1 (u')^2(1-x^2)^{(n-2)/2}\, dx}{\displaystyle\int_s^1 u^2(1-x^2)^{(n-1)/2}\, dx}$$

(by changing variables $x = x_n = \cos\theta$, in polar coordinates).

(b) ([FH]) The minimum decreases with dimension, since an n_1-dimensional configuration can be extended (without changing the homogeneities) to a higher $n_2 = n_1 + k$ dimension, so we scale properly and study the problem for large dimension.

Note that since $\alpha(\alpha + (n-2)) = \lambda$, then

$$\alpha = \frac{1}{2}\left[\sqrt{(n-2)^2 + 4\lambda} - (n-2)\right] = \frac{\lambda}{n-2} + o\Big(\frac{1}{n^2}\Big)$$

for n large, so it is enough to study $\tilde{\alpha}$ for $n \to +\infty$, with

$$\tilde{\alpha} = \frac{1}{n} \inf_{u(s_n)=0} \frac{\displaystyle\int_{s_n}^{1} (u')^2 (1-x^2)^{(n-2)/2}\, dx}{\displaystyle\int_{s_n}^{1} u^2 (1-x^2)^{(n-1)/2}\, dx}.$$

This suggests the change of variable $y = n^{1/2} x$. Thus, letting $h_n = n^{1/2} s_n$,

$$\tilde{\alpha} = \inf_{u(h_n)=0} \frac{\displaystyle\int_{h_n}^{n^{1/2}} (u')^2 \big(1 - \frac{y^2}{n}\big)^{(n-2)/2}\, dy}{\displaystyle\int_{h_n}^{n^{1/2}} u^2 \big(1 - \frac{y^2}{n}\big)^{(n-1)/2}\, dy}.$$

But $(1 - \frac{y^2}{n})^{(n-2)/2}$ converges in compact sets to $e^{-y^2/2}$, so, if we choose a convergent sequence of h_n such that $h_{n_j} \to h$ (h may be $\pm\infty$), then (see [FH]) $\tilde{\alpha}$ converges to the Gaussian eigenvalue:

$$\Lambda(h) = \inf_{u(h)=0} \frac{\displaystyle\int_{h}^{\infty} (u')^2 e^{-y^2/2}\, dy}{\displaystyle\int_{h}^{\infty} u^2 e^{-y^2/2}\, dy}. \tag{12.9}$$

Observe also that if $\omega_n = |\partial B_1|$ and $|\Gamma^*| = \gamma \omega_n$, $0 < \gamma < 1$, γ fixed, then

$$\gamma = \frac{1}{\sqrt{2\pi}} \int_h^{+\infty} \exp(-t^2/2)\, dt \tag{12.10}$$

by "passing" to the limit for $n \to +\infty$ in the relation

$$\gamma = \frac{|\Gamma^*|}{\omega_n} = \frac{\omega_{n-1}}{\omega_n} \int_{s_n}^{1} (1-x^2)^{(n-1)/2}\, dx\,.$$

Therefore, our problem is reduced to showing that if $\Gamma_1, \Gamma_2 \subset \partial B_1$, $\Gamma_1 \cap \Gamma_2 = \emptyset$ and $\Lambda(h_1), \Lambda(h_2)$ correspond to Γ_1 and Γ_2, respectively, then

$$\Lambda(h_1) + \Lambda(h_2) \geq 2. \tag{12.11}$$

Now $\Lambda = \Lambda(h)$ is the first Dirichlet eigenvalue of the one-dimensional Ornstein-Uhlenbeck operator

$$-\frac{d^2}{dx^2} + x\frac{d}{dx}$$

in the set $(h, +\infty)$. We analyze the properties of $\Lambda(h)$.

(c) ([BKP], [CK]) $\Lambda = \Lambda(h)$ *is convex.* Let u_h be the first normalized eigenfunction associated to $\Lambda(h)$, i.e.,

$$\begin{cases} \left(-\dfrac{d^2}{dx^2} + x\dfrac{d}{dx}\right) u_h(x) = \Lambda(h) u_h(x), & x > h, \\ u_h(h) = 0\ , \quad \dfrac{1}{\sqrt{2\pi}} \displaystyle\int_h^{+\infty} u_h^2(x) \exp\Big(-\frac{x^2}{2}\Big)\, dx = 1. \end{cases}$$

Put $v_h(x) = u_h(x) e^{-x^2/4}$. Then

$$-\frac{d^2}{dx^2} v_h(x) + \Big(\frac{1}{4}x^2 - \frac{1}{2}\Big) v_h(x) = \Lambda(h) v_h(x)$$

and $v_h(h) = 0$, $\int_h^{+\infty} v_h^2\, dx = 1$. This means that v_h is the first Dirichlet eigenfunction on $[h, +\infty)$ associated to the Hermite operator

$$-\frac{d^2}{dx^2} + \Big(\frac{1}{4}x^2 - \frac{1}{2}\Big)$$

and $\Lambda(h)$ is the corresponding eigenvalue.

Since $V(x) = \frac{1}{4}x^2 - \frac{1}{2}$ is convex and $\int_{\mathbb{R}} \exp(-tV(x))\, dx < \infty$ for every $t > 0$, from [BL] we deduce that $\Lambda(h)$ is a *convex function*.

(d) From [BKP], [CK] again, $\Lambda = \Lambda(h)$ is *strictly increasing*, $\Lambda(0) = 1$ and $\Lambda(h) \to 0$ as $h \to -\infty$, $\Lambda(h) \to +\infty$ as $h \to +\infty$. Moreover

$$\Lambda''(0) = \frac{4}{\pi}(1 - \ln 2). \tag{12.12}$$

In particular, if $h_1 + h_2 \geq 0$, then

$$\Lambda(h_1) + \Lambda(h_2) \geq 2\Lambda\left(\frac{h_1 + h_2}{2}\right) \geq 2\Lambda(0) = 2\ . \tag{12.13}$$

We can now go back to our optimization problem and in particular to inequality (12.11). Given $\Gamma_1, \Gamma_2 \subset \partial B_1$, $\Gamma_1 \cap \Gamma_2 = \emptyset$, let the numbers $\gamma_i = |\Gamma_i|/\omega_n$ and h_i be defined according to formula (12.10).

Observe that since $\Gamma_1 \cap \Gamma_2 = \emptyset$, then $\gamma_1 + \gamma_2 \leq 1$ and hence $h_1 + h_2 \geq 0$. Indeed, if $h_1 + h_2 < 0$ and, for instance, $h_1 < 0 < h_2$, we would have $\gamma_1 > \gamma(-h_2)$, $\gamma(-h_2) + \gamma_2 = 1$ (by the evenness of the Gaussian) so that $\gamma_1 + \gamma_2 > 1$, a contradiction.

From (12.13) we conclude that

$$\frac{J'(1)}{J(1)} \geq 2\Big[\Lambda(h_1) + \Lambda(h_2) - 2\Big] \geq 0 . \tag{12.14}$$

We summarize and refine the above discussion in the following important result ([ACF6]).

Theorem 12.3 (Monotonicity formula). *Let u_1, u_2 be nonnegative subharmonic functions in $C(B_1)$. Assume that $u_1 \cdot u_2 = 0$ and that $u_1(0) = u_2(0) = 0$. Set*

$$J(r) = \frac{1}{r^4} \int_{B_r} \frac{|\nabla u_1|^2}{|x|^{n-2}}\, dx \int_{B_r} \frac{|\nabla u_2|^2}{|x|^{n-2}}\, dx\ , \qquad 0 < r < 1\ .$$

Then $J(r)$ is finite and is an increasing function of r. Moreover,

$$J(r) \leq c(n)\|u_1\|^2_{L^2(B_1)} \cdot \|u_2\|^2_{L^2(B_1)}, \qquad 0 < r \leq \frac{1}{2}\ . \tag{12.15}$$

Proof. Observe that Δu_1 and Δu_2 are measures. Denote by v_m mollifiers of u_1. Then

$$\Delta v_m^2 = 2|\nabla v_m|^2 + 2v_m \Delta v_m \geq 2|\nabla v_m|^2$$

and therefore, if φ is a test function in $C_0^\infty(B_1)$,

$$\begin{aligned}\int_{B_1} u_1^2 \Delta\varphi &= \lim_{m\to+\infty} \int_{B_1} v_m^2 \Delta\varphi = \lim_{m\to+\infty} \int_{B_1} \Delta v_m^2 \varphi \\ &\geq \lim_{m\to+\infty} 2\int_{B_1} |\nabla v_m|^2 \varphi = 2\int_{B_1} |\nabla u_1|^2 \varphi\ .\end{aligned}$$

This means that $\Delta u_1^2 \geq 2|\nabla u_1|^2$ in the sense of measures and that $u_1 \in H^1_{\text{loc}}(B_1)$. Let now ψ be a cut-off function, $\psi \equiv 1$ in B_r, $\psi \equiv 0$ outside B_{2r}, $r < \frac{1}{2}$, and put

$$g_\varepsilon = \eta_\varepsilon * g$$

where $g(x) = |x|^{2-n}$ and η_ε is an approximation of the identity. We have

$$\begin{aligned}2\int_{B_{2r}} &\psi g_\varepsilon |\nabla u_1|^2\, dx \leq \int_{B_{2r}} \Delta(\psi g_\varepsilon) u_1^2\, dx \\ &= \int_{B_{2r}} u_1^2 \psi \Delta g_\varepsilon\, dx + 2\int_{B_{2r}\setminus B_r} u_1^2 \nabla\psi \cdot \nabla g_\varepsilon\, dx + \int_{B_{2r}\setminus B_r} u_1^2 \Delta\psi g_\varepsilon\, dx \\ &\leq c(n)\int_{B_{2r}} u_1^2 \psi \eta_\varepsilon\, dx + c(n) r^{-n} \int_{B_{2r}\setminus B_r} u_1^2\, dx.\end{aligned}$$

Letting $\varepsilon \to 0$, since $u_1(0) = 0$, we get

$$\int_{B_r} \frac{|\nabla u_1|^2}{|x|^{n-2}}\, dx \leq c(n) r^{-n} \int_{B_{2r}\setminus B_r} u_1^2\, dx\ . \tag{12.16}$$

An analogous inequality holds for u_2 and hence $\varphi(r)$ is finite for $0 < r < 1$.

Since

$$r \longmapsto r^{2-n} \int_{\partial B_r} |\nabla u_i|^2 \, d\sigma \qquad (i = 2, 2)$$

is in $L^1(0,1)$, we have

$$\frac{d}{dr} \int_{B_r} \frac{|\nabla u_i|^2}{|x|^{n-2}} \, dx = r^{2-n} \int_{\partial B_r} |\nabla u_i|^2 \, d\sigma \quad \text{a.e.}$$

It follows that

$$\begin{aligned} J'(r) = &-\frac{4}{r^5} \int_{B_r} |x|^{2-n} |\nabla u_1|^2 \, dx \cdot \int_{B_r} |x|^{2-n} |\nabla u_2|^2 \, dx \\ &+ \frac{1}{r^4} \int_{B_r} |x|^{2-n} |\nabla u_1|^2 \, dx \cdot \int_{\partial B_r} r^{2-n} |\nabla u_2|^2 \, d\sigma \\ &+ \frac{1}{r^4} \int_{\partial B_r} r^{2-n} |\nabla u_1|^2 \, d\sigma \cdot \int_{B_r} |x|^{2-n} |\nabla u_2|^2 \, dx \ . \end{aligned}$$

We want to show that $J'(r) \geq 0$ a.e. By rescaling, we can assume that $r = 1$ and we can conclude from (12.14) that $J'(1) \geq 0$.

Since $J(r)$ is increasing, from (12.16) we deduce, for $0 < r \leq \frac{1}{2}$,

$$J(r) \leq J\Big(\frac{1}{2}\Big) \leq c(n) \|u_1\|^2_{L^2(B_1)} \cdot \|u_2\|^2_{L^2(B_1)}. \qquad \square$$

A more precise result can be obtained by observing that (12.14) gives

$$J'(r) \geq \frac{2}{r} J(r) \Big[\Lambda(h_1) + \Lambda(h_2) - 2 \Big] \ .$$

Here $\Gamma_i = \Gamma_i(r)$ $(i = 1, 2)$ are the projections of $\partial B_r \cap \{u_i > 0\}$ on the unit sphere ∂B_1.

Corollary 12.4. *The strict inequality*

$$\Lambda(h_1) + \Lambda(h_2) - 2 > 0$$

holds unless $\Gamma_i(r)$ are both half spheres. In particular, if at least one of Γ_i digresses from being a hemisphere by an area of size $\varepsilon \omega_n$, then

$$\Lambda(h_1) + \Lambda(h_2) - 2 \geq c\varepsilon^2 \ . \tag{12.17}$$

As a consequence $r^{-c\varepsilon^2} J(r)$ is a nondecreasing function and

$$J(r) \leq J(1) r^{c\varepsilon^2} \ . \tag{12.18}$$

Proof. If $\Gamma \subset \partial B_1$ and $|\Gamma| = (\frac{1}{2} - \varepsilon)\omega_n$, then

$$\frac{1}{2} - \varepsilon = \frac{1}{\sqrt{2\pi}} \int_h^\infty \exp(-t^2/2)\, dt = \frac{1}{2} - \frac{1}{\sqrt{2\pi}} \int_0^h \exp(-t^2/2)\, dt \geq \frac{1}{2} - \frac{1}{\sqrt{2\pi}} h .$$

Hence $h \geq \sqrt{2\pi}\, \varepsilon$ and from (d), (12.13),

$$\Lambda(h_1) + \Lambda(h_2) - 2 \geq c\Lambda''(0)(h_1^2 + h_2^2) \geq c\varepsilon^2 . \qquad \square$$

12.3. Consequences and applications

The usefulness of the function $J(r)$ in free boundary problems comes from the fact that it gives a control of the linear behavior of the solution from both sides of the free boundary.

More precisely, one has the following consequence of the monotonicity formula.

Introduce the notation

$$J(u; R) = \left(\frac{1}{R^2} \int_{B_R} \frac{|\nabla u^+|^2}{|x|^{n-2}} dx \right) \left(\frac{1}{R^2} \int_{B_R} \frac{|\nabla u^-|^2}{|x|^{n-2}} dx \right) \equiv J_R^+(u) J_R^-(u) .$$

Lemma 12.5. *Let u, v be continuous functions in B_2 such that $u(0) = v(0)$ and outside their zero sets, $\Delta u = 0$, $v\Delta v \geq 0$. If, near the origin,*

$$u^\pm(x) \leq v^\pm(x) + o(|x|), \tag{12.19}$$

then

$$J(u, 0+) \leq J(v, 0+) .$$

Proof. If $J(u; 0+) = 0$ or $J(v; 0+) = \infty$, there is nothing to prove. Also, if $J(v; 0+) = 0$, then $J(u; 0+) = 0$, by (12.19) and (12.16). Therefore, suppose $J(u; 0+) > 0$ and $0 < J(v; 0+) < \infty$. From Corollary 12.4, in particular from (12.18), for every $R \leq 1$ (say), we deduce

$$|\Omega^\pm(u) \cap B_R| \geq cR^n .$$

Set $u_R(x) = \frac{1}{R} u(Rx)$ and define U_R by

$$U_R^+ = \frac{u_R^+}{\sqrt{J_R^+(u)}} , \qquad U_R^- = \frac{u_R^-}{\sqrt{J_R^-(u)}} .$$

Observe the following.

(a) Poincaré inequality gives

$$\frac{1}{s^2} \fint_{B_s} (u_R^\pm)^2 \leq c \fint_{B_s} |\nabla u_R^\pm|^2 \leq c J_s^\pm(u_R) = c J_{sR}^\pm(u) .$$

(b) From (a), $J(U_R;s)$ is well defined for $R,s \le 1$. In particular $J_1^{\pm}(U_R) = 1$ for every $R \le 1$. Moreover, along some subsequence $\{R_j\}$,

$$U_{R_j}^{\pm} \to U_0^{\pm}$$

uniformly in any compact subset of B_1.

(c) $J(U_0;s) = 1$ for every $s \le 1$. Indeed

$$\frac{J(U_{R_j};s)}{J(U_{R_j};t)} = \frac{J(u;sR_j)}{J(u;tR_j)} \to 1$$

as $R_j \to 0$ and therefore we get $J(U_0;s) = J(U_0;t) = 1$, since $J(U_{R_j};r) \to J(U_0;r)$.

(d) From (c) and Corollary 12.4 again, we deduce that in a suitable system of coordinates, $U_0^{\pm}$ are linear functions, that is,

$$U_0(x) = \alpha x_n^+ - \beta x_n^-$$

with $\alpha^2\beta^2 = 16/\omega_n^2$. Now define $V_R = V_R^+ - V_R^-$ by

$$V_R^+ = \frac{v_R^+}{\sqrt{J_R^+(u)}}, \qquad V_R^- = \frac{v_R^-}{\sqrt{J_R^-(u)}}.$$

Since $0 < J(v;0+) < \infty$, we also have $0 < J(V_R;0+) < \infty$. From (12.19) it follows that

$$V_R^{\pm}(x) \ge U_R^{\pm}(x) + o(|x|) .$$

Consider V_R^+ and let $a(x') = \inf\{x_n : V_R^+(x',x_n) > 0,\ x = (x',x_n) \in B_r\}$ and $\ell(x') = \sqrt{s^2 - |x'|^2} - a(x')$. Notice that $\Omega^+(V_R)$ is tangent to $\{x_n = 0\}$ since $\alpha^+\beta^+ > 0$, so that the point $(x', a(x')) \in B_s$ for $|x'| < s - o(s)$, if $s \le r$. We have

$$\begin{aligned}
\int_{B_s} |\nabla V_R^+|^2 &\ge \int_{B_s} |(V_R^+)_{x_n}|^2 \\
&\ge \int_{\{|x'|<s-o(s)\}} \frac{1}{\ell(x')} \left(\int_{a(x')<x_n<\sqrt{s^2-|x'|^2}} |(V_R^+)_{x_n}|\, dx_n \right)^2 dx' \\
&\ge \int_{\{|x'|<s-o(s)\}} \frac{1}{\ell(x')} \left(\max_{a(x')<x_n<\sqrt{s^2-|x'|^2}} V_R^+\, dx_n \right)^2 dx' \\
&\ge \int_{\{|x'|<s-o(s)\}} \frac{1}{\ell(x')} \left(\max_{a(x')<x_n<\sqrt{s^2-|x'|^2}} U_R^+ + o(s) \right)^2 dx' \equiv A_{s,R}^+ .
\end{aligned}$$

The same inequality holds for V_R^-. As a consequence, $J_s(V_R^{\pm}) \ge s^{-n} A_{s,R}^{\pm}$. Since $U_R \to U_0$ uniformly in compact sets, one has, first letting $R \to 0$ and then $s \to 0$,

$$J(V_R; s) = \frac{J(v; sR)}{J(u; R)} \to \frac{J(v; 0+)}{J(u; 0+)} \geq 1 .$$

□

Corollary 12.6. *Let u be continuous in B_1, $u(0) = 0$ and $u\Delta u \geq 0$ outside its zero set. If near the origin, for some $\alpha \geq 0$, $\beta \geq 0$,*

$$u^+(x) \geq \alpha x_n^+ + o(x),$$
$$u^-(x) \geq \beta x_n^+ + o(x),$$

then

$$\alpha^2\beta^2 \leq c(n) J(u; 0+)$$

where $c(n) = 16/\omega_n^2$, with equality when $u = \alpha x_n^+ - \beta x_n^-$.

As a consequence we prove the Lipschitz continuity of a viscosity solution to a free boundary problem, in a very general situation.

Theorem 12.7. *Let u be a viscosity solution of a free boundary problem in B_1 satisfying the following condition: near every $x_0 \in F(u)$ where*

$$u(x) = \alpha\langle x - x_0, \nu\rangle^+ - \beta\langle x - x_0, \nu\rangle^- + o(|x - x_0|) \tag{12.20}$$

($\alpha \geq 0$, $\beta \geq 0$), there are constants C_1, C_2 such that

$$\beta \leq 1 \quad \textit{implies} \quad \alpha \leq C_1$$

and

$$\alpha \leq 1 \quad \textit{implies} \quad \beta \leq C_2$$

with C_1, C_2 independent of x_0 (they may depend on u). Then u is Lipschitz continuous in $B_{1/2}$ with the Lipschitz constant bounded by $\max\{C_1, C_2, \|u\|^2_{L^\infty(B_1)}\}$.

Proof. It is enough to prove $|u(x)| \leq c\,\mathrm{dist}(x, F(u))$ in $B_{1/2}$. Let $x \in B_{1/2}$, $u(x) = \lambda > 0$ and $h = \mathrm{dist}(x, F(u))$, $x_0 \in \partial B_h(x) \cap F(u)$. Then, near x_0, u has the nontangential asymptotic behavior (12.20) for some $\alpha > 0$, $\beta \geq 0$.

Comparing u with a radially symmetric harmonic function in the ring $B_h(x)\backslash\bar{B}_{h/2}(x)$ vanishing on $\partial B_h(x)$ and equal to $c\lambda$ on $\partial B_{h/2}(x)$, we deduce, as at the beginning of Section 12.1, that

$$\lambda \leq c\alpha h .$$

Now if $\beta \leq 1$, then $\lambda \leq C_1 h$, while if $\beta > 1$, Lemma 12.5 for $u_1 = u^+$, $u_2 = u^-$ and x_0 as the origin gives

$$\alpha^2 \leq c(n)\beta^{-2}\|u\|^4_{L^\infty(B_1)} \leq c(n)\|u\|^4_{L^\infty(B_1)} .$$

Therefore

$$\lambda \le c(n)\|u\|^2_{L^\infty(B_1)} h,$$

which gives the Lipschitz continuity of u^+.

Analogously we deduce the Lipschitz continuity of u^-. □

Another useful application of the monotonicity formula and specifically of Corollary 12.4, formula (12.18), occurs for instance in one-phase problems, to rule out some degenerate situations such as two connected components of the positivity set touching at a free boundary point. This cannot occur in presence of nontrivial linear behavior together with positive density of the zero set. Precisely we have

Lemma 12.8. *Let $u \ge 0$ be a continuous function in B_1, harmonic in its positivity set $\Omega^+(u)$. Suppose Ω_1 and Ω_2 are two connected components of $\Omega^+(u)$ and $0 \in \partial\Omega_1$. If*

(i) *near 0, in Ω_1*

$$u(x) = \alpha x_n^+ + o(|x|)$$

with $\alpha > 0$,

(ii) *$\mathcal{C}\Omega^+(u)$ has positive density at 0,*

then either

(a) $0 \notin \partial\Omega_2$

or

(b) *near 0, in Ω_2,*

$$u(x) = o(|x|) \ .$$

Proof. If $0 \in \partial\Omega_2$, applying the monotonicity formula to $u_1 = u_{|\Omega_1}$ and $u_2 = u_{|\Omega_2}$, extended by zero outside Ω_1 and Ω_2, respectively, from (ii) and formula (12.18) we get

$$J(r) \le c \cdot r^\gamma \|u\|^4_{L^\infty(B_1)} \ .$$

If u_2 has a nontrivial linear behavior near 0, Corollary 12.6 gives a contradiction. □

Even in more general situations, the monotonicity formula can give topological information on the positivity set $\Omega^+(u)$.

Lemma 12.9. *Let $u \ge 0$ be continuous in B_1 and harmonic in $\Omega^+(u)$. Let Ω_1 be a connected component of $\Omega^+(u)$ and let $0 \in \partial\Omega_1$. Let $J(r)$ be as in Theorem 12.3 for $u_1 = u_{|\Omega_1}$ and $u_2 = u - u_1$. Then if $J(0+) > 0$, exactly two connected components Ω_1 and Ω_2 of $\Omega^+(u)$ are tangent at 0, and in a suitable system of coordinates,*

$$u(x) = \alpha x_1^+ + \beta x_1^- + o(|x|)$$

with $\alpha > 0$, $\beta > 0$.

Clearly, $J(0+) > 0$ forces a nontrivial linear behavior near 0 in any connected component of $\Omega^+(u)$ touching at 0; therefore there is no room for more than two connected components of the positivity set touching at 0. If $J(0+) = 0$, in general, nothing can be deduced regarding the topological structure of $\Omega^+(u)$ near 0.

We end this section with a quantitative version of Lemma 12.8: in one-phase problems, like those in Section 13.2, the combination of Lipschitz continuity, nondegeneracy and positive density of the zero set forces the free boundary to be nontangentially accessible, preventing in particular two connected components of $\Omega^+(u)$ to touch.

Let us briefly review the notion of nontangentially accessible domain (N.T.A. domain). Their importance stems from the fact they are the most general domains for which the Harnack and comparison theorems (interior and at the boundary) presented in Sections 11.2 and 11.3 are valid (see [JK]).

Given a domain Ω and a positive number M, we say that a ball $B_r \subset \Omega$ is *M-nontangential* if its distance from $\partial\Omega$ is comparable to its radius, that is,

$$M^{-1}r \leq \operatorname{dist}(B_r, \partial\Omega) \leq Mr .$$

Given $x, y \in \Omega$, an *M-Harnack chain from x to y* is a finite sequence of *M-nontangential* balls such that the first one contains x, the last one contains y and such that consecutive balls intersect. The number of balls of a chain is the *length* of the chain.

Let us define what a *Harnack chain condition* is.

Definition 12.1. Let Ω be an open set in $\mathbb{R}^n$; Ω satisfies an *M-Harnack chain condition* if, for every $\delta > 0$ and every couple of points $x, y \in \Omega$ such that $|x - y| \leq m\delta$ and $B_\delta(x), \beta_\delta(y) \subset \Omega$, there is an *$M$-Harnack chain from x to y* whose length depends on m but not on δ.

We now define N.T.A. domains.

Definition 12.2. A bounded open set $\Omega \subset \mathbb{R}^n$ is an N.T.A. domain if there exist M and r_0 such that the following hold.

(i) For every $Q \in \partial\Omega$ and every $r \leq r_0$, there is $y \in \Omega$ for which

$$M^{-1}r < |Q - y| < r \quad \text{and} \quad B_{r/M}(y) \subset \Omega .$$

(ii) An *M-Harnack chain condition* holds.

(iii) $\mathbb{R}^n \setminus \Omega$ has uniform positive density at every point, i.e., there exists $c_0 > 0$ such that, for $x \in \mathbb{R}^n \setminus \Omega$,

$$|B_r(x) \setminus \Omega| \geq c_0 r^n .$$

The following result holds.

Theorem 12.10. *Let $u \geq 0$ be a continuous function in B_1, harmonic in $\Omega^+(u)$. Assume that for each $x \in F(u) \cap B_{1/2}$ and $r \leq \frac{1}{3}$*

(i) *u is Lipschitz and nondegenerate, that is,*

$$\sup_{B_r(x)} u \geq cr,$$

(ii) *there exists $\eta > 0$ such that*

$$|B_r(x) \setminus \Omega^+(u)| \geq \eta r^n \ .$$

Then $\Omega^+(u) \cap B_{1/2}$ is an N.T.A. domain.

Proof. Since u is Lipschitz and nondegenerate, we only have to prove the Harnack chain condition. Suppose then that x, y are points in $\Omega^+(u)$ with $|x - y| \leq m\delta$ and

$$B_\delta(x) \subset \Omega^+(u) \ , \quad B_\delta(y) \subset \Omega^+(u)$$

for some $m > 0$ and $\delta > 0$.

Suppose $d(x, F(u)) \leq d(y, F(u)) = \delta_0$. If $\delta_0 \geq 2m\delta$, then $x \in B_{m\delta}(y) \subset \Omega^+(u)$ and one can easily find the Harnack balls chain.

Thus let $\delta_0 < 2m\delta$ and $y_0 \in F(u)$ be such that $|y_0 - y| = \delta_0$. Then, for $R \geq r_0 \equiv 4m\delta$, x and y belong to $B_{R/2}(y_0)$. Let $d = \frac{1}{2}\min\{u(x), u(y)\}$.

We will show that if $R > cr_0$, where c may depend on u but not on x and y, then the connected components A_x and A_y of $B_R(y_0) \cap \{u > d\}$ that contain x and y, respectively, are actually the same.

Assume $A_x \neq A_y$ and use the monotonicity formula for $u_1 = (u - d)^+_{|A_x}$ and $u_2 = (u - d)^+_{|A_y}$ (extended by zero outside A_x and A_y). Since (ii) holds, from Corollary 12.4 we get that if

$$\varphi(r) = \frac{1}{r^4} \int_{B_r(y_0) \cap A_x} \frac{|\nabla u_1|^2}{|z - y_0|^{n-2}} \, dz \cdot \int_{B_r(y_0) \cap A_y} \frac{|\nabla u_2|^2}{|z - y_0|^{n-2}} \, dz \ ,$$

for some positive $\beta = \beta(\eta)$ the function

$$r \longmapsto r^{-\beta} \varphi(r)$$

is nondecreasing.

Now, if L denotes the Lipschitz constant of u, we have

$$\varphi(r) \leq c(n) L^4 \equiv C \ .$$

Claim. *If $r \geq r_0$, then*

$$\varphi(r) \geq \bar{c} > 0$$

with $\bar{c}$ independent of r.

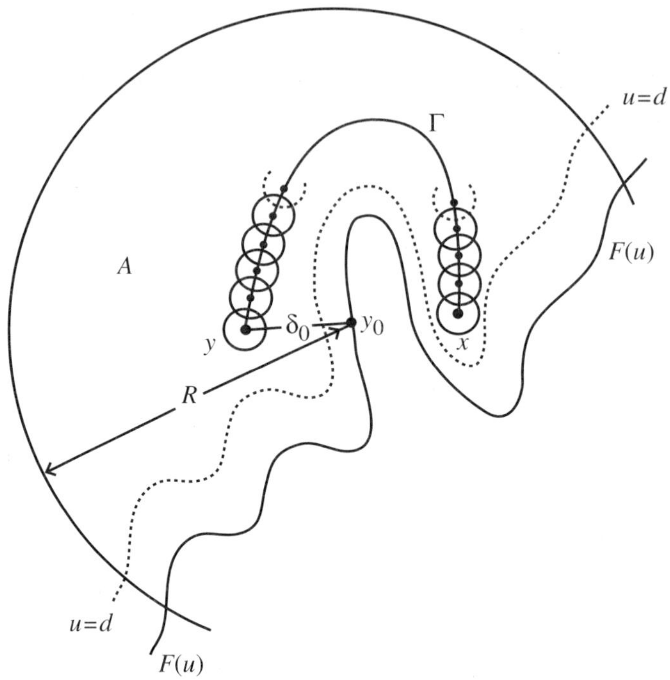

Figure 12.2. The Harnack chain condition for $\Omega^+(u)$ in Theorem 12.10

Using the claim, we have

$$\bar{c}r_0^{-\beta} \le r_0^{-\beta}\varphi(r_0) \le R^{-\beta}\varphi(R) \le cR^{-\beta}$$

so that

$$R \le cr_0 .$$

Thus, if $R > cr_0$, $A_x = A_y = A$. Since A is open and connected, there is a curve $\Gamma \subset A$ joining x and y. For each $z \in \Gamma$ we have, from (i), that

$$u(z) \ge d = \frac{1}{2}\min\{u(x), u(y)\} \ge \tilde{c}\delta .$$

Therefore, if $z \in \Gamma$ and $|p - z| \le \frac{\tilde{c}}{2L}$, we have

$$u(p) \ge u(z) - L|p - z| \ge \frac{\tilde{c}}{2L}\delta$$

so that if $\rho = \frac{\tilde{c}}{2L}\delta$, $B_\rho(z) \subset \Omega^+(u)$. Since

$$\Gamma \subset \bigcup_{z\in\Gamma} B_\rho(z),$$

we can find a sequence of points $z_1 = x,\ z_2, \ldots, z_{N-1},\ z_N = y$ of Γ such that $\Gamma \subset \bigcup_{j=1}^N B_\rho(z_j)$ and such that

$$\sum_{j=1}^N \chi_{B_\rho(z_j)}(y) \leq c(n) \ .$$

Moreover, since $\rho = \frac{\tilde{c}}{2L}\delta$, $r_0 = 4m\delta$ and $z_j \in B_{cr_0}(y_0)$, we can choose the length N independent on x, y and δ.

We now prove the claim. We show that

$$\int_{B_{r_0}(y_0)\cap A_x} \frac{|\nabla u_1|^2}{|z-y_0|^{n-2}}\, dz \geq c r_0^2 \ . \tag{12.21}$$

Since

$$2mL\delta \geq u(x) \geq c_0\delta \ ,$$

if $\sigma = c_0/4L$, then $B_{\sigma\delta}(x) \subset A_x$ (since $|x-y_0| \leq 3m\delta$) with

$$u(z) \geq \frac{3}{4}u(x) \ \text{ in } \ B_{\sigma\delta}(x) \ .$$

Let C be the convex hull of $B_{\sigma\delta}(x)$ and $\{y_0\}$. Every point $z \in C$ can be described by $\rho y'$ with $\rho = |z-y_0|$, $0 \leq \rho < \rho_1(y')$, and y' in a subset Σ of ∂B_1.

For each $y' \in \Sigma$, consider $\rho_0(y') = \sup\{\rho : u(\rho y') = d\}$ so that the segment $\rho y'$, $\rho_0(y') < \rho < \rho_1(y')$, is contained in A_x. Notice that $\rho_0(y') \geq c\delta$ and, since $\rho_1 y' \in \partial B_{\sigma\delta}(x)$,

$$u(\rho_1 y') \geq \frac{3}{4}u(x) \ .$$

Therefore, for each $y' \in \Sigma$,

$$\frac{1}{4}u(x) \leq u(\rho_1 y') - u(\rho_0 y') \leq \int_{\rho_0}^{\rho_1} |\nabla u(\rho y')|\, d\rho$$

which implies

$$\int_{\rho_0}^{\rho_1} |\nabla u(\rho y')|\rho\, d\rho \geq c\delta^2 \ .$$

Integrating this equality over Σ, we get

$$\int_C |\nabla u(z)| \cdot |z-y_0|^{2-n}\, dz \geq C\delta^2$$

and Schwarz inequality gives (12.21) with $C = C(m, L, n)$. In the same way one can prove that

$$\int_{B_{r_0}(y_0)\cap A_y} \frac{|\nabla u_1|^2}{|z-y_0|^{n-2}}\, dz \geq c r_0^2$$

and the claim follows. $\square$

12.4. A parabolic monotonicity formula

In this section we present two monotonicity formulas for pairs of disjointly supported subsolutions of the heat equation (one global in space and one local) ([C4]).

In the last section we show an application to a two-phase singular perturbation evolution problem.

The global formula

Theorem 12.11. *Let u_1 and u_2 be two disjointly supported, continuous and nonnegative subcaloric functions in the strip $R^n \times [-1, 0)$, i.e.,*

(a) $\Delta u_i - D_t u_i \geq 0$,

(b) $u_1 u_2 \equiv 0$,

(c) $u_1(0,0) = u_1(0,0) = 0$.

Assume that the u_i have moderate growth at infinity, for instance

$$\int_{B_R} u_i^2(x, -1)\, dx \leq C e^{\frac{|x|^2}{(4+\varepsilon)}}$$

for R large, some $\varepsilon > 0$ and let

$$\Gamma(x,t) = \frac{1}{t^{n/2}} e^{-\frac{|x|^2}{4t}} .$$

Then for $0 < t \leq 1$, the function

$$J(u_1, u_2, t) = \frac{1}{t^2} \left(\int_{R^n} \int_{-t}^{0} |\nabla u_1|^2 \Gamma(x, -s)\, dx\, ds \right) \times \left(\int_{R^n} \int_{-t}^{0} |\nabla u_2|^2 \Gamma(x, -s)\, dx\, ds \right)$$

is monotone increasing in t.

Remark. If the u_i are linear functions, i.e., $u_1 = \alpha x_n^+$, $u_2 = \beta x_n^-$, J is constant.

Proof. By scaling, it is enough to prove that

$$J'(u_1, u_2, 1) \geq 0.$$

Computing the derivative, we get, setting $J(t) = J(u_1, u_2, t) \equiv \frac{1}{t^2} I_1(t) \cdot I_2(t)$,

$$-2I_1 I_2 + I_1' I_2 + I_1 I_2', \text{ at } t = 1 .$$

Thus, we must prove that

$$\frac{I_1'}{I_1} + \frac{I_2'}{I_2} \geq 2 .$$

Using that $u_1(0,0) = u_2(0,0) = 0$, the fact that

$$\Delta u^2 - D_t u^2 = 2u(\Delta - D_t)u + 2|\nabla u|^2 \geq 2|\nabla u|^2$$

and that

$$[\Delta + D_t](\Gamma(x, -s)) = 0,$$

we may transform the I_i in spatial integrals at time $t = -1$, i.e., we must prove that

$$\frac{\int_{R^n} |\nabla u_1(x,-1)|^2 e^{-\frac{|x|^2}{4}} dx}{\int_{R^n} [u_1(x,-1)]^2 e^{-\frac{|x|^2}{4}} dx} + \frac{\int_{R^n} |\nabla u_2(x,-1)|^2 e^{-\frac{|x|^2}{4}} dx}{\int_{R^n} [u_2(x,-1)]^2 e^{-\frac{|x|^2}{4}} dx}$$

attains its minimum when u_i are a pair of linear functions (in which case all formulas are exact and J is constant).

Notice that in the process of integrating by parts, we need a growth control at infinity in space.

Since the u_i are subcaloric and nonnegative, a control at $t = -1$ suffices.

It is now easy to deduce the optimality estimate using once more the results in [BKP], [CK]. First, among all domains of given Gauss mass, the smallest eigenvalue is attained by a half space, i.e.,

$$\frac{\int_\Omega |\nabla u|^2 e^{-x^2} dx}{\int_\Omega u^2 e^{-x^2} dx}$$

is minimum for $\Omega = \{x_n \geq \alpha\}$ with appropriate α. Second, the first eigenvalue is a convex function of α, in particular

$$\lambda(\alpha) + \lambda(-\alpha) \geq 2\lambda(0)\ .$$

But $\lambda(0)$ is the eigenvalue corresponding to a linear function as eigenfunction. This completes the proof of the formula. □

The local formula

Next we give a local theorem.

Theorem 12.12. *Assume that u_i $(i = 1, 2)$ satisfy the hypotheses (a), (b), (c) of Theorem 12.11 and moreover that they are in L^2 of the unit cylinder $\mathcal{C}_1 = B_1(0) \times (-1, 0)$. Let φ be a cut-off function in x, i.e., $\varphi \equiv 0$ outside $B_{2/3}$, $\varphi \equiv 1$ in $B_{1/2}$ and smooth.*

Then if $w_i = u_i\varphi$ and $J(t) = J(w_1, w_2, t)$,

$$J(0^+) - J(t) \leq A e^{-c/t} \|u_1\|^2_{L^2(\mathcal{C}_1)} \|u_2\|^2_{L^2(\mathcal{C}_1)}$$

for some $C = C(n)$, $A = A(\|u_i\|_{L^2})$.

Remark. It follows from the proof that

$$I_i(t) \le \frac{1}{t^{n/2}} \|u_i\|^2_{L^2(B_1 \times \{t\})} + e^{-c/t} \|u_i\|^2_{L^2(\mathcal{C}_1)} \,.$$

In particular, say,

$$J(1/2) \le c(n) \|u_1\|^2_{L^2(\mathcal{C}_1)} \|u_2\|^2_{L^2(\mathcal{C}_1)} \,.$$

Proof. Since $u_i \in L^2(\mathcal{C}_1)$ and $u_i \ge 0$, $\Delta u_i - D_t u_i \ge 0$, we have

$$\nabla u_i \in L^2(\mathcal{C}_{2-\varepsilon}) \,.$$

We compute as before

$$J'(t) = -\frac{2}{t^3} I_1 I_2 + \frac{1}{t^2} I_1' I_2 + \frac{1}{t^2} I_1 I_2' \,,$$

but when we try to transform I_i into a boundary integral for time $-t$, we have to estimate both the error in I_i and in I_i' in order to control the product. The error in I_i comes from trying to estimate

$$\int_{\mathbb{R}^n} \int_{-t}^{0} |\nabla (u\varphi)|^2 \Gamma(x, -s) \, dx \, ds$$

from above by $\int_{\mathbb{R}^n} \int_{-t}^{0} 1/2(\Delta - D_t)(u\varphi)^2 \Gamma \, dx \, ds$.

Now

$$\begin{aligned} 1/2(\Delta - D_t)(u\varphi)^2 - |\nabla(u\varphi)|^2 &= u\varphi(\Delta - D_t)(u\varphi) \\ &\ge (u\varphi)[u\Delta\varphi + 2\nabla u \nabla \varphi] \,. \end{aligned}$$

All of these terms are supported outside the ball of radius $1/2$, where $\varphi \equiv 1$. Thus, since $|\Gamma(x,s)| \le ce^{-C/t}$ for $-t < s < 0$ and since $\|u_i\|^2_{H^1(\mathcal{C}_{2/3})} \le c\|u_i\|^2_{L^2(\mathcal{C}_1)}$,

$$I_i(t) \le \int_{\mathbb{R}^n} w_i^2(x, -t) \Gamma(x, t) \, dx + \|u_i\|^2_{L^2(\mathcal{C}_1)} e^{-C/t} \,.$$

I_i' introduces no error by itself, only as a factor in the error introduced by I_i.

So, for I_i' we only need a crude estimate from above. We use simply that $\sup_{|x|} \Gamma(x,t) \le t^{-n/2}$ to estimate

$$I_i'(t) \le Ct^{-n/2} \int_{B_{2/3}} [|\nabla u_i(x, -t)|^2 + (u_i(x, -t))^2] \, dx \,.$$

Putting it all together, we have (changing the constant in $e^{-C/t}$ to absorb all negative powers of t)

$$\begin{aligned} J'(t) \ge -Ce^{-C/t} \Big\{ &\|u_1(\cdot, -t)\|^2_{H^1(B_{2/3})} \cdot \|u_2\|^2_{L^2(\mathcal{C}_1)} \\ &+ \|u_2(\cdot, -t)\|^2_{H^1(B_{2/3})} \cdot \|u_1\|^2_{L^2(\mathcal{C}_1)} \Big\} \,. \end{aligned}$$

Therefore

$$J(0^+) - J(t) \leq Ae^{-C/t} \cdot \|u_1\|^2_{L^2(\mathcal{C}_1)} \|u_2\|^2_{L^2(\mathcal{C}_1)} \ .$$

□

12.5. A singular perturbation parabolic problem

Finally, we give an application to an equation appearing in combustion, [CV].

Theorem 12.13. *Let u be a solution in $\mathcal{C}_1 = B_1 \times (-1, 0)$ of*

$$\Delta u - u_t = \beta_\varepsilon(u) \ ,$$

where

(i) $0 \leq \beta_\varepsilon(u) \leq \frac{C}{\varepsilon}$,

(ii) *support of* $\beta_\varepsilon(u) = \{0 \leq u \leq \varepsilon\}$,

and assume that,

$$\|u\|_{L^\infty(\mathcal{C}_1)} \leq C \ .$$

Then

$$\|\nabla u\|_{L^\infty(\mathcal{C}_{1/2})} \leq \bar{C}$$

with $\bar{C}$ independent of ε.

Note: In [CV] this theorem is proved for nonnegative solutions.

Proof. We may apply Theorem 12.12 to

$$u_1 = (u - \lambda)^+ \ , \quad u_2 = (u - \lambda)^-$$

for any $\lambda < 0$.

Therefore on $\mathcal{C}_{3/4}$, $\|\nabla u\|_{L^\infty} \leq \bar{C}$ independent of ε (just from i) in the region $u \leq 0$.

Note that

$$\|u\|_{W^{2,p}(\mathcal{C}_{3/4})} \leq C(\varepsilon) \ ,$$

and therefore it is easy to compute

$$J(0^+) = C|\nabla u(0)|^4 \leq C\|u\|^4_{L^2(\mathcal{C}_1)}$$

whenever $u \leq 0$ at the point chosen as the origin. More precisely in the region $\{u < 2\varepsilon\}$,

$$|\nabla u| \leq \bar{C} \ .$$

Indeed consider (x_0, t_0) in such a region and rescale the problem

$$w(x,t) = \frac{1}{\varepsilon} u(\varepsilon(x - x_0) + x_0, \varepsilon^2(t - t_0) + t_0) \ .$$

This rescales an ε-parabolic neighborhood of (x_0, t_0) onto the unit cylinder around the origin. Let us bound w from below in the closure of $C = B_1 \times (-\sigma, 0)$ for $\sigma > 0$ small. Since on $w < 0$ we have

$$|\nabla w| = |\nabla u| \leq \bar{C} \ ,$$

on those sections of time on which $\bar{B}_{3/4} \cap \{w \geq 0\}$ is not empty, $w \geq -C$ (just by integrating along segments).

The family of times for which $\bar{B}_{3/4} \cap \{w \geq 0\}$ is empty forms an open set. Suppose that for some $t_0 \in (t_\alpha, t_\beta)$, $\inf_x w(x, t_0) = -M \ll -C$. Since $|\nabla w| = |\nabla u| \leq C$ on all of the slice,

$$w(x,t) \leq -\bar{M} \ll -C \ .$$

We now consider the barrier

$$\frac{\bar{M}}{2n}(|x|^2 - 1) + \bar{M}(t - t_0)$$

which controls w from above for $t_0 \leq t \leq t_\beta$ where w is negative and caloric.

Then $w(0, t_\beta) \leq -\frac{\bar{M}}{2n} + \bar{M}\sigma \leq -\frac{\bar{M}}{4} < -C$ if $\sigma = 1/4n$. This contradicts the fact that $w \geq -C$ on t_β.

But now with w being bounded from below, $(\Delta - D_t)w$ being bounded and

$$w(0,0) \leq 2,$$

the Harnack's inequality and standard a priori estimates imply that

$$|\nabla u(x_0, t_0)| = |\nabla w(0,0)| \leq \bar{C} \ .$$

Finally, we have to control $|\nabla u|$ on the domain $\Omega_\varepsilon = \{u > \varepsilon\}$. This follows from the standard Bernstein technique.

Let φ be a cut-off function for the 1/2 cylinder $\mathcal{C}_{1/2} = B_{1/2} \times (-1/2, 0)$ and consider in Ω_ε the function

$$\varphi^2 |\nabla u|^2 + \lambda u^2$$

for λ a large constant. Then

$$\begin{aligned}(\Delta - D_t)(\varphi^2 |\nabla u|^2 + \lambda u^2) &= [(\Delta - D_t)\varphi^2]|\nabla u|^2 + 2\varphi\varphi_i u_{ij} u_j \\ &\quad + \varphi^2 |D_{ij} u|^2 + 2\lambda |\nabla u|^2 \geq 0\end{aligned}$$

if λ is large. Since $\varphi^2 |\nabla u|^2 + \lambda u^2$ is under control on $\partial(\Omega_\varepsilon \cap \mathcal{C}_{1/2})$, the theorem is complete. □

Chapter 13

Boundary Behavior of Caloric Functions

13.1. Caloric functions in Lip(1, 1/2) domains

By a λ-caloric function (and if $\lambda = 1$, just caloric) we mean a nonnegative solution, u, of the heat equation

$$H_\lambda u = \Delta u - \lambda u_t = 0 \qquad (\lambda > 0) \tag{13.1}$$

in some domain $D \subset \mathbb{R}^{n+1}$ that vanishes locally on some distinguished part of ∂D. The term refers to the fact that, in some sense, along that part of ∂D, u resembles the fundamental solution of the heat equation in D.

In particular, we will consider the case when the domain D is the intersection of some $(n+1)$-dimensional cube Q with Ω, one side of the graph of a function, i.e., $\Omega = \{x_n > f(x', t)\}$, and the distinguished part of ∂D is precisely $\partial\Omega \cap Q$.

In this section, f will be a Lip(1, 1/2)-function, that is, for some positive constant L

$$|f(x', t) - f(y', s)| \le L(|x' - y'| + |t - s|^{1/2})$$

and

$$D = \{(x', x_n, t) : x_n > f(x', t),\ x_n < 8nL,\ |x'| < 2,\ |t| < 4\}\ .$$

We assume that $f(0,0) = 0$. The symbol $\partial_p D$ denotes the parabolic boundary of D:

$$\partial_p D = \underbrace{(\bar{D}_n\{t = -4\})}_{\text{bottom points}} \cup \underbrace{(\partial D \cap \{|t| < 4\})}_{\text{side points}} \equiv D_{-4} \cup S.$$

This kind of domain is *regular* for the Dirichlet problem.

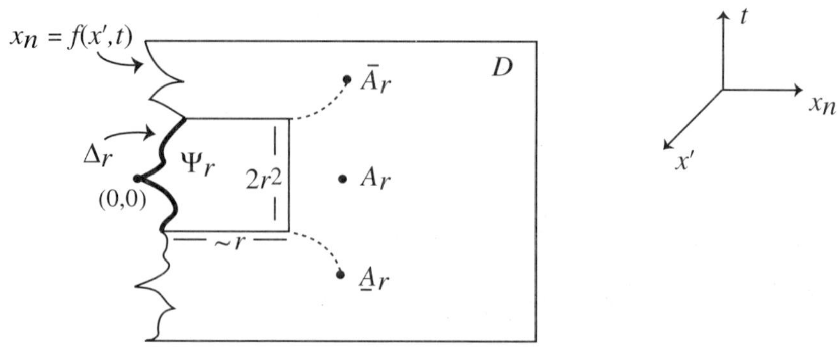

Figure 13.1. Parabolic box and disc

For $(\xi,\tau) \in$ graph(f) and $r > 0$, we introduce parabolic cubes, boxes and surface discs as follows:

$$Q_r(\xi,\tau) = \{(x,t) : |x'| < r,\ |x_n| < 4nLr,\ |t| < r^2\},$$
$$\Psi_n(\xi,\tau) = Q_r(\xi,\tau) \cap D,\ \Delta_r(\xi,\tau) = Q_r(\xi,\tau) \cap \partial D\ .$$

When $(\xi,\tau) = (0,0)$, we simply write Q_r, Ψ_r and Δ_r.

Notice that $D = \Psi_2$, $Q_2 \cap \partial D = Q_2 \cap \text{graph}(f) = \Delta_2$. Furthermore, we define, for $|\tau| \le 2$ and $r \le 1$,

- *inward* point $A_r(\xi,\tau) = (\xi', \xi_n + 6nLr, \tau)$,
  inward future point $\bar{A}_r(\xi,\tau) - (\xi', \xi_n + 6nLr, \tau + 2r^2)$,
  inward past point $\underline{A}_r(\xi,\tau) = (\xi', \xi_n - 6nLr, \tau - 2r^2)$,
- *parabolic distances* $\delta((x,t),(y,s)) = |x-y| + |t-s|^{1/2}$

and

$$\delta_{x,t} = \delta((x,t),(0,0)).$$

Accordingly, a caloric function in Ψ_r is a nonnegative solution of (13.1), vanishing on Δ_r.

Caloric functions in Lip$(1,1/2)$ domains enjoy a number of important properties that parallel those valid for harmonic functions in Lipschitz domains, such as a Carleson estimate (or boundary Harnack inequality) and a boundary comparison principle.

In these results, the presence of a time lag reflects irreversibility in time. However, for caloric functions vanishing on the lateral boundary S, a backward (in time) Harnack inequality holds up to S. This implies a doubling property for the caloric measure and the Hölder continuity of the quotient of two caloric functions vanishing on the same disc on S.

In the rest of this section we state precisely the above results.

The Green function for the domain D with pole at (y, s) is defined when $t > s$ by

$$G(x, t; y, s) = \Gamma(x - y, t - s) - V(x, t; y, s)$$

where

$$\Gamma(x, t) = [4\pi t]^{-n/2} \exp\left[-\frac{x^2}{4t}\right], \qquad t > 0,$$

is the fundamental solution of the heat equation and $V(\cdot, \cdot; y, s)$ is the solution of the problem

$$\begin{cases} Hu = 0 & \text{in } D, \\ u(\sigma, t) = \Gamma(\sigma - y, t - s) & \text{on } \partial_p D. \end{cases}$$

The *adjoint* Green function

$$G^*(y, s; x, t) = G(x, t; y, s)$$

corresponds to the adjoint operator $H^* = \Delta + \partial_t$.

For E a Borel subset of $\partial_p D$, the *caloric measure* $\omega^{(x,t)}(E) = \omega_D^{(x,t)}(E)$ evaluated at $(x, t) \in D$ is defined as the value at (x, t) of the solution to the problem

$$\begin{cases} Hu = 0 & \text{in } D, \\ u = \chi_E & \text{on } \partial_p D \end{cases}$$

where χ_E is the characteristic function of E. If $g \in C(\partial_p D)$, the solution of the problem

$$\begin{cases} Hu = 0 & \text{in } D, \\ u = g & \text{on } \partial_p D \end{cases}$$

is given by

$$u(x, t) = \int_{\partial_p D} g(\sigma) \omega^{(x,t)}(d\sigma) .$$

The (forward) Harnack inequality adapted to our case says:

Theorem 13.1 (Interior Harnack inequality). *Let u be a caloric function in D. Then there exists a positive constant $c = c(n, L)$ such that*

$$u(\underline{A}_r) \leq c u(\bar{A}_r)$$

for any $r \leq 1$.

Moreover ([K], [FGS])

Theorem 13.2 (Boundary Harnack principle or Carleson estimate). *Let u be caloric in Ψ_{2r}. Then there exists $c = c(n, L)$ and $\alpha = \alpha(n, L)$, $0 < \alpha \leq 1$, such that*

$$u(x, t) \leq c \left(\frac{\delta_{x,t}}{r}\right)^{\alpha} u(\bar{A}_r) \qquad \forall\ (x, t) \in \Psi_{r/2} .$$

Theorem 13.3 (Comparison). *Suppose u, v are positive caloric in $\Psi_{2r} \subset D$. Then there exists $c = c(n, L)$ such that*

$$\frac{u(x,t)}{v(x,t)} \geq c \frac{u(\underline{A}_r)}{v(\bar{A}_r)}$$

for every $(x, t) \in \Psi_{r/2}$.

If we consider caloric functions vanishing on the lateral part S of D, the above results can be refined as follows.

Theorem 13.4 (Backward Harnack inequality (B.H.I.)). *Let u be caloric in D, vanishing on S. Then*

$$u(\bar{A}_r) \leq c u(\underline{A}_r) \tag{13.2}$$

with $c = c(n, L)$.

Let us point out some consequences of (13.2).

Theorem 13.5 (Mutual continuity of caloric functions). *Let u, v be caloric in D, $u = v = 0$ on S. Then*

$$c^{-1} \frac{u(A_1)}{v(A_1)} \leq \frac{u(x,t)}{v(x,t)} \leq c \frac{u(A_1)}{v(A_1)} \quad in \ \Psi_1 \tag{13.3}$$

with $c = c(n, L)$. Moreover, $u/v \in C^{\alpha, \alpha/2}(\bar{\Psi}_1)$, $0 < \alpha < 1$.

In particular, notice that if $u > v$, applying (13.3) to $u - v$ and u, we get **Hopf principle:** In Ψ_1, $u \geq \gamma v$ with $\gamma > 1$ depending on $u(A_1)/v(A_1)$ (say).

Theorem 13.6 (Doubling property for the caloric measure). *Let $r \leq 1$. For every $(x, t) \in D \setminus \Psi_{3r/2}$, $t \geq r^2$,*

$$\omega^{(x,t)}(\Delta_r) \leq c \omega^{(x,t)}(\Delta_{r/2}) \tag{13.4}$$

with $c = c(n, L)$.

In view of the application to free boundary problems we actually need the following version of the backward Harnack inequality, valid for caloric functions vanishing just on a lateral disc ([ACS1]).

Theorem 13.7. *Let u be caloric in D, $m = u(\underline{A}_{3/2})$ and $M = \sup_D u$. Then there exists a constant $c = c(n, L, M/m)$ such that if $r \leq 1/2$,*

$$u(\bar{A}_r) \leq c u(\underline{A}_r) \ . \tag{13.5}$$

Proof. Let $B_1 = \{t = -1\} \cap \bar{D}$ and $\beta = \partial \Psi_1 \cap \{-\frac{2}{3} < t < -\frac{1}{2}; x_n = 4nL\}$.

Denote by $\omega_1^{(x,t)}$ the caloric measure in Ψ_1 and by S_1 the lateral boundary of Ψ_1. For $r \leq 1/2$,

$$u(\bar{A}_r) = \int_{B_1} u(\sigma)\omega_1^{\bar{A}_r}(d\sigma) + \int_{S_1} u(\sigma)\omega_1^{\bar{A}_r}(d\sigma) \equiv u_1(\bar{A}_r) + u_2(\bar{A}_r) \ .$$

Since u_1 vanishes on S_1, the B.H.I. gives

$$u_1(\bar{A}_r) \leq c u_1(\underline{A}_r) \ . \tag{13.6}$$

On the other hand, the doubling property of the caloric measure implies that

$$\omega_1^{(x,t)}(S_1) \leq c(n, L)\omega^{(x,t)}(\beta) \tag{13.7}$$

for all $(x, t) \in \Psi_{2/3}$. Therefore, by the maximum principle,

$$u_2(x, t) \leq c(n, L)M\omega^{(x,t)}(\beta) \tag{13.8}$$

in $\Psi_{2/3}$. Now, from Harnack's inequality, since $r \leq 1/2$,

$$u_2(\underline{A}_r) \geq \int_\beta u(\sigma)\omega^{\underline{A}_r}(d\sigma) \geq m\omega^{\underline{A}_r}(\beta) \ . \tag{13.9}$$

Since $\omega^{(x,t)}(\beta)$ is zero on the lateral part of $\Psi_{2/3}$, we have

$$\omega^{\bar{A}_r}(\beta) \leq c(n, L)\omega^{\underline{A}_r}(\beta)$$

so that, from (13.6)–(13.9) we obtain

$$\begin{aligned} u(\bar{A}_r) &\leq c u_1(\underline{A}_r) + cM\omega^{\bar{A}_r}(\beta) \leq c u_1(\underline{A}_r) + cM\omega^{\underline{A}_r}(\beta) \\ &\leq c u_1(\underline{A}_r) + c\frac{M}{m} u_2(\underline{A}_r) \leq c\left(1 + \frac{M}{m}\right) u(\underline{A}_r) \ . \end{aligned}$$

□

Corollary 13.8. *Let* u_1, u_2 *as in Theorem 13.7. Then in* $\Psi_{1/2}$,

$$c^{-1}\frac{u_1(A_{1/2})}{u_2(A_{1/2})} \leq \frac{u_1(x, t)}{u_2(x, t)} \leq c\frac{u_1(A_{1/2})}{u_2(A_{1/2})}$$

where $c = c(n, L, M_1/m_1, M_2/m_2)$. *Moreover* $u_1/u_2 \in C^\alpha(\bar{\psi}_{1/2})$ *for some* $0 < \alpha < 1$. *The* C^α *norm of* u_1/u_2 *and* α *depend only on* $n, L, \frac{m_1}{M_1}, \frac{m_2}{M_2}$.

Proof. From Theorems 13.3 and 13.7,

$$c^{-1}\frac{u_1(A_{1/2})}{u_2(A_{1/2})} \leq \frac{u_1(x, t)}{u_2(x, t)} \leq c\frac{u_1(A_{1/2})}{u_2(A_{1/2})} \tag{13.10}$$

with $c = c(n, L, \frac{m_i}{M_i})$. To prove that u_1/u_2 is Hölder continuous, we show inductively by renormalization that in $\Psi_{2^{-k}}$, $k \geq 0$,

$$\lambda_k u_2 \leq u_1 \leq \Lambda_k u_2 \tag{13.11}$$

with $\Lambda_k - \lambda_k \leq \gamma^k$ for some $\gamma < 1$, γ depending on $n, L, M_i/m_i$. For this purpose, suppose (13.11) holds in $\Psi_{2^{-k}}$ and renormalize by setting

$$u_k(x,t) = u_1(2^{-k-1}x, 4^{-k-1}t), \quad v_k(x,t) = u_2(2^{-k-1}x, 4^{-k-1}t) \ .$$

Then

$$\lambda_k \leq \frac{u_k}{v_k} \leq \Lambda_k \tag{13.12}$$

in Ψ_2. Consider the larger of the ratios

$$\frac{u_k(\underline{A}_{3/2}) - \lambda_k v_k(\underline{A}_{3/2})}{(\Lambda_k - \lambda_k)v_k(\underline{A}_{3/2})} \quad \text{or} \quad \frac{\Lambda_k v_k(\underline{A}_{3/2}) - u_k(\underline{A}_{3/2})}{(\Lambda_k - \lambda_k)v_k(\underline{A}_{3/2})} \ .$$

To fix ideas, assume the larger is the first one. Then

$$u_k(\underline{A}_{3/2}) - \lambda_k v_k(\underline{A}_{3/2}) \geq \frac{1}{2}(\Lambda_k - \lambda_k)v_k(\underline{A}_{3/2}) \ .$$

Let $m_k = u_k(\underline{A}_{3/2}) - \lambda_k v_k(\underline{A}_{3/2})$ and let

$$M_k = \sup_{\Psi_2}(u_k - \lambda_k v_k) \ .$$

From the boundary Harnack principle,

$$\begin{aligned} M_k &\leq c[u_k(\bar{A}_{3/2}) - \lambda_k v_k(\bar{A}_{3/2})] \\ &\leq c(\Lambda_k - \lambda_k)v_k(\bar{A}_{3/2}) \end{aligned}$$

since $u_k(\bar{A}_{3/2}) \leq \Lambda_k v_k(\bar{A}_{3/2})$. Thus

$$\frac{M_k}{m_k} \leq 2c\frac{v_k(\bar{A}_{3/2})}{v_k(\underline{A}_{3/2})} \leq c\left(n, L, \frac{M_2}{m_2}\right)$$

by Theorem 13.7.

With the ratio M_k/m_k under control, we apply (13.10) to $w_k = u_k - \lambda_k v_k$ and $z_k = (\Lambda_k - \lambda_k)v_k$. We get, in Ψ_1,

$$\frac{w_k}{z_k} \geq c\frac{w_k(A_{1/2})}{z_k(A_{1/2})} \geq c\frac{w_k(\underline{A}_{3/2})}{z_k(\bar{A}_{3/2})} \geq c\left(n, L, \frac{M_2}{m_2}\right) \ .$$

Thus, in Ψ_1

$$\lambda_{k+1} \equiv \lambda_k + c(\Lambda_k - \lambda_k) \leq \frac{u_k}{v_k} \leq \Lambda_k \ .$$

Letting $\Lambda_{k+1} = \Lambda_k$, we have

$$\Lambda_{k+1} - \lambda_{k+1} \leq (\Lambda_k - \lambda_k)(1-c) \equiv \gamma(\Lambda_k - \lambda_k) \leq \gamma^{k+1} \ . \qquad \square$$

13.2. Caloric functions in Lipschitz domains

In this section we examine the boundary behavior of caloric functions in domains D above the graph of a Lipschitz function $x_n = f(x', t)$, that is,

$$|f(x',t) - f(y',s)| \leq L(|x - x'| + |t - s|) .$$

Lipschitz continuity in time versus Lipschitz continuity in space is the proper homogeneity balance in studying the phase transition relations of the form $F(u_\nu^+, u_\nu^-, V_\nu)$ considered in Chapters 8–10.

The main result is that, in a neighborhood of the graph of f, the time derivative of a caloric function u is controlled from above by its spatial gradient. This amounts to saying that there the level sets of u are uniformly Lipschitz surfaces in space and time w.r.t. the e_n axis and there exists a space-time cone of directions $\Gamma(e_n, \theta)$ along which u is monotone increasing. Thus, the situation is perfectly analogous to the elliptic one.

We start with those results analogous to Theorem 11.10. Clearly all the results of Section 13.1 remain valid in the context of Lipschitz domains. We keep the same notation. In particular $M = \sup_D u$ and $m = u(\underline{A}_{3/2})$.

We introduce also the elliptic distances

$$d((x,t),(y,s)) = (|x-t|^2 + |t-s|^2)^{1/2},$$
$$d_{x,t} = \inf\{d((x,t),(y,s)) : (y,s) \in \operatorname{graph}(f)\} .$$

Lemma 13.9. *Let u be caloric in D. If $D_{e_n} u \geq 0$ in Ψ_2, then*

$$c^{-1}\frac{u(x,t)}{d_{x,t}} \leq D_{e_n}u(x,t) \leq c\frac{u(x,t)}{d_{x,t}}$$

in Ψ_1, with $c = c(n, L, M/m)$.

Proof. Let $(x,t) \in \Psi_1$ and choose $(\xi, \tau) \in \operatorname{graph}(f)$ such that $(x,t) = A_r(\xi,\tau)$, for some $r > 0$. Now

$$u(\underline{A}_r(\xi,\tau)) - u(\xi + \delta e_n, \tau - 2r^2) = \int_\delta^{nLr} D_{e_n} u(\xi + s e_n, \tau - 2r^2)\, ds .$$

By the parabolic Carleson estimate (Theorem 13.2)

$$u(\xi + \delta e_n, \tau - 2r^2) \leq c\left(\frac{\delta}{r}\right)^\alpha u(\bar{A}_r(\xi,\tau)) .$$

Choosing δ small enough and using Theorem 13.7, we get

$$\frac{1}{2}u(A_r(\xi,\tau)) \leq crD_{e_n}u(A_r(\xi,\tau)) .$$

On the other hand, by Schauder and Harnack inequalities and Theorem 13.7 again,

$$crD_{e_n}u(A_r(\xi,\tau)) \leq cu(\bar{A}_r(\xi,\tau)) \leq cu(A_r(\xi,\tau)) .$$

Since $r \sim d_{x,t}$, the proof is complete. □

Lemma 13.10. *Let u be caloric in D. Then there exists $d = d(n, L, M/m) > 0$ such that in $\Psi_1 \cap \{d_{x,t} < d\}$,*

$$D_{e_n} u \geq 0 \ .$$

Proof. Let $\Sigma = \partial_p \Psi_{3/2} \setminus \mathrm{graph}(f)$ and set $u_1 = u$, $u_2 = c\omega^{(x,t)}_{\Psi_{3/2}}(\Sigma)$, with c chosen such that

$$u_1(A_1) = u_2(A_2) \ .$$

By Corollary 13.8,

$$c^{-1} \leq \frac{u_1}{u_2} \leq c \tag{13.13}$$

in Ψ_1, with $c = c(n, L, M/m)$. By comparing u_2 with its e_n-translations, we obtain

$$D_{e_n} u_2 > 0$$

in Ψ_1. Since $u_1/u_2 \in C^{\alpha,\alpha/2}(\bar{\psi}_1)$ by Corollary 13.8, for all $(x,t), (y,s) \in \psi_1$ we can write

$$\left| \frac{u_1(x,t)}{u_2(x,t)} - \frac{u_1(y,s)}{u_2(y,s)} \right| \leq c(|x-y| + |t-s|^{1/2})^\alpha \ . \tag{13.14}$$

Fix $(y,s) \in \psi_1$. Then if (x,t) varies in a parabolic cylindrical neighborhood of (y,s) of radius $\frac{1}{2} d_{y,s}$, we have, from (13.14), Harnack's inequality and Lemma 13.9,

$$\begin{aligned} \left| u_1(x,t) - \frac{u_1(y,s)}{u_2(y,s)} u_2(x,t) \right| &\leq c u_2(x,t)\, d^\alpha_{y,s} \leq c u_2(y,s)\, d^\alpha_{y,s} \\ &\leq c D_{e_n} u_2(y,s)\, d^{\alpha+1}_{y,s} \ . \end{aligned}$$

Now, by interior Schauder estimates, in a parabolic neighborhood of (y,s) of radius $\frac{1}{4} d_{y,s}$,

$$\left| D_{e_n} u_1(x,t) - \frac{u_1(y,s)}{u_2(y,s)} D_{e_n} u_2(x,t) \right| \leq c D_{e_n} u_2(y,s)\, d^\alpha_{y,s}$$

and therefore, from (13.13),

$$\begin{aligned} D_{e_n} u_1(y,s) &\geq \left[\frac{u_1(y,s)}{u_2(y,s)} - c\, d^\alpha_{y,s} \right] D_{e_n} u_2(y,s) \\ &\geq (c - c\, d^\alpha_{y,s}) D_{e_n} u(y,s) \end{aligned}$$

which is positive if $d_{y,s}$ is small enough depending on $n, L, M/m$. □

We now establish a control for the derivatives of u along admissible directions involving a time component. First we need some L^2-estimates for u_t and the spatial gradient ∇u.

Lemma 13.11. *Let u be caloric in D. Then $\nabla u, u_t \in L^2(\Psi_1)$ and*

$$\|\nabla u\|_{L^2(\Psi_1)}, \|u_t\|_{L^2(\Psi_1)} \leq c\|u\|_{L^2(\Psi_{3/2})}$$

with $c = c(n, L)$.

Proof. By approximation, it is enough to prove the lemma assuming D is a smooth domain (and thus u smooth in $\bar{D}$), as long as the estimates under consideration depend only on the Lipschitz character of ∂D.

The estimate

$$\|\nabla u\|_{L^2(\Psi_1)} \leq c\|u\|_{L^2(\Psi_1)}$$

is a standard energy inequality and follows from the fact that when extended by zero outside D, across graph(f), u is a subsolution of the heat equation. It is then enough to multiply the equation $\Delta u = u_t$ by $\varphi^2 u$, where φ^2 is a cut-off function in space, independent of time, and to integrate by parts.

We now want to prove

$$\|u_t\|_{L^2(\Psi_1)} \leq c\|u\|_{L^2(\Psi_1)} \cdot$$

Let η^2, $\eta = \eta(x,t)$, be a cut-off function such that $\eta \equiv 1$ in Q_1 and $\eta \equiv 0$ outside of $\Psi_{3/2}$. Now

$$\int_D \{H^*(\eta^2 u)u_{x_n} + (Hu_{x_n})\eta^2 u\}\, dx\, dt = \int_{\Delta_{3/2}} \eta^2 u_{\nu_x} u_{x_n}\, d\sigma_x\, dt \tag{13.15}$$

where $H^* = \Delta + \partial_t$, $\Delta_{3/2} = \text{graph}(f) \cap Q_{3/2}$, ν_x is the space interior normal to D and $d\sigma_x$ denotes surface measure in space. On the other hand, $Hu_{x_n} = 0$ and

$$H^*(\eta^2 u) = 2\eta^2 u_t + 2\nabla\eta^2 \cdot \nabla u + uH^*(\eta^2)\ .$$

Therefore, using Schwarz inequality and the L^2-estimate for ∇u, from (13.15) we get

(13.16)

$$\int_{\Delta_{3/2}} \eta^2 u_{\nu_x} u_{x_n}\, d\sigma_x\, dt \leq c\left\{\frac{1}{\varepsilon}\int_{\Psi_{3/2}} u^2\, dx\, dt + \varepsilon \int_{\Psi_{3/2}} \eta^2 (u_t)^2\, dx\, dt\right\}$$

for any $\varepsilon > 0$, with $c = c(n, L)$.

We want to control the L^2 norm of u_t by the left-hand side of (13.16). To this purpose, multiply the heat equation by $\eta^2 u_t$ and integrate by parts;

we get

$$\int_{\Psi_{3/2}} \eta^2(u_t)^2\,dx\,dt = \int_{\Psi_{3/2}} \eta^2 u_t \Delta u\,dx\,dt$$
$$= -\int_{\Delta_{3/2}} \eta^2 u_t u_{\nu_x}\,d\sigma_x\,dt - \int_{\Psi_{3/2}} 2\eta u_t \nabla u\cdot\nabla\eta\,dx\,dt$$
$$-\frac{1}{2}\int_{\Psi_{3/2}} \eta^2(|\nabla u|^2)_t\,dx\,dt\ .$$

Another integration by parts gives

$$\int_{\Psi_{3/2}} \eta^2(|\nabla u|^2)_t\,dx\,dt = -\int_{\Delta_{3/2}} \eta^2|\nabla u|^2\nu_t\,d\sigma - \int_{\Psi_{3/2}} 2\eta\eta_t|\nabla u|^2\,dx\,dt$$

where ν_t denotes the t-component of the interior normal to $\Psi_{3/2}$ and $d\sigma$ is surface measure in space and time.

Now, along $\Delta_{3/2}$, due to the Lipschitz character of D, we have

$$|\nabla u| = u_{\nu_x},\qquad |u_t| \le c(n,L)u_{x_n}\ .$$

Therefore

$$\int_{\Delta_{3/2}} \eta^2|u_t|u_{\nu_x}\,d\sigma_x\,dt \le c(n,L)\int_{\Delta_{3/2}} \eta^2 u_{x_n}u_{\nu_x}\,d\sigma_x\,dt$$

and

$$\left|\int_{\Delta_{3/2}} \eta^2|\nabla u|^2\nu_t\,d\sigma_x\,dt\right| \le c(n,L)\int_{\Delta_{3/2}} \eta^2 u_{x_n}u_{\nu_x}\,d\sigma_x\,dt\ .$$

As a consequence, from (13.16),

$$\int_{\Psi_{3/2}} \eta^2(u_t)^2\,dx\,dt \le c(n,L)\left\{\frac{1}{\varepsilon}\int_{\Psi_{3/2}} u^2\,dx\,dt + \varepsilon\int_{\Psi_{3/2}} \eta^2(u_t)^2\,dx\,dt\right\}\ .$$

The proof is complete if we choose ε small enough. □

Lemma 13.12. *Let u be caloric in D. Then for any direction entering into D, i.e., $\mu = \alpha e_n + \beta e_{n+1}$, $\alpha^2+\beta^2 = 1$, $\alpha > 0$, such that $0 < \tan^{-1}(\frac{|\beta|}{\alpha}) < \frac{1}{2}\cot^{-1}(L)$,*

$$D_\mu u \ge 0$$

in $\Psi_1 \cap \{d_{x,t} < d_0\}$, with $d_0 = d_0(n, L, M/m, \|u\|_{L^2(\Psi_{3/2})})$.

Proof. Take $\bar\mu = \bar\alpha e_n + \bar\beta e_{n+1}$ with $\bar\alpha^2+\bar\beta^2 = 1$, $\bar\alpha > 0$ and $0 < \tan^{-1}(\frac{|\bar\beta|}{1+\bar\alpha}) < \frac{1}{2}\cot^{-1}(L)$. For $0 < h < h_0$ small, set $p = (x,t)$ and

$$w_h^-(p) = \left[\frac{u(p+h\bar\mu) - u(p)}{h}\right]^-$$

for all $p \in \Psi_{5/3}$, where the superscript denotes the negative part. Note that w_h^- is subcaloric, nonnegative in $\Psi_{5/3}$ and $w_h^-(p) = 0$ on $\Delta_{5/3}$. Extend w_h^- to all of $Q_{5/3}$ by setting $w_h^- = 0$ in $Q_{3/2} \backslash D$. Let Q^* be a subcube of $Q_{5/3}$, chosen so that $\partial_p Q^* \subset Q_{5/3} \setminus \bar{Q}_{3/2}$ and $\|u_t\|_{L^2(\partial_p Q^*)}$, $\|\nabla u\|_{L^2(\partial_p Q^*)} \leq c\|\nabla u\|_{L^2(\Psi_{5/3})}$ for some (universal) big constant c. Denote by ω_*^p the caloric measure in Q^* evaluated at p and define

$$z_h(p) = \int_{\partial_p Q^*} w_h^- \, d\omega_*^p .$$

Then if H^n denotes the n-dimensional Hausdorff measure on $\partial_p Q^*$,

$$\begin{aligned} z_h(p) &\leq \|w_h^-\|_{L^2(\partial_p Q^*)} \|d\omega_*^p / dH^n\|_{L^2(\partial_p Q^*)} \\ &\leq c(\|\nabla u\|_{L^2(\partial_p Q^*)} + \|u_t\|_{L^2(\partial_p Q^*)}) \|d\omega_*^p / dH^n\|_{L^2(\partial_p Q^*)} \end{aligned} \tag{13.17}$$

with $c = c(n, L)$, since

$$w_h^-(p) \leq M_{\bar{\mu}}(\nabla u)(p)$$

where $M_{\bar{\mu}}$ is the Hardy-Littlewood maximal function along the $\bar{\mu}$-line through p. Now consider

$$v_h(p) = \int_{\partial_p (Q^* \cap \Psi_{5/3})} w_h^- \, d\omega^p$$

where ω^p is the caloric measure in $Q^* \cap \Psi_{5/3}$.

Observe that v_h is caloric in $Q^* \cap \Psi_{5/3}$. By the maximum principle, we see that

$$w_h^- \leq v_h \leq z_h \tag{13.18}$$

in $Q^* \cap \Psi_{5/3}$. From Theorem 13.3 we have

$$\frac{v_h}{u} \leq c(n, L) \frac{v_h(\bar{A}_1)}{u(\underline{A}_1)} \tag{13.19}$$

in Ψ_1. On the other hand,

$$\left\| \frac{d\omega_*^{\bar{A}_1}}{dH^n} \right\|_{L^2(\partial_p Q^*)} \leq c(n)$$

so that (13.17) and Lemma 13.11 give

$$z_h(\bar{A}_1) \leq c(n, L) \max_{\bar{\Psi}_{5/3}} u . \tag{13.20}$$

From the Carleson estimate and B.H.I.

$$\max_{\bar{\Psi}_{5/3}} u \leq c(n, L, M/m) u(\underline{A}_1)$$

and therefore, from (13.18), (13.19) and (13.20),

$$w_h^- \leq c(n, L, M/m) u$$

in Ψ_1. Since the last estimate is independent of h, we obtain

$$(D_{\bar{\mu}}u)^- \le c(n, L, M/m)u \tag{13.21}$$

in Ψ_1. Now choose $\mu = \alpha e_n + \beta e_{n+1}$ with

$$\alpha = \sqrt{1+\bar{\alpha}}/\sqrt{2}\,, \quad \beta = \bar{\beta}/\sqrt{2(1+\bar{\alpha})}\,.$$

Then $\alpha^2 + \beta^2 = 1$, $\tan^{-1}(\frac{|\beta|}{\alpha}) = \tan^{-1}(\frac{|\bar{\beta}|}{1+\bar{\alpha}}) < \frac{1}{2}\cot^{-1}(L)$.

By Lemma 13.9 and (13.21), in Ψ_1,

$$\begin{aligned} D_\mu u(p) &= \frac{1}{\sqrt{2(1+\bar{\alpha})}}[D_{e_n}u(p) + D_{\bar{\mu}}u(p)] \\ &\ge \frac{1}{\sqrt{2(1+\bar{\alpha})}}\left(\frac{c}{d_p} - c\right)u(p)\,. \end{aligned}$$

Thus for d_p small enough, $D_\mu u(p) \ge 0$ and the proof is complete. □

Lemmas 13.10 and 13.12 remain valid if we replace the e_n-directional derivative by any spatial derivative along a direction making an angle less than $\frac{1}{2}\cot^{-1}(L)$ with the e_n-axis. This allows us, using Lemma 13.12, to obtain a whole cone of directions in *space and time* along which a caloric function is monotone increasing. Thus we have

Theorem 13.13. *Let u be caloric in D. There exists*

$$d_0 = d_0(n, L, M/m, \|u\|_{L^2(\Psi_{3/2})})$$

such that, in $\Psi_1 \cap \{d_{x,t} < d_0\}$, u is monotonically increasing along every direction $\tau \in \Gamma(e_n, \theta)$, where $\theta < \frac{1}{2}\cot^{-1}(L)$.

Theorem 13.13 has a number of interesting and useful consequences. For instance some of the results in this section become invariant under *elliptic scaling*.

Indeed, let us introduce elliptic cubes and boxes:

$$\begin{aligned} &Q_r^*(\xi,\tau) = \{(x,t) : |x'| < r,\ |x_n| < 4nLr,\ |t| < r\}, \\ &\Psi_r^*(\xi,\tau) = Q_r^*(\xi,\tau) \cap D\,. \end{aligned}$$

We have

Corollary 13.14. *Let u be caloric in Ψ_{2r}^*, monotone increasing along every $\sigma \in \Gamma(e_n, \theta)$. Then there exist positive constants c, C depending only on n, L, such that, in Ψ_r^*, for $r \le 1$,*

$$u(x,t) \le c\left(\frac{d_{x,t}}{r}\right)^{\alpha} u(A_r), \tag{13.22}$$

$$c^{-1}\frac{u(x,t)}{d_{x,t}} \le |\nabla u(x,t)| \le c\frac{u(x,t)}{d_{x,t}}\,, \quad |u_t| \le c\frac{u(x,t)}{d_{x,t}}\,. \tag{13.23}$$

Proof. The parabolic Carleson estimate, Theorem 13.2, gives that if $(\xi,\tau)\in Q_r^*\cap\partial D$, for some $0<\beta<1$,

$$u(x,t)\le c\left(\frac{d_{x,t}}{r}\right)^{\beta}u(\bar{A}_r(\xi,\tau))$$

for every $(x,t)\in\Psi_r(\xi,\tau)$. Now observe that if $\sigma\in\Gamma(e_n,\theta)$, $\sigma=\alpha e_n-\beta e_t$, $\alpha^2+\beta^2=1$, $\beta>0$, then

$$u(\bar{A}_r(\xi,\tau))\le u(\bar{A}_r(\xi,\tau)+s\sigma)\qquad\left(0<r\le\frac{r}{2}\right).$$

Choose (say) $s=r/2$ and use Harnack's inequality to get

$$u(\bar{A}_r(\xi,\tau))\le cu(A_r)\ .$$

Since $(\xi,\tau)\in Q_r^*\cap\partial D$ is arbitrary, (13.22) follows.

To prove (13.23), it is enough to notice that the existence of a cone of monotonicity in space and time implies

$$|u_t|\le c|\nabla u|\ ,\quad |\nabla u|\le c|D_{e_n}u|$$

for some $c=(c,n,\theta)$. Then (13.23) follows as in Lemma 13.9 using the *elliptic Carleson estimate* (13.22), where $c=c(n,L)$. □

In turn, the estimate (13.23) implies that at each level of time, a caloric function u is "almost a harmonic function". Precisely, we have

Lemma 13.15. *Let u be caloric in D, monotone increasing along every $\sigma\in\Gamma(e_n,\theta)$. Then there exist $\varepsilon>0$, depending only on n,L and $d>0$, depending only on n,L and (say) $u(A_1)$ such that*

$$w_+=u+u^{1+\varepsilon}\ ,\quad w_-=u-u^{1+\varepsilon}\tag{13.23bis}$$

are subharmonic and superharmonic in

$$\Psi_1\cap\{d_{x,t}<d\}\cap\{t=\bar{t}\}\ .$$

Proof. It is enough to consider $\bar{t}=0$. From (13.23),

$$|\nabla u(x,0)|\approx\frac{u(x,0)}{d_{x,0}}\quad\text{and}\quad|\Delta u(x,0)|=|u_t(x,0)|\le c_1\frac{u(x,0)}{d_{x,0}}\tag{13.24}$$

where $c=c(n,L)$. By Harnack's inequality and using the cone of monotonicity, we have, for a large $N=N(n,L)$,

$$u(x,0)\ge cu(A_1)\cdot d_{x,0}^N\tag{13.25}$$

with $c=c(n,L)$. Choosing $\varepsilon<1/N$, we get

$$\Delta w_+=\Delta u(1+(1+\varepsilon)u^{\varepsilon})+\varepsilon(1+\varepsilon)u^{\varepsilon-1}|\nabla u|^2$$

and, using (13.24), (13.25),

$$\Delta w_+(x,0) \geq \left(-c_1 + c_2 \cdot u(A_1)^\varepsilon d_{x,0}^{\varepsilon N-1}\right) \frac{u(x,0)}{d_{x,0}} .$$

Thus if $d_{x,0}$ is small enough, w_+ is subharmonic. A similar argument gives w_- superharmonic. □

An important consequence of the subharmonicity of w_+ is that in typical phase transition problems, we can apply the monotonicity formula of Theorem 12.3. In fact, we have

Corollary 13.16. *Let u_1 be a λ_1-caloric function in Ψ_1 and let u_2 be a λ_2-caloric function in $Q_1 \setminus \bar{\Psi}_1$. Suppose u_1 and u_2 satisfy the hypotheses of Lemma 13.15. Then for any $r \leq r_0$, r_0 depending only on $n, L, \lambda_1, \lambda_2$, $u_1(A_1), u_2(A_1)$, and any n-dimensional ball B'_r centered at 0, on the hyperplane $t = 0$,*

$$⨍_{B'_r \cap \Psi_1} |\nabla u_1|^2 \cdot ⨍_{B'_r \setminus \Psi_1} |\nabla u_2|^2 \leq c \|u_1\|^2_{L^\infty(B'_1 \cap \Psi_1)} \|u_2\|^2_{L^\infty(B'_2 \setminus \Psi_1)}$$

where c depends only on $n, L, \lambda_1, \lambda_2$.

Proof. Choose ε, d_1 (for u_1) and d_2 (for u_2) as in Lemma 13.15 and let $r_0 \leq \min\{d_1, d_2\}$. Now consider $u_1 + (u_1)^{1+\varepsilon}$ and call w_1 its extension by zero in $Q_1 \setminus \bar{\Psi}_1$. Similarly, call w_2 the extension by zero in Ψ_1 of $u_2 + (u_2)^{1+\varepsilon}$.

Then w_1, w_2 fall into the hypotheses of the monotonicity formula for $r \leq r_0$. Then

$$\frac{1}{r^4} \int_{B'_r} \frac{|\nabla w_1|^2}{|x|^{n-2}}\, dx \cdot \int_{B'_r} \frac{|\nabla w_2|^2}{|x|^{n-2}}\, dx \leq c(n) \|w_1\|^2_{L^\infty(B'_1)} \|w_2\|^2_{L^\infty(B'_1)} \cdot$$

Due to the Lipschitz character of Ψ_1,

$$\frac{1}{r^2} \int_{B'_r} \frac{|\nabla w_i|^2}{|x|^{n-2}}\, dx \geq c(n, L) ⨍_{B'_r \cap \{w_t > 0\}} |\nabla w_i|^2\, dx \qquad (i = 1, 2)$$

and the proof can be easily completed. □

13.3. Asymptotic behavior near the zero set

13.3.1. Lipschitz domains. The object of this section is a study of the behavior of a caloric function near a part of the boundary where it vanishes.

In the next two lemmas D is a Lipschitz domain as in Section 13.2. Then by Lemma 13.15, u is "almost" harmonic at each time level near $Q_1 \cap \partial D$ and its boundary behavior is like that in Lemma 11.17 and Remark 11.18.

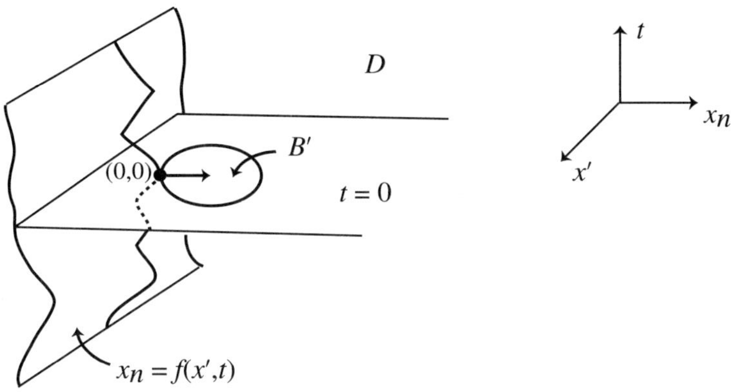

Figure 13.2

Lemma 13.17. *Let u be caloric in D, monotone increasing along every $\sigma \in \Gamma(e_n, \theta)$. Suppose there is an n-dimensional ball $B' \subset D \cap \{t = 0\}$ (resp. $B' \subset \mathcal{C}D \cap \{t = 0\}$) such that $B' \cap \partial D = \{(0, 0)\}$.*

Then near $x = 0$, on the hyperplane $t = 0$, if ν denotes the inward (resp. outward) normal unit vector to ∂B at $(0, 0)$,

$$u(x, 0) = \alpha \langle x, \nu \rangle^+ + o(|x|) \tag{13.26}$$

for some $\alpha \in (0, \infty]$ (resp. $\alpha \in [0, \infty)$). When $\alpha = \infty$, we mean that u grows faster than any linear function.

Proof. Let w_+ and w_- be as in Lemma 13.15. Since $u^{1+\varepsilon} = o(u)$, (13.26) will follow if we prove that

$$w_-(x, 0) = \alpha \langle x, \nu \rangle^+ + o(|x|)$$

when $B' \subset D \cap \{t = 0\}$, and

$$w_+(x, 0) = \alpha \langle x, \nu \rangle^+ + o(|x|)$$

when $B' \subset \mathcal{C}D \cap \{t = 0\}$.

Suppose $B' \subset D \cap \{t = 0\}$ and let $r > 0$ be the radius of B'. With no loss of generality we may assume that $\nu = e_n$. For $\delta < \frac{r}{n-1}$ and $a = \frac{r}{2(n-1)}$, define in $B'_\delta(0) \cap B'$ the function

$$\varphi(x) = x_n + \frac{1}{2a} x_n^2 - \frac{1}{2a(n-1)} |x'|^2 \, .$$

Then $\Delta \varphi = 0$ and $B'_\delta(0) \cap \{\varphi > 0\} \subset B'_\delta(0) \cap B'$. Moreover, it is enough to prove (13.26) in $B_\delta(0) \cap \{\varphi > 0\}$.

Let

$$\alpha_k = \sup\{m : w_-(x, 0) \geq m\varphi(x), \ \forall \ x \in B_{\delta/2^k}(0) \cap \{\varphi > 0\}\}$$

and, since α_k is nondecreasing, set

$$\alpha = \sup \alpha_k = \lim_{k\to\infty} \alpha_k \ .$$

Notice that $\alpha > 0$. If $\alpha = \infty$, then we are done. If $\alpha < \infty$, then

$$w_-(x,0) \geq \alpha\varphi(x) + o(|x|) = \alpha x_n + o(|x|) \tag{13.27}$$

in $B_\delta(0) \cap \{\varphi > 0\}$. We want to show that equality holds in (13.27).

Suppose not. Then there exists a sequence of points $\{x^k\}$ with $x^k \to 0$, such that

$$w_-(x^k,0) - \alpha\varphi(x^k) \geq \eta|x^k|$$

for some $\eta > 0$. Since w_- is monotone increasing along $\Gamma(e_n,\theta)$, with no loss of generality we may suppose that $d_{x^k,0} \approx |x^k| = \delta/2^k$ and that $k > k_0$, with k_0 large, to be chosen later. Let h be harmonic in $B'_{\delta 2^{-k_0}}(0) \cap D$ with $h = w_+$ on $\partial(B'_{\delta 2^{-k_0}} \cap D)$. Then

$$h \geq w_+ \geq w_- \quad \text{in} \ \ B'_{\delta 2^{-k_0}} \cap D$$

and, *a fortiori*, since $\alpha_k \leq \alpha_{k+1} \leq \alpha$,

$$\begin{aligned} &h(x^k) - \alpha_{k+1}\varphi(x^k) \geq \eta|x^k|, \\ &h(x) - \alpha_{k+1}\varphi(x) \geq 0 \ \text{ in } \ B'_{\delta 2^{-k+1}} \cap \{\varphi > 0\} \ . \end{aligned} \tag{13.28}$$

By Harnack's inequality, on a fixed (independent of k) portion of $\partial B'_{\delta 2^{-k}}(0)$, we have

$$h(x) - \alpha_{k+1}\varphi(x) \geq c\eta|x^k| \ . \tag{13.29}$$

Rescale by setting

$$w_k(x) = 2^{k-1}h(2^{-k+1}x) - \alpha_{k+1}2^{k-1}\varphi(2^{-k+1}x) \equiv h_k(x) - \alpha_{k+1}\varphi_k(x) \ .$$

Then w_k is harmonic and positive in $B'_{2\delta}(0)$ and from (13.29), on a fixed portion Γ_δ of $\partial B'_\delta(0)$,

$$w_k \geq c\eta \ .$$

Let w_k^x be the harmonic measure in $B'_\delta(0) \cap \{\varphi_k > 0\}$. Then in $B'_{\delta/2}(0) \cap \{\varphi_k > 0\}$, by the comparison principle for harmonic functions,

$$w_k(x) \geq c\eta\omega_k^x(\Gamma_\delta) \geq c(\delta)\eta\varphi_k(x) \ .$$

Rescaling back, we obtain

$$h(x) - \alpha_{k+1}\varphi(x) \geq c(\delta)\eta\varphi(x) \tag{13.30}$$

in $B'_{\delta 2^{-k-1}}(0) \cap \{\varphi > 0\}$.

Now choose k_0 and c, depending only on ε, n, L, such that

$$w_+ \leq (1 + c^{-k_0})w_-$$

in $B'_{\delta 2^{-k_0}}(0) \cap \{\varphi > 0\}$. Then by the maximum principle, since w_- is superharmonic, in this set we have

$$h \leq (1 + c^{-k_0}) w_-$$

and from (13.30)

$$w_-(x) \geq \frac{\alpha_{k+1} + c(\delta)}{1 + c^{-k_0}} \varphi(x) \tag{13.31}$$

in $B'_{\delta 2^{-k-1}}(0) \cap \{\varphi > 0\}$, a contradiction to the definition of α_{k+1}, if k_0 is large.

In the case $B' \subset \mathcal{C}D \cap \{t = 0\}$, choose δ and a as above and define

$$\tilde{\varphi}(x) = x_n - \frac{1}{2a} x_n^2 + \frac{1}{2a(n-1)} |x'|^2 .$$

Then $\tilde{\varphi}$ is harmonic and $B'_\delta(0) \cap \mathcal{C}B \subset B'_\delta(0) \cap \{\tilde{\varphi} > 0\}$. Moreover, set

$$\tilde{w}_+(x) = \begin{cases} w_+(x, 0) & x \in D \cap B'_\delta(0), \\ 0 & x \in \mathcal{C}D \cap B'_\delta(0). \end{cases}$$

Then $\tilde{w}_+$ is subharmonic in $B'_\delta(0)$. Now define

$$\alpha_k = \inf\{m : \tilde{w}_+ \leq m\tilde{\varphi}, \quad \forall\, x \in B'_{\delta 2^{-k}}(0) \cap \{\tilde{\varphi} > 0\}\}$$

and, since α_k is nonincreasing, set

$$\alpha = \inf \alpha_k = \lim_{k \to \infty} \alpha_k .$$

We have $0 \leq \alpha < \infty$ and, in $B'_\delta(0) \cap \{\tilde{\varphi} > 0\}$,

$$\tilde{w}_+(x) \leq \alpha \tilde{\varphi}(x) + o(|x|) = \alpha x_n^+ + o(|x|) . \tag{13.32}$$

To show that equality holds in (13.32), one can proceed as before. □

We now prove an asymptotic estimate in space and time at regular points of the zero set.

Lemma 13.18. *Let u be caloric in D, monotone increasing along every $\sigma \in \Gamma(e_n, \theta)$. Assume that there is an $(n+1)$-dimensional ball B such that $\bar{B} \cap \partial D = \{(0,0)\}$, and*

(i) $B \subset D$. *Then, near $(0,0)$, for $t \leq 0$*

$$u(x,t) \geq [\beta t + \alpha \langle x, \nu \rangle]^+ + o(d(x,t)) \tag{13.33}$$

for some $\beta \in \mathbb{R}$, $\alpha \in (0, \infty]$, where ν is the spatial unit inward normal to $\partial B \cap \{t = 0\}$ and $d(x,t) = \sqrt{|x|^2 + t^2}$.

(ii) $B \subset {}^C D$. *Then, near* $(0,0)$, *for* $t \le 0$

$$u(x,t) \le [\beta t + \alpha\langle x, \nu\rangle]^+ + o(d(x,t)) \tag{13.34}$$

for some $\beta \in \mathbb{R}$, $\alpha \in [0,\infty)$, *where* ν *is the outward unit normal to* ∂B *at* $(0,0)$.

Moreover, equality holds in (13.33), (13.34) *when* $t = 0$.

Proof. First assume $B \subset D$ and, without loss of generality, that $\nu = e_n$. Define

$$\psi(x,t) = \alpha x_n + \beta t + \frac{\beta}{2}x_n^2 - c_1 t\left[t + \frac{|x|^2}{n}\right] - c_2\left[\frac{|x'|^2}{n-1} - x_n^2\right] - \frac{c_1}{12n}\sum_{j=1}^n x_j^4\,,$$

where α is as in Lemma 13.17 if $\alpha < \infty$ and otherwise it is any $\alpha > 0$; where $\beta \in \mathbb{R}$ is chosen so that the level surface $\{\psi(x,t) = 0\}$ is tangent to B at $(0,0)$.

The function ψ is a solution of the heat equation $H\psi = 0$. Moreover one can choose $c_1 \gg 1$ and $c_2 > 0$, $\varepsilon > 0$ so that $\{\psi(x,t) > 0\} \cap B_\varepsilon(0,0) \subset B$. The conclusion of the lemma will follow if we can prove that, near $(0,0)$, for $t \le 0$,

$$u(x,t) \ge \psi(x,t) + o(d(x,t)) \tag{13.35}$$

in $\{\psi(x,t) > 0\} \cap B_\varepsilon(0,0)$.

Define

$$\psi_\delta(x,t) = \psi(x,t) - \delta\left(x_n + \frac{\beta}{2\alpha}x_n^2 + \frac{\beta}{\alpha}t\right)$$

and, for $h > 0$,

$$D_{h,\delta} = \{\psi_\delta(x,t) > 0\} \cap \{|x_n| < \sqrt{h},\ -h < t < 0\}\,.$$

Then (13.35) follows if we show that for every small $\delta > 0$ and for a small $h > 0$

$$u(x,t) \ge \psi_\delta(x,t) + o(d(x,t)) + o(h)$$

in $D_{h,\delta}$.

Take $0 < \delta_1 < \delta$ and observe the following.

(a) On $\partial D_{h,\delta_1} \cap \{\psi_{\delta_1}(x,t) = 0\}$, we have clearly

$$u \ge \psi_{\delta_1}\,.$$

(b) On $\partial D_{h,\delta_1} \cap \{x_n \ge h\cot\theta,\ t = -h\}$, we have, by the cone of monotonicity,

$$u(x,-h) - \psi_{\delta_1}(x,-h) \ge u(x', x_n - h\cot\theta, 0) - \psi_{\delta_1}(x,-h)$$

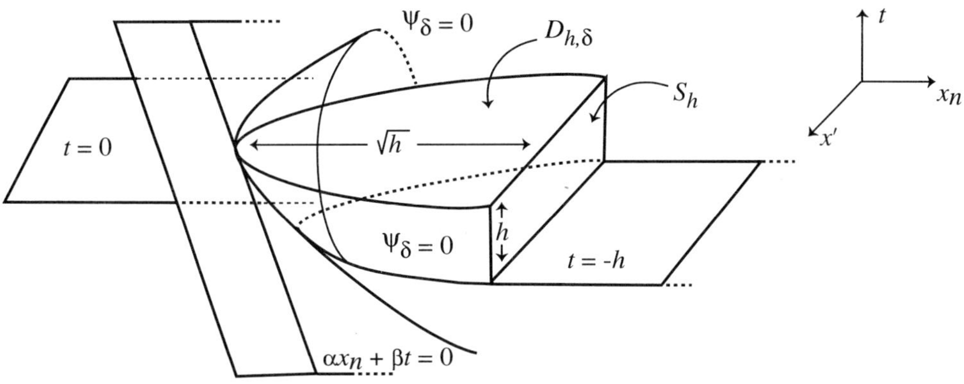

Figure 13.3

and by Lemma 13.17

$$\geq \alpha(x_n - h\cot\theta) + o(|x|) - \psi_{\delta_1}(x, -h)$$
$$\geq -ch + o(|x|) + o(h) \equiv -ch + o(x, h).$$

(c) On $\partial D_{h,\delta_1} \cap \{x_n < h\cot\theta,\ t = -h\}$, $u \geq 0$ and $\psi_{\delta_1} \leq ch + o(|x|) + o(h)$ so that

$$u(x, -h) - \psi_{\delta_1}(x, -h) \geq -ch + o(|x|) + o(h) \equiv -ch + o(x, h).$$

(d) On $S_h = \partial D_{h,\delta_1} \cap \{x_n = \sqrt{h}\}$, we have, by the cone of monotonicity,

$$u(x', \sqrt{h}, t) \geq u(x', \sqrt{h} - (t + h)\cot\theta, -h)\ .$$

For h small, $\sqrt{h} - (t + h)\cot\theta \geq h\cot\theta$, so that, from (b),

$$u(x', \sqrt{h}, t) \geq -ch + o(|x|) + o(h) \equiv -ch + o(x, h)\ .$$

Now set $g(x,t) = -\bar{c}h\omega_{h,\delta_1}^{(x,t)}$, where $\omega_{h,\delta_1}^{(x,t)}$ denotes the caloric measure in D_{h,δ_1} of the set $\partial D_{h,\delta_1} \setminus \{\psi_{\delta_1} = 0\}$. Choosing $\bar{c}$ large enough, by the maximum principle, we can write

$$u - \psi_{\delta_1} \geq g \quad \text{in} \quad D_{h,\delta_1}\ .$$

By *a priori* estimates, in $D_{h/2,\delta_1}$,

$$|g_t|, |\nabla g| \leq c\sqrt{h}$$

and therefore, in the same set,

$$g(x,t) \geq -c\sqrt{h}\, x_n\ .$$

Thus, in $D_{h/2,\delta_1}$,

$$u(x,t) - \psi_{\delta_1}(x,t) \geq -c\sqrt{h}\, x_n\ . \tag{13.36}$$

Let now h be so small that

$$\delta_1 + c\sqrt{h} < \delta \ .$$

Then, from (13.36),

$$\begin{aligned} u(x,t) &\geq \psi_\delta(x,t) - \frac{\beta}{\alpha} t(\delta_1 - \delta) + o(d(x,t)) \\ &\geq \psi_\delta(x,t) + o(d(x,t)) + o(h) \ . \end{aligned}$$

In case (ii), set

$$\tilde{\psi}(x,t) = -\psi(-x,t)$$

and observe that $B_\varepsilon(0,0) \cap \mathcal{C}B \subset B_\varepsilon(0,0) \cap \{\tilde{\psi} > 0\}$.

Now, if we extend u by zero across ∂D, we can proceed as above to prove that, near $(0,0)$, for $t \leq 0$

$$u(x,t) \leq \tilde{\psi}(x,t) + o(d(x,t))$$

in $B_\varepsilon(0,0) \cap \{\tilde{\psi} > 0\}$.

The conclusion follows easily. □

Remark. One can prove that, actually, the conclusions of Lemma 13.18 hold near $(0,0)$ also for $t \geq 0$. See, anyway, the next lemma.

13.3.2. General domains. We now drop the hypotheses that D is a Lipschitz domain and suppose that it is merely an open set. Thus, let u be caloric in D, vanishing (say) on $F = \partial D \cap Q_1$ and assume $(0,0) \in F$ and that it is a regular point.

At this level of generality asymptotic inequalities like (13.33) and (13.34) still hold, restricting $(x,t) \in B$ in the case (i) and $(x,t) \in \mathcal{C}B$ in the case (ii). On the other hand, when D is Lipschitz, equality holds for $t = 0$. In general, we can only achieve equality along paraboloids coming from the past. To show this, we need (13.33) and (13.34) to hold also for $t > 0$, near $(0,0)$. Precisely, we have

Lemma 13.19. *Let u be caloric in the open set D, vanishing on $F = \partial D \cap Q_1$. Suppose $(0,0) \in F$ and that there is an $(n+1)$-dimensional ball B such that $\bar{B} \cap F = \{(0,0)\}$. Assume that the tangent plane to B at $(0,0)$ is given by*

$$\bar{\beta} t + \bar{\alpha}\langle x, \nu\rangle = 0$$

for some spatial unit vector ν and some real numbers $\bar{\alpha}$, $\bar{\beta}$, $\bar{\alpha} > 0$ ($-\bar{\beta}/\bar{\alpha}$ finite). Then either u grows more than any linear function or the following hold.

(a) *The asymptotic behavior in (i) and (ii) of Lemma 13.18 hold near $(0,0)$ in B (with α^+ finite) and in $\mathcal{C}B$, respectively.*

(b) *Equality holds in* (13.33), (13.34) *along any paraboloids of the form* $t = -\gamma\langle x, \nu\rangle^2$, $\gamma > 0$.

Proof. Assume first $B \subset D$ and $\nu = e_n$. Take ψ as in Lemma 13.18. Let $C_k = B_{2^{-k}}(0,0) \cap (-4^{-k}, 4^{-k})$ and

$$D_k = C_k \cap \{\psi > 0\} .$$

For $2^{-k} < \varepsilon$, set

$$m_k = \sup\{m : u(x,t) \geq m\psi(x,t),\ \forall\ (x,t) \in D_k\} .$$

Clearly, $m_k > 0$ for each k and m_k is nondecreasing. Define $m_\infty = \sup m_k$.

If $m_\infty = \infty$, then u grows more than any linear function near $(0,0)$. On the other hand, if m_∞ is finite, we have

$$u(x,t) \geq m_\infty\psi(x,t) + o(d(x,t)) \text{ in } B_\varepsilon \cap \{\psi > 0\} .$$

Setting $\bar{\alpha} = m_\infty\alpha$, $\bar{\beta} = m_\infty\beta$, we have (13.33), with $\alpha = \bar{\alpha}$, $\beta = \bar{\beta}$.

If (ii) holds, let

$$\tilde{\psi}(x,t) = -\psi(-x,t)$$

and extend u by zero outside D. Moreover, let

$$\tilde{D}_k = C_k \cap \{\tilde{\psi} > 0\} .$$

If c_1, c_2, and ε are suitably chosen, we have $B_\varepsilon(0,0) \cap \mathcal{C}B \subset B_\varepsilon(0,0) \cap \{\tilde{\psi} > 0\}$. Define

$$m_k = \inf\{m : u(x,t) \leq m\tilde{\psi}(x,t),\ \forall\ (x,t) \in \tilde{D}_k\} .$$

Then m_k is nonincreasing and if we set $m_\infty = \inf m_k$, we have

$$u(x,t) \leq m_\infty\tilde{\psi}(x,t) + o(d(x,t))$$

in $B_\varepsilon(0,0) \cap \{\tilde{\psi} > 0\}$. If we set $\bar{\alpha} = m_\infty\alpha$, $\bar{\beta} = m_\infty\beta$, we obtain (13.34), again with $\alpha = \bar{\alpha}$ and $\beta = \bar{\beta}$.

Now take a paraboloid $t = -\gamma x_n^2$, $\gamma > 0$, and assume that equality does not hold in (13.33). Then there exists a sequence of points $\{p_k = (x_k, t_k)\}$ with $t_k = -\gamma(x_k)_n^2$ such that $p_n \to (0,0)$ and

$$u(p_k) - m_\infty\psi(p_k) \geq \eta\psi(p_k)$$

for some $\eta > 0$ and k large enough. Let $C_j = B_{2^{-j}} \times (-4^{-j}, 4^{-j})$. We can find a subsequence $\{j_k\}$ such that $p_k \in C_{j_k} \setminus \bar{C}_{j_{k+1}}$ and $t_k < -4^{-j_{k+1}}2$, $2^{-j_{k+1}} < |x_k| < 2^{-j_k - 1}$. Then with m_k as in Lemma 13.18, since $m_{j_{k+1}} < m_\infty$,

$$u(p_k) - m_{j_{k+1}}\psi(p_k) \geq \eta\psi(p_k) .$$

Perform the dilation $x \mapsto 2^{-j_k}x$, $t \mapsto 4^{-j_k}t$ and set

$$\tilde{u}(x,t) = 2^{j_k}u(2^{-j_k}x, 4^{-j_k}t) , \quad \tilde{\psi}(x,t) = 2^{j_k}\psi(2^{-j_k}x, 4^{-j_k}t) .$$

C_{j_k} transforms into $\tilde{C}_1$ and $\tilde{C}_{j_{k+1}}$ transforms into $\tilde{C}_{k_0}$, for some k_0. In $\tilde{Q}_1 = \tilde{C}_1 \cap \{\tilde{\psi} > 0\}$ we have

$$\tilde{u}(p) - m_{j_{k+1}}\tilde{\psi}(p) \geq 0$$

and

$$\tilde{u}(\tilde{p}) - m_{j_{k+1}}\tilde{\psi}(\tilde{p}) \geq \eta\tilde{\psi}(\tilde{p}) \tag{13.37}$$

where $\tilde{p} \in \tilde{Q}_1 \setminus \tilde{Q}_{1/2}$ is the dilated p_k. Let $\tilde{p} = (\tilde{p}', \tilde{x}_n, \tilde{t})$ and introduce the set

$$\Sigma = \left\{(x', x_n, t) : |x'| < b,\ x_n = \tilde{x}_n + b,\ \tilde{t} + b < t < \tilde{t} + 2b\right\}$$

where b is positive and small enough so that $\Sigma \subset \tilde{Q}_1$, $\tilde{t} + 3b < -4^{-k_0}$.

From Harnack's inequality and (13.37), on Σ

$$\tilde{u}(p) - m_{j_{k+1}}\tilde{\psi}(p) \geq c\eta\tilde{\psi}(p) \qquad (c = c(n,b))\ .$$

Now let ω^p be the caloric measure in $\tilde{Q}_1 \cap \{x_n < \tilde{x}_n + b\}$. Then in $\tilde{Q}_{k_0} = \tilde{C}_{k_0} \cap \{\tilde{\psi} > 0\}$

$$\begin{aligned}\tilde{u}(p) - m_{j_k}\tilde{\psi}(\tilde{p}) &\geq \int_\Sigma [\tilde{u} - m_{j_{k+1}}\tilde{\psi}]\, d\omega^p \\ &\geq c\eta\omega^p(\Sigma)\end{aligned}$$

with $c = (n, b)$. On the other hand, since $\omega^p(\Sigma)$ vanishes on $\tilde{C}_{k_0} \cap \{\tilde{\psi} = 0\}$, by the comparison Theorem 13.3,

$$\frac{\omega^p(\Sigma)}{\tilde{\psi}(p)} \geq c\frac{\omega^{\bar{A}}(\Sigma)}{\tilde{\psi}(\underline{A})} \geq c(n,b)$$

in $\tilde{Q}_{k_0}$, where $\bar{A} = (0, 2^{-k_0+1}, 4^{-k_0+1})$, $\underline{A} = (0, -2^{-k_0+1}, -4^{-k_0+1})$. Rescaling back, we obtain

$$u(p) - m_{j_{k+1}}\psi(p) \geq c\eta\psi(p)$$

in $C_{j_{k+1}}$, which contradicts the definition of $m_{j_{k+1}}$.

In the case where the ball B touches ∂D at $(0,0)$ from the other side, the proof is similar. □

13.4. ε-monotonicity and full monotonicity

The notion of ε-monotonicity can be clearly extended to function $u = u(x,t)$: given $\varepsilon > 0$, u is ε-monotone in the direction τ if

$$u(p + \lambda\tau) \geq u(p) \qquad (p = (x,t))$$

for every $\lambda \geq \varepsilon$.

Suppose one of the following conditions holds.

(i) u is uniformly close to a two-plane function of the type $ax_n^+ - bx_n^-$, $a, b > 0$.

(ii) u is trapped between two strictly monotone functions (i.e., with derivatives along some (space-time) direction strictly positive).

(iii) u is uniformly close to a strictly monotone function along some spatial direction and with small oscillation in time.

Then it is easy to check that, for a suitable $\varepsilon > 0$, u is ε-monotone in a cone of (space-time) directions.

Situations like (i), (ii) or (iii) occur, for instance, in studying asymptotic limits (as $t \to +\infty$ or $t \to 0^+$) or blow-up sequences $\frac{1}{\lambda}u(\lambda x, \lambda t)$ as $\lambda \to 0$. Two examples of applications of these notions are given in Section 10.8.

We denote by the symbols $\Gamma_x(e, \theta^x)$ and $\Gamma_t(\nu, \theta^t)$ with $\nu \in \operatorname{span}\{e_n, e_t\}$, respectively, a spatial circular cone of opening θ^x and axis e and a two-dimensional space-time cone of opening θ^t and axis ν. Also, we recall, $\alpha(\tau_1, \tau_2)$ denotes the angle between the two directions τ_1 and τ_2.

In this section we show that if a nonnegative caloric function u is ε-monotone in a cylinder $Q_{\sqrt{\varepsilon M}} = B'_{\sqrt{\varepsilon M}} \times (-\varepsilon M, \varepsilon M)$ along every direction $\tau \in \Gamma_x(e_n, \theta^x) \cup \Gamma_t(\nu, \theta^t)$, then in half the cylinder, u is fully monotone along the directions of slightly smaller cones. Precisely, the main result of this section follows.

Theorem 13.20. *Let u be a nonnegative caloric function in $Q_{\sqrt{\varepsilon M}}$, ε-monotone along every $\tau \in \Gamma_x(e_n, \theta^x) \cup \Gamma_t(\nu, \theta^t)$, $\nu \in \operatorname{span}\{e_n, e_t\}$. Then there exists $c > 0$ such that*

$$D_\tau u(0,0) \geq 0$$

for every $\tau \in \Gamma_x(e_n, \theta^x - c\varepsilon) \cup \Gamma_t(\nu, \theta^t - c\sqrt{\varepsilon})$.

We divide the proof into four steps.

Step 1. Control of u_t by u from below for the solution of $\Delta u + h u_{x_n} - u_t = 0$.

Step 2. ε-monotonicity implies monotonicity for directions $\tau = \alpha e_n + \beta e_t$, $|\beta| \geq \beta_0 > 0$.

Step 3. Monotonicity along $\tau \in \Gamma_t(\nu, \theta_0)$ and ε-monotonicity along a spacial direction e implies monotonicity along $\bar{e} = e + c\varepsilon\nu$.

Step 4. ε-monotonicity along $\tau = \alpha e_n + \beta e_t$, $\beta \neq 0$, implies monotonicity along $\tau_\varepsilon = \tau + c|\alpha|\sqrt{\varepsilon}\, e_n$.

Lemma 13.21 (Step 1). *Let u be a nonnegative solution of $\mathcal{L}u = \Delta u + h u_{x_n} - u_t = 0$ in $B'_2(0) \times (-2, 0)$ where h is a real constant. Then there exists a constant $\eta > 0$, $\eta = \eta(n, h)$, such that*

$$u_t(x, 0) \geq -c(\eta) u(x, t)$$

for every $x \in B'_{1/2}(0)$ and every t, $-\eta \leq t \leq 0$.

Proof. Let G denote the Green's function in $B'_1 \times (-1, 0]$ for the operator $\mathcal{L}$. Then if ν denotes the exterior normal to B'_1,

$$\begin{aligned} u_t(x,0) &= \int_{B'_1} G(x,0;y,-1)u_t(y,-1)\,dy \\ &\quad - \int_{-1}^{0}\int_{\partial B'_1} \partial_\nu G(x,0;y,t)u_t(y,t)\,d\sigma_y\,dt \\ &= \int_{B'_1} G(x,0;y,-1)(\Delta u + h u_{x_n})(y,-1)\,dy \\ &\quad + \int_{-1}^{0}\int_{\partial B'_1} \partial_t\partial_\nu G(x,0;y,t)u(y,t)\,dt\,d\sigma_y \\ &\quad - \int_{\partial B'_1}\int_{-1}^{0} \partial_t(\partial_\nu G(x,0;y,t)u(y,t))\,dt\,d\sigma_y \\ &= \int_{B'_1} \Delta G(x,0;y,-1)u(y,-1)\,dy \\ &\quad - \int_{B'_1} hG_{y_n}(x,0;y,-1)u(y,-1)\,dy \\ &\quad + \int_{-1}^{0}\int_{\partial B'_1} \partial_t\partial_\nu G(x,0;y,t)u(y,t)\,d\sigma_y\,dt\ . \end{aligned}$$

Now there exists $\eta > 0$ such that $\partial_t\partial_\nu G(x,0;y,t)$ is nonnegative in $\partial B'_1 \times [-\eta, 0)$. In fact, perform the change of variables $y_n = -ht + \tilde{y}_n$, $y_i = \tilde{y}_i$, $i = 1, \ldots, n-1$, $t = s$, and set

$$v(y,s) = u(y', -hs + y_n, s)\ .$$

The function v satisfies the heat equation in the tilted cylinder

$$\tilde{C} = \left\{(y,s) : |y'|^2 + (-hs + y_n)^2 \le 4,\ -2 \le s \le 0\right\}\ .$$

If $\tilde{G}$ denotes the corresponding Green's function, we have

$$G(x,0;y', -hs + y_n, 0) = \tilde{G}(x,0;y,s)$$

and therefore what we want to prove is that $(\partial_s - h\partial_{y_n})\tilde{G}_\nu(x,0;y,s)$ is nonnegative on the lateral side of $\tilde{C}$ for $|s|$ small.

Let $\Gamma(x,0;y,s)$ be the fundamental solution of the heat equation. Then for $s < 0$,

$$\partial_s\Gamma(x,0;y,s) = \left[2\pi n(-4\pi s)^{-\frac{n}{2}-1} - (-4\pi s)^{-\frac{n}{2}-2}\left[\frac{(x-y)^2}{4}\right]\right] \cdot \Gamma(x,0;y,s),$$

$$\partial_{y_n}\Gamma(x,0;y,s) = \frac{y_n - x_n}{2}(-4\pi s)^{-\frac{n}{2}-1}\Gamma(x,0;y,s)\ ,$$

and therefore $(\partial_s - h\partial_{y_n})\Gamma(x,0;y,s) < 0$ for $|x-y| \geq 1/4$ and $|s|$ small enough.

Since $\tilde{G}(x,0;y,s) = \Gamma(x,0;y,s) + w_{(x,0)}(y,s)$ where w at $s=0$ is zero of infinite order, we conclude that, for η small enough, $\tilde{G}$ is strictly decreasing along the direction $e_s - he_n$ on the lateral part of the cylinder

$$\tilde{C}_x = \left\{|y'-x'|^2 + (-hs + y_n - x_n)^2 \leq \tfrac{1}{16},\ s \in [-\eta, 0)\right\} .$$

Since $\tilde{G}$ vanishes for $s = 0$ in the set $B_1' - B_{1/4}'(x)$ and on the lateral part of $\tilde{C}_1$, we conclude that $\tilde{G}_\nu$ is increasing along $e_s - he_n$ on the set $\partial B_1' \times [-\eta, 0)$. Going back to the original variables, we have $\partial_t G_\nu > 0$ on $\partial B_1' \times [-\eta, 0)$.

By the fact that G is smooth away from the pole and by the Harnack inequality, the proof is complete. □

Lemma 13.22 (Step 2)**.** *Let u be a nonnegative caloric function in $Q_{\sqrt{\varepsilon M}} = B'_{\sqrt{\varepsilon M}} \times (-\varepsilon M, \varepsilon M)$. Suppose that u is ε-monotone in the direction $\tau = \alpha e_n + \beta e_t$ with $\alpha^2 + \beta^2 = 1$ and $|\beta| \geq 1/2$. Then if M is large enough,*

$$D_\tau u(0,0) \geq 0 .$$

Proof. Suppose first that $\alpha = 0$ and $\beta = 1$. For any λ, $1 < \lambda < \frac{1}{4}\sqrt{M}$, define in $Q_{\sqrt{\varepsilon M}/2}$

$$w_\lambda(x,t) = u(x, t + \lambda\varepsilon) - u(x,t) ;$$

then

$$D_t w_\lambda(0, -\lambda\varepsilon) = D_t u(0,0) - D_t u(0, -\lambda\varepsilon)$$

or

$$\int_1^2 D_t w_\lambda(0, -\lambda\varepsilon)\, d\lambda = D_t u(0,0) - \int_1^2 D_t u(0, -\lambda\varepsilon)\, d\lambda .$$

Therefore

$$\begin{aligned}\int_1^2 D_t w_\lambda(0, -\lambda\varepsilon)\, d\lambda &= D_t u(0,0) + \int_1^2 \frac{1}{\varepsilon} D_\lambda u(0, -\lambda\varepsilon)\, d\lambda \\ &= D_t u(0,0) - \frac{1}{\varepsilon} w_1(0, -2\varepsilon) .\end{aligned}$$

If $1 \leq \lambda \leq \sqrt{M}/10$, by Harnack's inequality, $w_\lambda(0,y) \leq c\lambda w_1(0,y)$ in $Q_{\sqrt{\varepsilon M}/8}$, so that, from Lemma 13.22,

$$D_t w_\lambda(0, -\lambda\varepsilon) \geq -\frac{C(\eta)}{\varepsilon M} w_\lambda(0, -\varepsilon M\eta) \geq -\frac{C(\eta)}{\varepsilon M} \lambda w_1(0, -2\varepsilon)$$

if M is chosen large enough. Hence

$$D_t u(0,0) \geq \frac{1}{\varepsilon}\left(1 - \frac{C}{M}\right) ,$$

which is positive if M is large.

Suppose, now, that $\alpha > 0$, $\beta \geq 1/2$. Perform the transformation $y_i = x_i$, $i = 1, \ldots, n-1$, $y_n = \frac{\alpha}{\beta} t - x_n$, and $s = t$ and set

$$v(y, s) = u(x, t) \; .$$

Then v is $\varepsilon\beta$-monotone in the e_s direction and

$$Lv - v_s = \Delta v - \frac{\alpha}{\beta} v_{y_n} - v_s = 0 \;\text{ in }\; B'_{\sqrt{\varepsilon M}/2} \times \left(-\frac{\beta}{\alpha}\varepsilon M, \frac{\beta}{\alpha}\varepsilon M \right) \cap Q_{\sqrt{\varepsilon M}} \; .$$

Then as above, $D_\tau u(0,0) \geq 0$. The other cases are treated in a similar fashion. □

From Lemma 13.22 we infer that if a nonnegative caloric function is ε-monotone in a cylinder $Q_{\sqrt{\varepsilon M}}$ along all the directions of a two-dimensional space-time cone $\Gamma_t(\nu, \theta_0^t)$, then it is fully monotone along the same directions in a smaller concentric cylinder. Therefore in this smaller cylinder we have an estimate of the type $|u_t| \leq c|\nabla_x u|$, with $c = c(n, M, \theta_0^t)$. This observation will be used in the following lemma:

Lemma 13.23 (Step 3). *Let u be a nonnegative caloric function in $Q_{\varepsilon M} := B'_{\varepsilon M} \times (-\varepsilon^2 M^2, \varepsilon^2 M^2)$. Suppose that u is monotone in $\Gamma_t(e_n, \theta_0)$ for some θ_0 and ε-monotone along a space direction e. Then if $M = M(n, \theta_0)$ is large enough and εM is small (< 1), there exists a $c = c(n, \theta_0)$ such that $D_{\bar{e}} u(0,0) \geq 0$ where $\bar{e} = e + c\varepsilon M e_n$.*

Proof. For any $1 \leq \lambda < M/2$ and $(x,t) \in Q_{\varepsilon M/2}$ define

$$w_\lambda(x,t) = u(x + \lambda\varepsilon e, t) - u(x,t) \; .$$

Using the Harnack inequality in $Q_{\varepsilon M/4}$, we have

$$w_\lambda(x,t) \leq w_{[\lambda]+2}(x,t) \leq C\lambda w_1(x, t + rM^2)$$

for some fixed $r > 0$. Since

$$D_e w_\lambda(0,0) = D_e u(\lambda\varepsilon e, 0) - D_e u(0,0) \; ,$$

by the above inequality and Schauder estimates, we obtain

$$\begin{aligned} w_1(\varepsilon e, 0) &= u(2\varepsilon e, 0) - u(\varepsilon e, 0) \\ &= \int_1^2 \varepsilon D_e w_\lambda(0,0)\, d\lambda + \varepsilon D_e u(0,0) \\ &\leq \frac{C}{M} \int_1^2 w_\lambda(0, r\varepsilon^2 M^2)\, d\lambda + \varepsilon D_e u(0,0) \\ &\leq \frac{C'}{M} w_1(\varepsilon e, 2\varepsilon^2 M^2) + \varepsilon D_e u(0,0) \; . \end{aligned}$$

Now,

$$\begin{aligned}w_1(\varepsilon e, 2\varepsilon^2 M^2) &= w_1(\varepsilon e, 2\varepsilon^2 M^2) - w_1(\varepsilon e, 0) + w_1(\varepsilon e, 0)\\ &\le \max|D_t w_1| 2\varepsilon^2 M^2 + w_1(\varepsilon e, 0)\\ &\le C' D_{e_n} u(0,0)\cdot 2\varepsilon^2 M^2 + w_1(\varepsilon e, 0)\ .\end{aligned}$$

Therefore, if M is large enough

$$D_e u(0,0) + c\varepsilon M D_{e_n} u(0,0) \ge w_1(\varepsilon e, 0)\left(1 - \frac{c}{M}\right)\frac{1}{\varepsilon} \ge 0\ .$$

□

Lemma 13.24 (Step 4). *Suppose u is a nonnegative caloric function in*

$$Q_{\sqrt{\varepsilon M}} = B'_{\sqrt{\varepsilon M}} \times (-\varepsilon M, \varepsilon M)$$

that is (fully) monotonically increasing along a cone $\Gamma_t(e_n, \theta_0)$ and ε-monotone along a direction $\tau = \alpha e + \beta e_t$ where $\alpha^2 + \beta^2 = 1$, $\beta \ne 0$, and e a spatial direction. Then if $M = M(n, \theta_0)$ is large enough and ε small enough,

$$D_{\tau_\varepsilon} u(0,0) \ge 0$$

where $\tau_\varepsilon = \tau + c|\alpha|\sqrt{\varepsilon} e_n$ for some $c = c(n,\theta)$.

Proof. Suppose $\beta > 0$ ($\beta < \frac{1}{2}$). Let $1 \le \lambda \le \frac{1}{2}\sqrt{M}$ and set

$$w_\lambda(p) = u(p) - u(p - \lambda\varepsilon\tau) \qquad (p = (x,t))\ .$$

Since

$$D_\lambda u(-\lambda\varepsilon\tau) = -\varepsilon D_\tau u(-\lambda\varepsilon\tau)\ ,$$

we have

$$\begin{aligned}w_1(-\varepsilon\tau) = u(-\varepsilon\tau) - u(-2\varepsilon\tau) &= -\int_1^2 D_\lambda u(-\lambda\varepsilon\tau)\, d\lambda\\ &= \int_1^2 \varepsilon D_\tau u(\lambda\varepsilon\tau)\, d\lambda = \varepsilon D_\tau u(0) - \varepsilon\int_1^2 D_\tau w_\lambda(0)\, d\lambda\\ &= \varepsilon D_\tau u(0) - \varepsilon\int_1^2 \alpha D_e w_\lambda(0)\, d\lambda - \varepsilon\int_1^2 \beta D_t w_\lambda(0)\, d\lambda\ .\end{aligned}$$

By Schauder estimates in $Q_{\sqrt{\varepsilon M}/4}$ and Harnack's inequality,

$$|D_e w_\lambda(0)| \le \frac{C}{\sqrt{\varepsilon M}} w_\lambda(\varepsilon M e_t) \le \frac{C\lambda}{\sqrt{\varepsilon M}} w_1(-\alpha\varepsilon e + 2\varepsilon M e_t)$$

and

$$w_1(-\alpha\varepsilon e + 2\varepsilon M e_t) \le C\varepsilon M D_{e_n} u(0) + w_1(-\varepsilon\tau)\ .$$

On the other hand, by Lemma 13.21

$$D_t w_\lambda(0) \ge -\frac{c(\eta)}{M\varepsilon} w_\lambda(-\varepsilon\bar{t}e_t)\ , \qquad -M\eta \le \bar{t} \le 0\ .$$

Choosing M large enough and $\bar{t}$ of order 1, by Harnack's inequality

$$D_t w_\lambda(0) \geq -\frac{c(\eta)}{M\varepsilon} w_1(-\varepsilon\tau) .$$

Collecting all our estimates, we have

$$\begin{aligned} w_1(-\varepsilon\tau) &\leq \varepsilon D_\tau u(0) + \frac{C}{M}\beta w_1(-\varepsilon\tau) + c|\alpha|\sqrt{\frac{\varepsilon}{M}}\, w_1(-\varepsilon\tau) \\ &\quad + c|\alpha|\varepsilon\sqrt{\varepsilon}\sqrt{M} D_{e_n} u(0) \end{aligned}$$

or

$$D_{\tau_\varepsilon} u(0) \geq w_1(-\varepsilon\tau)\left[1 - \frac{c\beta}{M} - c|\alpha|\frac{\sqrt{\varepsilon}}{\sqrt{M}}\right]\frac{1}{\varepsilon}$$

which is positive if M is large enough. The case $\beta < 0$ can be treated in a similar fashion. □

Remark. Note that if $\alpha = 0$ in the above lemma, then $\tau_\varepsilon = \tau$.

13.5. An excursion on caloric measure

Although caloric measure in a Lip(1, 1/2) domain D has the doubling properties expressed in Theorem 13.6, it is not absolutely continuous with respect to the measure $dH^{n-1} \times dt$ on the parabolic boundary $\partial_p D$ (see [KW]). On the other hand, if D is a Lipschitz cylinder, that is, $D = \Omega \times (0, T)$ where $\Omega \subset \mathbb{R}^n$ is a Lipschitz domain, then the two measures are mutually absolutely continuous and a theorem similar to Lemma 11.9 holds ([FS]). It turns out that the same kind of result holds in Lip(1, 1) domains also.

Theorem 13.25. *Let $D = \Psi_1$ be a Lipschitz domain on one side of a graph $A = \{x_n = f(x', t)\}$, with $|x'| < 1$, $|t| < 1$. Let $P_0 = (x_0, t_0) \in D$, fixed, denote by ω^{P_0} the caloric measure in D at P_0 and denote by σ the surface measure on $\partial_p D$. Let $K_0 = \frac{d\omega^{P_0}}{d\sigma}$. Then on A*

(a) $\omega^{p_0} \ll \sigma$, $\sigma \ll \omega^{P_0}$,

(b) *$K_0 \in L^2(A)$ and, for every parabolic surface disc $\Delta_r \subset A \cap \{-1 + \eta \leq t \leq t_0 - \eta\}$*

$$\left(\fint_{\Delta_r} K_0^2\, d\sigma\right)^{1/2} \leq c(n, \mathrm{Lip}(f), P_0, \eta) \fint_{\Delta_r} K_0\, d\sigma,$$

(c) *$K_0 \in A_\infty(d\sigma)$ on A.*

Proof. We can assume that D is smooth, as long as all the estimates depend only on $n, \mathrm{Lip}(D), P_0$. Let $(0,0) \in \mathrm{graph}(f)$ and $\Delta_r = \Delta_r(0,0)$. From [FGS] we have for $r \leq t_0/2$ and $A_r = A_r(0,0)$

$$\omega^{P_0}(\Delta_r) \sim r^n G(P_0, A_r), \tag{13.38}$$

that is, on Δ_r,

$$K_0(y,s) \sim G_{y_n}(P_0; y, s) \ .$$

Since $H^* G_{y_n}(P_0; y, s) = 0$ away from P_0, we have, from the adjoint version of (13.15),

$$\int_{\Delta_r} K_0^2 \, d\sigma \leq c(n, \mathrm{Lip}(f), P_0) \int_D H(\varphi^2 G) G_{y_n} \, dy \, ds \tag{13.39}$$

where $\varphi = \varphi(y,s)$ is a cut-off function such that $\varphi \equiv 1$ in Q_r and $\varphi \equiv 0$ outside $\psi_{3r/2}$.

Since

$$H(\varphi^2 G) = 2\nabla\varphi^2 \cdot \nabla G + GH(\varphi^2),$$

we have, using Schwarz inequality and Lemma 13.11,

$$\begin{aligned} \left| \int_D H(\varphi^2 G) G_{y_n} \, dy \, ds \right| &\leq c(n, \mathrm{Lip}(f), P_0) \cdot r^{-3} \int_{\psi_{3/2}} G^2 \, dy \, ds \\ &\leq c(n, \mathrm{Lip}(f), P_0) r^{n-1} G^2(P_0, A_r) \end{aligned}$$

by the boundary Harnack principle.

From (13.38) and (13.39), we get

$$\int_{\Delta_r} K_0^2 \, d\sigma \leq c r^{-n-1} (\omega^{P_0}(\Delta_r))^2$$

from which (a) and (b) follow easily. Now (c) is a consequence of the results in [CF]. □

Not only caloric and surface measure are mutually absolutely continuous on D.

Another important consequence of Lemma 13.15 is that caloric, harmonic and surface measure are mutually absolutely continuous on ∂D at each time level. More precisely, let $G = G(x,t;y,s)$ be the Green's function for the domain D and set $\tilde{G}(x) = G(e_n, 1, x, 0)$. Moreover, let $g = g(x)$ be the Green's function for the Laplace operator in $D \cap \{t = 0\}$ with pole at e_n. Let $G_\pm = \tilde{G} \pm (\tilde{G})^{1+\varepsilon}$ and denote by $P(G_\pm)$ the harmonic replacements of $G_\pm$ in D, respectively. From the adjoint version of Lemma 13.15 and the comparison theorem for harmonic functions, it follows that, for suitable d_1 and constants c, C, depending only on n, L, and the values of $\tilde{G}$ and g at $x = \frac{1}{2}e_n$ (say), we have

$$P(G_+) \geq cg \quad \text{and} \quad P(G_-) \leq Cg$$

in $\Psi_1 \cap \{d_{x,0} < d_1\} \cap \{t = 0\}$.

By the maximum principle, lowering d_1 if necessary, we can write

$$P(G_+) \leq CP(G_-) \leq CG_- \leq CG$$

and

$$P(G_-) \geq cP(G_+) \geq cG_+ \geq cG$$

in $\Psi_1 \cap \{d_{x,0} < d_1\} \cap \{t = 0\}$. Therefore in this set

$$\tilde{G}/g \sim c\,. \tag{13.40}$$

It follows that, basically, the normal derivative, $\tilde{G}_\nu$, of $\tilde{G}$ on $\Delta_1 \cap \{t = 0\}$ determines simultaneously the densities of caloric and harmonic measure.

Therefore, they are mutually absolutely continuous at each level time between -1 and 1. From (13.40) their relative density is controlled from above and below by a constant depending only on $n, L, \bar{G}(\frac{1}{2}e_n)$, $g(\frac{1}{2}e_n)$.

Bibliography

The following references correspond mainly to the original articles where the main theorems in this book appear. There is, of course, an enormous body of literature concerning these problems in the form of books, lecture notes and articles, but we felt that with the current electronic search capabilities it was not worthwhile to try to compile a more or less complete bibliography.

[AC] H. W. Alt and L. A. Caffarelli, Existence and regularity for a minimum problem with free boundary, *J. Reine Angew. Math.* **325** (1981), 105–144.

[ACF1] H. W. Alt, L. A. Caffarelli, and A. Friedman, Axially symmetric jet flows, *Arch. Rational Mech. Anal.* **81** (1983), 97–149.

[ACF2] ________, Asymmetric jet flows, *Comm. Pure Appl. Math.* **35** (1982), 29–68.

[ACF3] ________, Jet flows with gravity, *J. Reine Angew. Math.* **331** (1982), 58–103.

[ACF4] ________, Jet flows with two fluids, I. One free boundary, *Indiana University Math. J.*

[ACF5] ________, Jet flows with two fluids, II. Two free boundaries, *Indiana University Math. J.*

[ACF6] ________, Variational problems with two phases and their free boundaries, *Trans. Amer. Math. Soc.* **282**, No. 2 (1984), 431–461.

[AthC] I. Athanasopoulos and L. A. Caffarelli, A theorem of real analysis and its applications to free boundary problems, *Comm. Pure Appl. Math.* **38**, No. 5 (1985), 499–502.

[ACS1] I. Athanasopoulos, L. A. Caffarelli, and S. Salsa, Caloric functions in Lipschitz domains and the regularity of solutions to phase transition problems, *Ann. of Math.* **143** (1996), 413–434.

[ACS2] ________, Regularity of the free boundary in parabolic phase transition problems, *Acta Math* **176** (1996), 245–282.

[ACS3] ________, Phase Transition Problems of parabolic type: Flat Free Boundaries are smooth, *Comm. Pure Appl. Math.*, Vol. L1 (1998), 77–112.

[BKP] W. Beckner, C. Kenig, and J. Pipher, unpublished.

[BL] H. J. Brascamp and E. H. Lieb, On extensions of the Brunn-Minkowski and Prèkopa-Leindler Theorems, including inequalities for log concave functions, and with an application to the diffusion equation, *J. Funct. Anal.* **22** (1976), 366–389.

[BuL] J. D. Buckmaster and G. S. S. Ludford, *Theory of Laminar Flames*, Cambridge University Press, Cambridge, 1982.

[C1] L. A. Caffarelli, A Harnack inequality approach to the regularity of free boundaries. Part I, Lipschitz free boundaries are $C^{1,\alpha}$, *Revista Math. Iberoamericana* **3** (1987), 139–162.

[C2] ________, A Harnack inequality approach to the regularity of free boundaries. Part II, Flat free boundaries are Lipschitz, *Comm. Pure Appl. Math.* **42** (1989), 55–78.

[C3] ________, A Harnack inequality approach to the regularity of free boundaries. Part III, Existence Theory, Compactness and Dependence on X, *Ann. S.N.S. di Pisa*, IV, vol. XV (1988).

[C4] ________, A monotonicity formula for heat functions in disjoint domains. *Boundary Value Problems for P.D.E. and applications* (J.L. Lions, ed.), Masson, Paris, 1993, 53–60.

[C5] ________, The Obstacle Problem Revisited. *Fourier Anal.* **4**, No. 4–5 (1998), 383–402, MR 1658612 (2000b: 49004)

[CE] L. A. Caffarelli and L. C. Evans, Continuity of the temperature in the two-phase Stefan problems, *Arch. Rational Mech. Anal.* **81** (1983), 199–220.

[CFK] L. A. Caffarelli, E. Fabes, and C. Kenig, Completely singular harmonic measures, *Indiana Univ. Math. J.* **30**, No. 6 (1981), 917–924.

[CFMS] L. A. Caffarelli, E. Fabes, S. Mortola, and S. Salsa, Boundary behavior of nonnegative solutions of elliptic operators in divergence form, *Indiana Univ. Math. J.* **30**, No. 4 (1981), 621–640.

[CK] L. A. Caffarelli and C. Kenig, Gradient estimates for variable coefficients parabolic equations and singular perturbation problems, *Amer. J. Math.* **120** (1998), 391–439.

[CLW1] L. A. Caffarelli, C. Lederman, and N. Wolanski, Uniform estimates and limits for a two phase parabolic singular perturbation problem.

[CLW2] ________, Pointwise and viscosity solutions for the limit of a two phase parabolic singular perturbation problem, *Indiana Univ. Math. J.* **46**, no. 3 (1997).

[CV] L. A. Caffarelli and J. L. Vazquez, A free boundary problem for the heat equation arising in flame propagation, *Trans. Amer. Math. Soc.* **347** (1995), 411–441.

[CF] R. Coifman and C. Fefferman, Weighted norm inequalities for maximal functions and singular integrals, *Studia Math.* **51** (1974), 241–250.

[CL] M. Crandall and P. L. Lions, Viscosity solutions of Hamilton-Jacobi equations, *Trans. Amer. Math. Soc.* **277** (1983), 1–42.

[D] B. Dahlberg, On estimates of harmonic measures, *Arch. Rational Mech. Anal.* **65** (1977), 272–288.

[Du] G. Duvaut, Résolution d'un problème de Stefan (Fusion d'un bloc de glace à zero degré), *C. R. Acad. Sci. Paris* **276** (1973), 1461–1463.

[FGS] E. B. Fabes, N. Garofalo, and S. Salsa, Comparison theorems for temperatures in non-cyclindrical domains, *Atti Accad. Naz. Lincei, Read.* Ser. 8, **78** (1984), 1–12.

[FS] E. Fabes and S. Salsa, Estimates of caloric measure and the initial-Dirichlet problem for the heat equation in Lipschitz cylinders, *Trans. Amer. Math. Soc.* **279** (1983), 635–650.

[F1] A. Friedman, The Stefan problem in several space variables, *Trans. Amer. Math. Soc.* **133** (1968), 51–87.

[F2] A. Friedman, *Variational Problems and Free Boundary Problems*, Wiley, New York, 1982.

[FH] S. Friedland and W. K. Hayman, Eigenvalue inequalities for the Dirichlet problem on spheres and the growth of subharmonic functions, *Comm. Math. Helv.* **51** (1979), 133–161.

[G] E. Giusti, *Minimal Surfaces and Functions of Bounded Variation*, Monographs in Mathematics, 1984.

[JK] D. Jerison and C. Kenig, Boundary behaviour of harmonic functions in nontangentially accessible domains, *Adv. in Math.* **46**, No. 1 (1982), 80–147.

[KA] L. Kamenomostskaja, On Stefan's problem, *Math. Sbornik* **53** (95) (1965), 485–514.

[KW] R. Kaufman and J. M. Wu, Singularity of parabolic measures, *Compositio Math.* **40**, No. 2 (1980), 243–250.

[K] J. T. Kemper, Temperatures in several variables: Kernel functions, representation and parabolic boundary values, *Trans. Amer. Math. Soc.* **167** (1972), 243–262.

[LSW] W. Littman, G. Stampacchia, and H. Weinberger, Regular points for elliptic equations with discontinuous coefficients, *Ann. Scuola Norm. Sup. di Pisa* (3) **17** (1963), 43–77.

[MM] L. Modica and S. Mortola, Construction of a singular elliptic-harmonic measure, *Manuscripta Math.* **33** (1980), 81–98.

[N] R. H. Nochetto, A class of non-degenerate two-phase Stefan problems in several space variables, *Comm. Partial Differential Equations* **12** (1987), 21–45.

[S] E. Sperner, Zum symmetrisierung von Functionen auf Sphären, *Math. Z.* (1973), 317–327.

[W] K.-O. Widman, Inequalities for the Green function and boundary continuity of the gradient of solutions of elliptic differential equations, *Math. Scand.* **21** (1967), 17–37.

Index